ENERGY METABOLISM

STUDIES IN THE AGRICULTURAL AND FOOD SCIENCES

STUDIES IN THE AGRICULTURAL AND FOOD SCIENCES

Energy Metabolism

Proceedings of the Eighth Symposium on Energy Metabolism held at Churchill College, Cambridge, September, 1979

LAURENCE E. MOUNT

Agricultural Research Council,
Institute of Animal Physiology,
Babraham, Cambridge

European Association for Animal Production,
EAAP Publication No. 26

BUTTERWORTHS
LONDON - BOSTON
Sydney - Wellington - Durban - Toronto

United Kingdom London	Butterworth & Co (Publishers) Ltd 88 Kingsway, WC2B 6AB
Australia Sydney	Butterworths Pty Ltd 586 Pacific Highway, Chatswood, NSW 2067 Also at Melbourne, Brisbane, Adelaide and Perth
Canada Toronto	Butterworth & Co (Canada) Ltd 2265 Midland Avenue, Scarborough, Ontario, M1P 3S1
New Zealand Wellington	Butterworths of New Zealand Ltd T & W Young Building, 77–85 Customhouse Quay, 1, CPO Box 472
South Africa Durban	Butterworth & Co (South Africa) (Pty) Ltd 152–154 Gale Street
USA Boston	Butterworth (Publishers) Inc 10 Tower Office Park, Woburn, Massachusetts 01801

First published 1980

ISBN 0 408 10641 7

British Library Cataloguing in Publication Data

Symposium on Energy Metabolism, *8th, Cambridge, 1979*
Energy metabolism. – (European Association for Animal Production. EAAP publications; no. 26). – (Studies in the agricultural and food sciences).
1. Livestock – Congresses
2. Energy metabolism – Congresses
I. Title II. Mount, Laurence Edward
III. Series IV. Series
636.089'2'39 SF5 80–40265

ISBN 0–408–10641–7

Typeset by Scribe Design, Chatham, Kent
Printed and bound by Redwood Burn Ltd., Trowbridge & Esher

The Energy Symposium dedicates this book to Grete Thorbek

PREFACE

This book provides a compilation of current work on the energy metabolism of animals, with emphasis on animals of agricultural importance. The chapters are written by authoritative workers in the field and are grouped in six parts. Discussions by leaders in the field deal with those topics that are of particular interest in each part.

Each chapter is based on a selected paper presented at the Eighth Symposium on Energy Metabolism held at Cambridge in September 1979. The Energy Symposium is not an isolated symposium. It is an international body of research scientists that has been in existence with a continuing organization since 1958 when the first meeting was held in Copenhagen at the instigation of a number of interested workers.

In connection with the origins and development of the Energy Symposium, it is a great pleasure to refer to Dr Grete Thorbek, to whom this volume is dedicated, and who was until 1979 the General Secretary. Throughout her career Dr Thorbek's name has always been strongly associated with the study of energy metabolism, and the first meeting of the Symposium that took place in Copenhagen in 1958 owed a great deal to her initiative. The succeeding meetings in The Netherlands (1961), Scotland (1964), Poland (1967), Switzerland (1970), Federal Republic of Germany (1973), France (1976) and now in Cambridge in 1979, have all benefited extensively from her continuing enthusiasm for the subject of energy metabolism, her contributions to that field of work and the very large part that she has played in the organization of meetings. She is succeeded as General Secretary by Professor K.H. Menke of Hohenheim University.

In addition to Dr Thorbek (General Secretary), the Symposium Committee prior to the 1979 Symposium included K.H. Menke (Federal Republic of Germany), P.W. Moe (USA), L.E. Mount (UK), M. Vermorel (France) and D.M. Walker (Australia).

The members of the Local Committee that was responsible for the organization of the 1979 Symposium were G. Bull, W.H. Close, H.N. Cook, E.L. Miller and L.E. Mount (Convener), with the help during the meetings of M.F. Fuller, Maria McDonnell, Vivien Masters, Jadwiga Necel and A.J.F. Webster.

Each of the six topics on which the meetings were based was in the hands of its own chairman, discussion leader and two scientific secretaries:

Topic No.	*Chairman*	*Discussion Leader*	*Scientific Secretaries*
1	Y. Henry	A.J.H. van Es	H.F. Tyrrell M. Vermorel
2	D.M. Walker	K.L. Blaxter	W.H. Close C. Wenk
3	P.W. Moe	R.L. Baldwin	L.S. Bull N.E. Smith
4	J.B. Ludvigsen	H.S. Bayley	M.F. Fuller J. Guillaume
5	K.H. Menke	W.N. Garrett	D.J. Farrell M.W.A. Verstegen
6	E.F. Annison	A.J.F. Webster	F. Berschauer C.W. Holmes

The financial assistance that was provided by the following organizations is gratefully acknowledged:

Agricultural Research Council,
The British Council,
European Association for Animal Production,
University of Cambridge.

Finally, I should like to thank Dr B.A. Cross, FRS, the Director of the Institute of Animal Physiology, for making the facilities of the Institute available to the Symposium; Butterworths, the publishers, for their help and cooperation; and Jadwiga Necel for her valuable secretarial work in the preparation of this book.

Babraham,
Cambridge,
1979

L.E. Mount

CONTENTS

PART 1

FEED EVALUATION

1

ENERGY UTILIZATION BY GROWING CATTLE AS DETERMINED IN 72 COMPARATIVE SLAUGHTER EXPERIMENTS

W.N. GARRETT
Animal Science Department, University of California, Davis, California 95616, USA

Summary

Fasting heat production determined in trials involving 708 steers and 341 heifers averaged 319 ± 3 kJ/$W^{3/4}$ per day (steers, 315 ± 4; heifers, 324 ± 6). The curvilinear relationships between metabolizable energy (ME; MJ/kg) and net energy values for maintenance (NE_m) or for gain (NE_g) were:

$$NE_m = 1.373\ ME - 0.0330\ ME^2 + 0.0006\ ME^3 - 4.676$$

and

$$NE_g = 1.415\ ME - 0.0415\ ME^2 + 0.0007\ ME^3 - 6.92$$

The data examined (1843 steers with hormonal implants and 861 control steers) indicated that the energy value of the gain for cattle receiving the implant was about 4.5% lower. The energy value of gain (MJ/kg) of the British breed steers (Y=0.280 ± 0.001 $W^{3/4}$; n=2704) was 17% lower than comparable heifers (Y=0.339 ± 0.002 $W^{3/4}$; n=735) but 35% higher than a small sample of Charolais steers (Y=0.183 ± 0.005 $W^{3/4}$; n=52).

Introduction

The comparative slaughter feeding-trial technique has been used on a routine basis for most nutritional experiments with growing cattle conducted at the University of California, Davis, since 1960. The data collected during these experiments have been made accessible for summary and computer-assisted analyses.

Materials and methods

The estimation of body composition using carcass density (Garrett and Hinman, 1969) provided the means to determine the energy content of the empty body of growing cattle. In all experiments an initial slaughter group established the beginning body composition. Energy retention was calculated as the difference between the estimated initial energy content and the final energy content determined on each animal at the termination of a relatively long feeding period (100–200 days). Digestion trials were conducted on most diets fed at about 1.1 times the maintenance level. The ME values were estimated from digestible energy by the factor 0.82. Animal weight in all instances is empty body weight unless otherwise specified.

The estimates of fasting heat production (FHP) were obtained by the regression of log heat production/$W^{3/4}$ against ME/$W^{3/4}$. This approach (Garrett, Meyer and Lofgreen, 1959; Lofgreen and Garrett, 1968) was adopted for routine use of the comparative slaughter technique with cattle because of the physiological and economic difficulties associated with keeping animals on a plane of nutrition which results in negative energy retention over long periods of time.

Maintenance requirements were determined by iteration using the equation describing the relationship between heat production (HP), and ME intake, when HP = ME. The NE_m and the efficiency of ME use for maintenance (k_m) were calculated from FHP and feed or ME required for maintenance. In experiments in which there were no animals fed at or near the maintenance level FHP was assumed to be 0.322 $W^{3/4}$ MJ/day.

The NE_g and the efficiency of ME use for gain (k_g) were determined by the appropriate calculation from energy retention above maintenance, feed available for gain and ME available for gain.

Results and Discussion

ESTIMATED FASTING HEAT PRODUCTION

The information from 17 experiments with steers and nine experiments with heifers was suitable (animals fed a range of intakes from maintenance to *ad lib.*) for estimating the FHP by extrapolation to zero of the regression log HP/$W^{3/4}$ vs ME/$W^{3/4}$. The results of this analysis are shown in *Table 1.1.*

Table 1.1 ESTIMATED FASTING HEAT PRODUCTION OF STEERS AND HEIFERS FED VARIOUS DIETS

Variable	*No. of animals*	*Regression equation*[a]		*r*	*FHP*[b]/*$W^{3/4}$*
		Slope ± *s.e.*	*Intercept*		
Steers	708	0.000375 ± 0.000005	2.4984 ± 0.0053	0.94	315 ± 4
Heifers	341	0.000368 ± 0.000008	2.5108 ± 0.0079	0.93	324 ± 6
All	1049	0.000372 ± 0.000004	2.5032 ± 0.0044	0.94	319 ± 3

[a] Log heat production/$W^{3/4}$ (Y) vs metabolizable energy intake/$W^{3/4}$ (X)
[b] kJ/day

The estimates of FHP are consistent with previous determinations (Garrett, 1970; Lofgreen and Garrett, 1968) using this same technique. The difference between steers and heifers is not significant (until a level of probability of about 20%), but it may indicate that the castrated male has a very slightly lower fasting metabolic rate than the intact female. Blaxter (1962) suggested this as a possibility after reviewing determinations of FHP of sheep.

RELATIONSHIPS BETWEEN METABOLIZABLE ENERGY AND NET ENERGY

The data from all animals (2766) fed at or near *ad lib* were used to determine the relationships between ME (MJ/kg) and NE_m or NE_g (MJ/kg) as well as k_m or k_g. The data were not uniformly distributed across the range of ME (MJ/kg) usually encountered in practical situations (1% < 8 MJ/kg; 22% > 8 MJ/kg but < 11 MJ/kg; 65% > 11 MJ/kg but < 12 MJ/kg; 12% > 12 MJ/kg). The analysis used a stepwise polynomial-regression technique to include only those variables (X to X^4) that significantly improved the regression. A curvilinear relationship (*Table 1.2*) between ME and NE_m and NE_g accounted for approximately 2% more of the variation between the variables than a linear regression, but extrapolation beyond the effective range of the data (ME of 9–12 MJ/kg) gave

Table 1.2 RELATIONSHIPS BETWEEN METABOLIZABLE ENERGY CONCENTRATION AND NET ENERGY VALUES OR THE EFFICIENCY OF ME USE

Relationship	*r*	SE	*ME, MJ/kg (X)*			
			9	*10*	*11*	*12*
Net energy, maintenance (Y)[a]						
Curvilinear (derived)	0.92	0.40	5.49	6.49	7.38	8.17
Curvilinear (extended)	(0.99)[e]	(0.03)[e]	5.44	6.35	7.23	8.08
Efficiency of ME use (%), maintenance (Y′)[b]						
Curvilinear (derived)	0.51	3.7	61.1	65.0	67.5	68.5
Curvilinear (extended)	(0.97)[e]	(1.6)[e]	59.8	63.1	65.7	67.7
Net energy, gain (Z)[c]						
Curvilinear (derived)	0.56	1.05	2.92	3.94	4.72	5.29
Curvilinear (extended)	(0.99)[e]	(0.03)[e]	2.96	3.78	4.55	5.29
Efficiency of ME use, gain (Z′)[d]						
Curvilinear (derived)	0.37	9.2	32.1	39.1	43.3	44.6
Curvilinear (extended)	(0.98)[e]	(1.8)[e]	31.8	36.8	40.9	44.2

[a] Derived (n = 2766), $Y = 1.27X - 0.00008X^4 - 5.41$; extended $Y = 1.373X - 0.0330X^2 + 0.0006X^3 - 4.676$
[b] Derived, $Y' = 17.6X - 0.72X^2 - 39$; extended, $Y' = 12.2X - 0.619X^2 + 0.0103X^3 - 7.35$
[c] Derived (n = 2766), $Z' = 3.160X - 0.1130X^2 - 16.36$; extended $Z = 1.415X - 0.0415X^2 + 0.0007X^3 - 6.92$
[d] Derived (n = 2766), $Z' = 33.57X - 1.40X^2 - 156.6$; extended $Z' = 18.2X - 0.90X^2 + 0.0145X^3 - 69.7$
[e] The extended equation was estimated from averages; therefore, these values have little statistical meaning

improbable values for NE_m and NE_g. In order to have some ability to predict NE_m and NE_g values for individual feedstuffs, whose ME concentrations were lower or higher than that for the mixed diets fed in these trials, an additional equation was synthesized. This extended-range equation was developed using means for NE_m and NE_g predicted from the derived equation (ME concentration between about 8.5 and 12.5 MJ), with NE_m and NE_g outside the range estimated from net-energy values calculated for individual feedstuffs with high (fat) or low (straw) concentrations of ME. The synthetic equations are, therefore, based on limited experimental data outside the ME range of 8.5–12.5 MJ/kg and need to be verified and modified when additional information is available.

Relationships were also developed between ME (MJ/kg) and the efficiency of ME use for maintenance (k_m) or for growth (k_g). The derived and 'extended' regressions were also curvilinear. The efficiency of ME use for maintenance and for gain tended to reach near-maximum values at 15 MJ/kg ME. The data reported by Blaxter and Boyne (1978) also show this trend.

ENERGY VALUE OF WEIGHT GAIN

Relationships between energy deposition and weight gains are necessary to use a net-energy system in formulating diets for specific rates of gain or to determine the rate of gain of an animal receiving a known amount of a specific feed or feed mixture. *Table 1.3* is a summary of the information obtained from these data relative to the relationship between scaled energy retention and weight gain. It is immediately obvious that the scaled energy content of the gain is influenced by sex, class and breed and to a lesser extent by the use of hormonal adjuvants.

Table 1.3 RELATIONSHIPS BETWEEN DAILY ENERGY GAIN (EG, Y) AND WEIGHT GAIN (X) WITH A DERIVED EXAMPLE

Group	*No. of animals*	*Equation*	*EG, MJ/day for a 350 kg animal, 1.0 kg gain*
Observed weight basis:			
Steers, not implanted	861	$Y=0.260 \pm 0.002X\ (W^{3/4})$	21.0
Implanted steers	1843	$Y=0.241 \pm 0.008X\ (W^{3/4})$	19.5
Heifers, not implanted	405	$Y=0.296 \pm 0.003X\ (W^{3/4})$	24.0
Implanted heifers	330	$Y=0.283 \pm 0.003X\ (W^{3/4})$	22.9
Empty weight basis:			
Steers, not implanted	861	$Y=0.289 \pm 0.002X\ (W^{3/4})$	23.4
Implanted steers	1843	$Y=0.276 \pm 0.001X\ (W^{3/4})$	22.3
Heifers, not implanted	405	$Y=0.348 \pm 0.003X\ (W^{3/4})$	28.1
Implanted heifers	330	$Y=0.332 \pm 0.003X\ (W^{3/4})$	26.9
Charolais steers	52	$Y=0.183 \pm 0.005X\ (W^{3/4})$	14.8

These relatively simple expressions (based on the standard error of prediction) describe the data about as well as the quadratic expressions used by Lofgreen and Garrett (1968) but have the inherent implication that the energy content of the gain is independent of rate of gain for an animal of a particular size. As a generalization this may be approximately correct, but is not completely satisfying since there are reports in the literature (MAFF, DAFS and DANI, 1975; Garrett, 1978) showing that the composition of the gain is influenced by the rate of gain made over approximately the same incremental change in body weight. Further analysis of the data may result in relationships between animal weight, weight gain and energy gain which can account for the fact that rate of gain does influence the composition of the gain.

Acknowledgements

The contributions of Dr G.P. Lofgreen in supplying some of the data and Mr N. Hinman to the computer-assisted analyses are gratefully acknowledged.

References

BLAXTER, K.L. (1962). *The Energy Metabolism of Ruminants*, 1st edn. London: Hutchinson

BLAXTER, K.L. and BOYNE, A.W. (1978). *J. agric. Sci., Camb.*, **99**, 47–68

GARRETT, W.N. (1970). In *Energy Metabolism of Farm Animals*, pp. 101–104. Eds. A. Schürch and C. Wenk. EAAP Publication No. 13. Zurich: Juris Druck and Verlag

GARRETT, W.N. (1978). In *Protein Metabolism and Nutrition*, pp. 115–118. Ed. S. Tamminga. EAAP Publication No. 22. The Netherlands: Wageningen

GARRETT, W.N. and HINMAN, N. (1969). *J. Anim. Sci.*, **28**, 1–5

GARRETT, W.N., MEYER, J.H. and LOFGREEN, G.P. (1959). *J. Anim. Sci.*, **18**, 528–547

LOFGREEN, G.P. and GARRETT, W.N. (1968). *J. Anim. Sci.*, **27**, 793–806

MAFF, DAFS and DANI (1975). Energy allowances and feeding systems for ruminants. *Technical Bulletin No. 33.* London: HMSO

2

INFLUENCE OF THE METABOLIZABLE ENERGY CONTENT OF THE DIET ON ENERGY UTILIZATION FOR GROWTH IN BULLS AND HEIFERS

Y. GEAY, J. ROBELIN and M. VERMOREL
I.N.R.A.–C.R.Z.V. de Theix, 63110 Beaumont, France

Summary

Long grass hay or wafers of dehydrated lucerne, associated with two different proportions of concentrate, were fed at the same levels of ME to four groups of 10 bulls and 10 heifers. The concentrate diets were also fed *ad lib.* to two groups of five bulls and five heifers. The average gain in various body components was assessed by the slaughter technique.

For the same ME intake the daily gain of live weight, protein fat and energy was lowest for the diets with the least metabolizability, but the decrease was less in bulls than in heifers. An inverse relationship occurred between k_{pf} and the amount of energy retained as protein. Both the above changes appeared to be independent of roughage processing.

Introduction

It is well known that the efficiency of metabolizable energy (ME) utilization for fat deposition (k_f) drops when the crude fibre content of the diet increases or when its metabolizability (q) decreases in adult ruminants. However, a first experiment (Geay, Robelin and Jarrige, 1976) showed that the efficiency of ME utilization for protein and fat deposition (k_{pf}) in young bulls was much less than k_f, and it decreased only slightly with the metabolizability. Therefore, a second experiment, the main features of which are presented in this paper, was undertaken to determine whether these characteristics of k_{pf} could be related to the high protein gain in bulls or to the forage processing previously used (wafers). The complete results are published in detail elsewhere (Geay *et al.*, 1980).

In this trial two diets differing in q value (various proportions of roughage) were given in the same quantities of ME to fattening bulls and heifers. Half the animals were fed long-grass hay (H) and the others were fed wafers of dehydrated lucerne (L). To determine the growth potential of cattle and the influence of feeding level on energy utilization, the diets with the highest q content were also offered *ad lib.*

Materials and methods

The diets were constituted by mixing forages (F) and concentrates (C) in the following proportions: 80 F/20 C and 20 F/80 C. On a dry matter basis, the hay contained 8.6% of crude protein and 33.3% of crude fibre, while the lucerne contained 19.2% of crude protein and 25.8% of crude fibre. The concentrates were a pelleted mixture of maize, barley, soybean meal and urea.

Digestibility of each diet and urinary losses were measured twice in four bulls at the beginning and end of the experiment by total collection of excreta for seven days. Methane energy losses were measured on three animals from each group in respiration chambers. The q values of the four diets (80 H/20 C, 20 H/80 C, 80 L/20 C, 20 L/80 C) were 0.49, 0.64, 0.54 and 0.69 respectively.

These diets were fed to 50 bulls and 50 heifers, Charolais x Salers. Charolais x Salers cattle were chosen for their high protein retention potential. For each sex four groups of 10 animals were formed and each group received one of the four diets. The 80% forage diets were offered *ad lib.* In addition, two groups of five animals, similar to the previous ones, were formed and received the 20% hay or 20% lucerne diets *ad lib.* The animals were tethered on chip litter and fed individually twice a day.

At the beginning of the trial, five additional and representative animals of each sex were slaughtered. Their whole body was separated into its main components (muscles, bones and fatty tissues) which were minced and analysed separately (energy, protein, lipids and ash content). At the end of the experiment the chemical composition was estimated from the anatomical composition of the 11th rib cut and the weight of the fifth quarter's fatty tissues, using the relationship established by Robelin and Geay (1978).

Results and discussion

Over the same length of fattening period, whatever the sex or roughage, the two groups of cattle fed a limited quantity of the richest diets (q = 0.64 or 0.69) consumed *the same quantity of ME* as the corresponding groups fed the poorest diets (q = 0.49 or 0.54) *ad lib.* However, the former had greater live weight, protein, lipid and energy gain, and thus a greater overall efficiency of ME utilization for maintenance and growth (k_{mf}) than the latter.

Maintenance requirements were estimated to be 326 kJ net energy (NE)/kg $W^{0.75}$, that is from 435–468 kJ ME according to q (van Es, 1975). For the same ME intake the k_{pf} ranged from 0.24 to 0.37 in bulls and from 0.31 to 0.44 in heifers. Taking into account all the animals the k_{pf} also increased with q, but significantly less ($P < 0.05$) in bulls than in heifers and much less than k_f in the adult ruminant according to the Blaxter equation (1974; *Figure 2.1*). The variation of k_{pf} with q was less affected by roughage processing than by the animal type.

For the high concentrate diets and for the same metabolizability, live weight, protein, lipid and energy gain increased with ME intake, although the increases in energy retention were smaller in bulls than in heifers for both forages (*Figure 2.1*). However, the k_{pf} on the wafered lucerne diets was smaller than that on the long-hay diets in spite of a higher metabolizability. Similarly, k_{pf} in bulls was significantly lower ($P < 0.05$) than in heifers. Furthermore, it was smaller in the Charolais X Salers bulls than in the earlier maturing Salers bulls (Geay, Robelin and Jarrige, 1976).

These large discrepancies in k_{pf} between bulls and heifers may be due to two causes, the relative importance of which could not be determined:

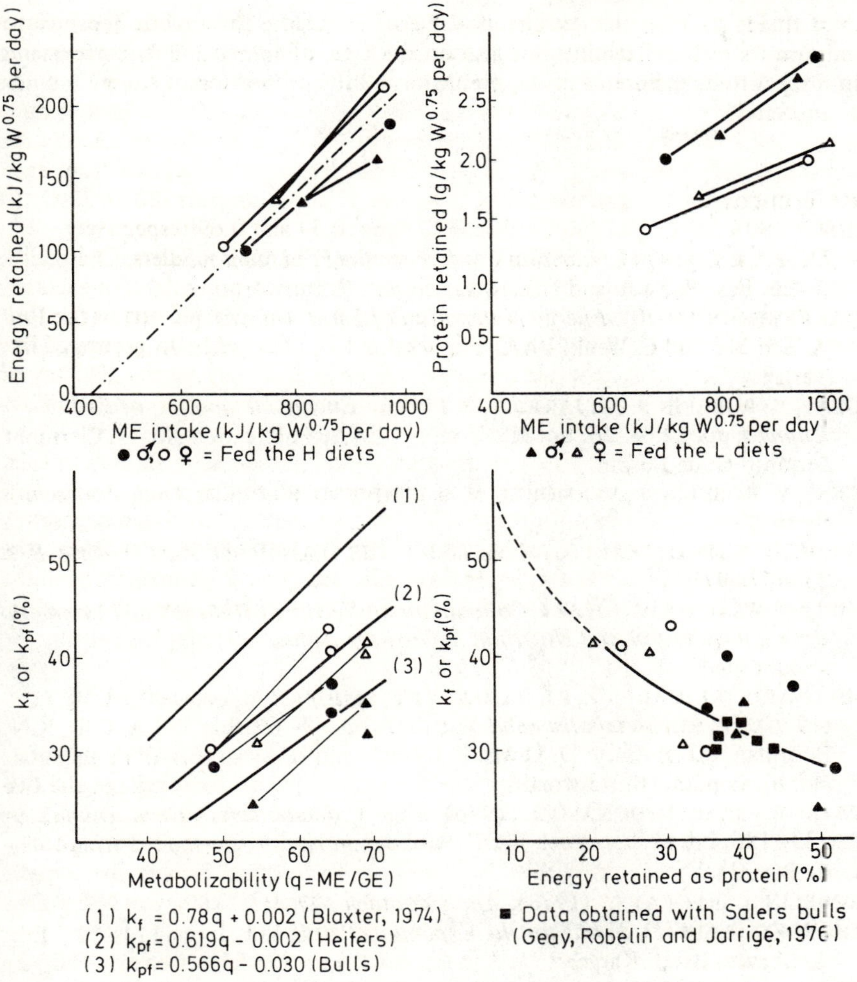

Figure 2.1 The retentions of energy and protein in bulls and heifers and efficiencies of energy utilization

1. A too low value for the maintenance requirement of young bulls in relation to their rapid growth (Graham, Searle and Griffiths, 1974) and their considerable protein synthesis (Millward *et al.*, 1976). Bulls have higher maintenance requirements than steers (Webster, 1976), which have the same requirements as heifers (Garrett, 1970). However, to obtain the same k_f value as that proposed by Blaxter (1974) the maintenance requirements of bulls would have to be increased by 15–45% and that of heifers by 9–16%.
2. A higher proportion of energy retained as protein (44% versus 28.9%), which could be responsible for a lower ME efficiency as has been shown in sheep (Ørskov and McDonald, 1976) and in pigs (Thorbek, 1972). A regular and curvilinear decrease in k_{pf} occurs when the proportion of energy retained as protein (x) increases (*Figure 2.1*). When x = o, k_{pf} = 0.65, so that the energy cost of lipid deposition amounts to 60.5 MJ ME/g, a value similar to the 71 MJ proposed by Kielanowski (1976).

If this is so, then the net energy value of feedstuffs for growth depends not only on its metabolizability but also on the type of animal and its performance in respect to its maintenance requirements and the proportion of energy retained as protein.

References

BLAXTER, K.L. (1974). *Nutrition Conference for Feed Manufacturers: 7*, pp. 3–26. Eds. H. Swan and D. Lewis. London: Butterworths

GARRETT, W.N. (1970). *Energy Metabolism of Farm Animals*, pp. 101–104. Eds. A. Schürch and C. Wenk. EAAP Publication No. 13. Zurich: Juris Druck and Verlag

GEAY, Y., ROBELIN, J. and JARRIGE, R. (1976). *Energy Metabolism of Farm Animals*, pp. 225–228. Ed. M. Vermorel. EAAP Publication No. 19. Clermont Ferrand: G. de Bussac

GEAY, Y., ROBELIN, J., VERMOREL, M. and THIVEND, P. (1980). *Annls Zootech.* In press

GRAHAM, N. Mc. C., SEARLE, T.W. and GRIFFITHS, D.A. (1974). *Aust. J. agric. Res.*, **25**, 957–971

KIELANOWSKI, J. (1976). *EEC Colloquium on Feeding Efficiency and Interaction Between Genotype and Nutrition in Growing Animals*, Theix, France, unpublished

MILLWARD, D.J., GARLICK, P.J., JAMES, W.P.T., SENDER, P.M. and WATERLOW, J.C. (1976). *Protein Metabolism and Nutrition*, pp. 49–69. Eds. D.J.A. Cole, K.N. Boorman, P.J. Buttery, D. Lewis, R.J. Neale and H. Swan. EAAP Publication No. 16. London: Butterworths

ØRSKOV, E.R. and McDONALD, I. (1976). *Energy Metabolism of Farm Animals*, pp. 133–136. Ed. M. Vermorel. EAAP Publication No. 19. Clermont Ferrand: G. de Bussac

ROBELIN, J. and GEAY, Y. (1978). *Annls Zootech.*, **27**, 159–167

THORBEK, G. (1972). *Foods for the Expanding World*, Vol. 3, pp. 114–121. Ed. A. Chavez. Basel: Karger

VAN ES, A.J.H. (1975). *Live Stk Prod. Sci.*, **2**, 95–107

WEBSTER, A.J.F., SMITH, J.S. and MOLLISON, G. (1976). *Energy Metabolism of Farm Animals*, pp. 221–224. Ed. M. Vermorel. EAAP Publication No. 19. Clermont Ferrand: G. de Bussac

3

THE EVALUATION OF THE ENERGY IN FORAGE BY THE COMPARATIVE SLAUGHTER TECHNIQUE

D.J. THOMSON and S.B. CAMMELL
The Grassland Research Institute, Hurley, Maidenhead, Berkshire SL6 5LR, England

Summary

The efficiency of utilization of energy for growth (k_f) on a range of pelleted red clover (*Trifolium pratense*) and pelleted perennial ryegrass (*Lolium perenne*) diets was determined using worm-free lambs. Energy retention was measured by the comparative slaughter technique. Lambs were fed at several planes of nutrition above maintenance (including *ad lib.*) and efficiency was calculated by regression analysis of scaled (kg $LW^{0.75}$) values of intake and retention. The results indicated separate relationships between ME (metabolizable energy) required for maintenance, ME per unit dry matter (DM) or *in vitro* digestibility and k_f for the pelleted red clover and pelleted grass diets. All k_f values were lower than previously published calorimetric estimates of k_f for pelleted diets.

Materials and methods

Forage plant breeders have supplied new tetraploid varieties of red clover (*Trifolium pratense*) with improved disease resistance characteristics and with annual yields of DM 30% higher than in previously available varieties. Little was known about the nutritional value of these new plant materials. The efficiency of utilization of ME for growth and fattening (k_f) of a range of red clover and perennial ryegrass (*Lolium perenne*) diets was determined. The forages were conserved by high temperature dehydration and were ground (3 mm screen) and pelleted.

Energy retention (ER) was measured by the comparative slaughter technique (CST) using worm-free lambs which were reared artificially from birth and weaned at five weeks of age (Thomson and Cammell, 1979). The lambs were fed at 2, 3 or 4 planes of nutrition above maintenance (occasionally including *ad lib.*) for periods of 70–100 days between eight and 20 weeks of age. Energy digestibility (DE) was measured on each lamb during the course of individual experiments and the ME content of the diets was derived using urine energy and a predicted value of methane (Blaxter and Clapperton, 1965) or from DE × 0.81. The initial energy content of the lambs was estimated from fasted (18 h) live weight (LW) using prediction equations obtained from sample groups of lambs slaughtered at the beginning of experiments. The final energy content was obtained from the summated energy contribution of carcass, gut, wool, skin and blood and the ER obtained was the difference between the initial and the final content. The daily ER and ME intake values were scaled (kg $LW^{0.75}$) and k_f calculated by regression analysis of the scaled values. Resolution by linear regression was obtained in all instances with the exception of two red clover and one ryegrass diet from which positive intercepts were calculated. For these diets exponentials were fitted and the mean of the slopes generated at each individual point was taken as the k_f value for that diet.

Results and discussion

Values for ME at the maintenance level of feeding (q) were obtained by extrapolation from the linear regression of ME and plane of nutrition (defined as multiples of maintenance) using a maintenance energy requirement value of 340 kJ ME/$kg^{0.75}$ per d (Thomson, Fenlon and Cammell, 1979). The decline in ME per unit (multiple of maintenance) change in level of feeding was on average 0.028 for pelleted red clover, compared with 0.108 for pelleted ryegrass.

The estimation of initial total body energy content from fasted LW was based on linear regression equations obtained from sample groups (15–20 lambs) slaughtered on day one of the experiments. Of the eight separate initial slaughter group equations examined, two sets, each of two equations, were able to be pooled, but some individual experiment regressions were distinct and preferred for the estimation of initial body energy content. The initial energy content of lambs was estimated from fasted LW with, on average, an accuracy (RSD) of ± 10 – 12 MJ. Individual regression equations did have RSDs as low as ± 4.8 MJ (23 lambs); smaller groups, or factors influencing gut fill (and therefore fasted LW) or the energy content of body weight (birth weight, sex, check at weaning and rate of gain from birth), led to higher errors associated with the estimation of initial energy content.

Knowledge of the errors associated with the estimation of the initial energy content, the small increments of retained energy, and the ER at *ad lib.*, where between-animal variation appeared to be higher, influenced both design and replication for later estimates of k_f. Live-weight gains of less than 60–80 g/d at the lowest levels of feeding were avoided, and animal replication was increased at the highest (*ad lib.*) and lowest compared with intermediate planes of nutrition

Table 3.1 VALUES FOR THE EFFICIENCY OF UTILIZATION OF METABOLIZABLE ENERGY (ME) FOR GROWTH AND FATTENING (k_f) FOR PELLETED RED CLOVER AND PELLETED RYEGRASS IN RELATION TO THE ME CONCENTRATION IN THE DRY MATTER (M/D)

M/D (MJ/kg DM)	*9*	*10*	*11*	*12*
Pelleted red clover	0.32	0.42	0.51	–
Pelleted ryegrass	–	0.37	0.40	0.42

The k_f values for the pelleted red clover and ryegrass diets were related to *in vivo* ME at q, M/D (ME concentration in the dry matter, MJ/kg DM) or *in vitro* digestibility (Tilley and Terry, 1963). The k_f values for the two categories of forage at a range of M/D values are given in *Table 3.1*.

The results indicated different relationships for the two forages with the k_f values being higher for the red clover than the ryegrass. Additional information is required for a more precise definition of the response in k_f with change in digestibility. The values, determined by CST using young lambs, were substantially lower than those reported by Blaxter (1974) which were measured by respiration calorimetry using mature adult sheep. The information on body composition and the energy and protein contained in the carcass were used to calculate the yields of edible energy and protein per unit area for red clover and ryegrass forages.

References

BLAXTER, K.L. (1974). *Nutrition Conference for Feed Manufacturers:* 7, pp. 3–25. Eds. H. Swan and D. Lewis. London: Butterworths

BLAXTER, K.L. and CLAPPERTON, J.L. (1965). *Br. J. Nutr.,* **19**, 511–522

THOMSON, D.J. and CAMMELL, S.B. (1979). *Br. J. Nutr.,* **41**, 297–310

THOMSON, D.J., FENLON, J.S. and CAMMELL, S.B. (1979). *Br. J. Nutr.,* **41**, 223–229

TILLEY, J.M.A. and TERRY, R.A. (1963). *J. Br. Grassld Soc.,* **18**, 104–111

4

ENERGY UTILIZATION IN SHEEP FED GRASS SILAGE AND HAY

F. SUNDSTØL and A. EKERN
Department of Animal Nutrition, Agricultural University of Norway, Institute of Animal Nutrition, 1432, Ås-NLH, Norway
P. LINGVALL, E. LINDGREN and J. BERTILSSON
Department of Animal Husbandry, Swedish University of Agricultural Sciences, Uppsala, Sweden

Summary

The content of metabolizable energy (ME) and the utilization of the ME in silage and hay were measured in two series of respiration experiments. The content of ME for grass clover cut at ear emergence and fed at maintenance was 10.4–10.8 MJ/kg dry matter (DM) for silage and 10.0–10.5 MJ/kg DM for hay. When fed at the production level the figures were 11.3 and 11.2 respectively. The utilization of ME for maintenance (k_m) was calculated from both series of experiments including a series of starvation experiments. The following k_m values were obtained (n = 17):

Grass silage	0.57
Hay	0.60
Silage + concentrate	0.75
Hay + concentrate	0.65

From the present investigation it can be concluded that the favourable effect of silage as compared to hay found in milk production experiments can hardly be explained fully by a better utilization of the feed energy.

Introduction

In the Nordic countries cattle, sheep and goats are normally stall-fed on conserved forages for 7–9 months of the year. In Norway more than 70% of the harvested grass is conserved as direct-cut silage, while in Sweden haymaking is still the dominant method for grass conservation.

Materials and methods

The forages used in this investigation were produced in Sweden. They were used in experiments with high-yielding dairy cows and in digestibility experiments with sheep. Part of the forage was transported to the Agricultural University of Norway and used in energy-balance experiments.

EXPERIMENT 1

The first series of respiration experiments were performed in 1975–76 with barn-dried hay and formic acid silage from a grass-clover ley cut in 1975 at Kungsängen. The experiments were carried out at about the maintenance level of feeding and the following plan was applied:

1.	Early-cut silage, cut at ear emergence	(a) without concentrate
		(b) with concentrate
2.	Late-cut silage, cut 10 days later than 1	(a) without concentrate
		(b) with concentrate
3.	Early-cut hay, cut at the same time as 1	(a) without concentrate
		(b) with concentrate

All forages were fed alone and in combination with concentrate (50:50 on a DM basis) making a 3 × 2 factorial plan with two wethers on each diet.

EXPERIMENT 2

The forages used in this experiment were produced at Röbäcksdalen Experimental Station in 1976. The grass was cut at ear emergence and conserved as hay and direct-cut formic acid silage. The following plan was used in the digestion and energy balance experiments:

1.	Grass silage	(a) without concentrate	(i) maintenance level of feeding
			(ii) production level of feeding
		(b) with concentrate	(i) maintenance level of feeding
			(ii) production level of feeding
2.	Hay	(a) without concentrate	(i) maintenance level of feeding
			(ii) production level of feeding
		(b) with concentrate	(i) maintenance level of feeding
			(ii) production level of feeding

Three animals were used for each treatment in the 2 × 2 × 2 factorial plan. Fasting metabolism was determined in the six young rams at the start, in the middle and at the end of the experiments.

Results

EXPERIMENT 1

The chemical composition of the feeds used in Experiment 1 is shown in *Table 4.1.* In the early-cut silage the DM contained more protein and crude fibre than late-cut silage and early-cut hay. The higher content of crude fibre in early-cut silage was unexpected and does not correspond with Swedish analyses of the same feeds showing highest crude fibre in late-cut silage.

The amount of feed supplied was calculated from energy requirements given by the Agricultural Research Council (1965), and the content of ME was estimated from digestion coefficients obtained for the same feeds determined in

Table 4.1 CHEMICAL COMPOSITION OF FEEDS (EXPERIMENTS 1 AND 2)

	Experiment 1				*Experiment 2*		
	Grass silage, early cut	*Grass silage, late cut*	*Hay, early cut*	*Concentrate*	*Grass silage*	*Hay*	*Concentrate*
Dry matter (%)	25.1	29.0	91.4	90.6	19.3	88.0	88.9
In 100 g DM:							
Ash (g)	6.8	5.9	7.8	4.0	7.7	8.1	2.1
Crude protein (g)	13.2	11.0	11.8	21.9	15.3	15.0	13.6
Ether extract (g)	5.2	4.5	2.8	3.9	4.7	2.6	2.6
Crude fibre (g)	32.7	28.3	28.2	8.2	29.6	27.5	4.9
N-free extracts (g)	42.1	50.3	49.4	62.0	42.7	46.8	76.8
Energy (MJ)	1.86	1.84	1.86	1.92	1.94	1.87	1.87
Organic acids (g)	6.7	6.4			10.8		
NH_3-N (g/100 g N)	4.8	4.4			3.6		
pH	3.8	3.7			3.6		

sheep in Sweden. The latter was 11.1 MJ/kg DM for early-cut silage, 9.92 MJ/kg DM for late-cut silage and 10.63 MJ/kg DM for early-cut hay.

The energy metabolism measured in Experiment 1 is presented in *Table 4.2*. The digestibility of energy was similar for early-cut silage and hay, whereas the digestibility of late-cut silage was significantly lower. The same tendency was found for mixed diets of the forages and concentrate. No difference in energy losses in urine and methane between the forages was observed, but the energy in methane as a percentage of energy intake (IE) tended to be higher when

Table 4.2 ENERGY METABOLISM IN WETHERS FED TWO CUTS OF CLOVER-GRASS SILAGE AND ONE CUT OF HAY. BOTH TYPES OF FORAGES FED ALONE AND IN COMBINATION WITH CONCENTRATE (50:50 ON A DM BASIS; EXPERIMENT 1)

	Forage (fed alone)			*Forage plus concentrate*		
	Grass silage, early cut	*Grass silage, late cut*	*Hay, early cut*	*Grass silage, early cut*	*Grass silage, late cut*	*Hay, early cut*
Number of observations	2	2	2	2	2	2
Live weight (kg)	71.0	72.0	68.8	70.5	66.0	65.3
Forage (g DM/d)	723.0	800.0	759.0	361.0	400.0	379.0
Concentrate (g DM/d)	–	–	–	356.0	406.0	374.0
Energy intake (IE; MJ/d)	13.43	14.72	14.09	13.56	15.17	14.25
Faeces energy	3.70	4.93	4.07	3.53	3.40	3.13
(FE; MJ/d (% of IE))	(27.6)	(33.5)	(28.9)	(23.0)	(24.1)	(22.1)
Digestible energy	9.73	9.79	10.02	10.03	11.77	11.12
(DE; MJ/d (% of IE))	(72.4)	(66.5)	(71.1)	(77.0)	(75.8)	(77.9)
Urine energy	0.76	0.69	0.80	0.75	0.73	0.65
(UE; MJ/d (% of IE))	(5.7)	(4.7)	(5.7)	(5.5)	(4.9)	(4.5)
Methane energy	1.13	1.21	1.23	1.28	1.46	1.31
(CE; MJ/d (% of IE))	(9.7)	(8.4)	(8.4)	(9.4)	(10.7)	(10.1)
Metabolizable energy	7.84	7.89	7.99	8.00	9.58	9.16
(ME; MJ/d (% of IE))	(57.0)	(53.6)	(57.0)	(62.0)	(60.3)	(63.2)
Heat production (H; MJ/d)	8.99	9.09	8.23	8.17	10.00	9.56
Energy retention (MJ/d)	–1.15	–1.20	–0.24	–0.17	–0.42	–0.40

concentrate was fed. The content of ME was lower for late-cut silage (9.9 MJ/kg DM) than for early-cut silage (10.8 MJ/kg DM) and early-cut hay (10.5 MJ/kg DM).

EXPERIMENT 2

In Experiment 2 the chemical composition and digestibility were not known when the respiration experiments started. This contributed to a feed intake which deviated somewhat from that planned, which again resulted in slightly positive or negative energy retention even at the production level of feeding. As in Experiment 1 there was no difference in energy digestibility between early-cut silage and hay. Nor was there any difference between silage and hay with regard to energy losses in urine or methane (as percentages of IE). However, urinary losses tended to be higher and methane losses lower when the forages were fed alone in contrast to when fed with a concentrate. At the maintenance level of feeding the content of ME/kg DM was 10.4 MJ and 10.0 MJ for silage and hay respectively, whereas the corresponding figures at the production level of feeding were 11.3 MJ and 11.2 MJ.

Discussion

From the regression equation between retained energy per $W^{0.75}$ and ME/$W^{0.75}$ the following coefficients for utilization of the ME were obtained for the early-cut forages of Experiments 1 and 2:

Starvation – maintenance	(k_m), hay	0.60 (n = 17)
Starvation – maintenance	(k_m), silage	0.57 (n = 17)
Starvation – maintenance	(k_m), hay + concentrate	0.65 (n = 17)
Starvation – maintenance	(k_m), silage + concentrate	0.75 (n = 17)
Maintenance – production	(k_f), hay	0.41 (n = 8)
Maintenance – production	(k_f), silage	0.27 (n = 8)
Maintenance – production	(k_f), hay + concentrate	0.37 (n = 8)
Maintenance – production	(k_f), silage + concentrate	0.50 (n = 8)

From the results obtained no significant differences between corresponding hay and silage diets were found with regard to the utilization of ME for maintenance and for fattening. For the production level of feeding the intake was generally too low and the number of observations small, which caused great variation in results obtained.

This is in accordance with results previously obtained by van Es and Nijkamp (1969), van der Honing *et al.* (1973) and Ekern and Sundstøl (1974). Lactation studies with dairy cows in Sweden showed that grass silage gave 5–15% higher yield of 4% fat-corrected milk than did hay cut at the same time (Lingvall, Knutsson and Bertilsson, 1978).

References

AGRICULTURAL RESEARCH COUNCIL (1965). *The Nutrient Requirement of Farm Livestock, No. 2: Ruminants.* HMSO: London

EKERN, A. and SUNDSTØL, F. (1974). *Energy Metabolism of Farm Animals, 6 Symposium, Stuttgart*, pp. 221–224. Eds. K.H. Menke, H.-J. Lantzsch and J.R. Reichl. EAAP Publication No. 14. Dokumentationsstelle: Universität Hohenheim

VAN ES, A.J.H., and NIJKAMP, H.J. (1969). In *Energy Metabolism of Farm Animals, 4 Symposium, Jablonna*, pp. 209–212. Eds. K.L. Blaxter, D. Thorbek and J. Kielanowski. EAAP Publication No. 12. Newcastle: Oriel Press

VAN DER HONING, Y., VAN ES, A.J.H., NIJKAMP, H.J. and TERLUIN, R. (1973). *Z. Tierphysiol., Tierernähr. Futtermittelk.*, **31**, 149–158

LINGVALL, P., KNUTSSON, P.-G. and BERTILSSON, J. (1978). European Association for Animal Production (EAAP), 29th Annual Meeting, Stockholm

5

CALCULATION OF ENERGY VALUE OF FEEDSTUFFS FOR MILK PRODUCTION: COMPARISON OF VARIOUS FEED EVALUATION SYSTEMS

Y. VAN DER HONING and A. STEG
Research Institute for Livestock Feeding and Nutrition, Postbox 160, 8200 AD Lelystad, The Netherlands

Summary

Based on a given chemical composition and the same digestibility coefficients per feedstuff, the energy values were calculated for 32 feedstuffs according to eight feed evaluation systems for milk production. Comparison of feeding values between systems were made directly and relative to the values of barley or a standard ration. Production potential, for maintenance and milk, of a given amount of a standard ration in the various systems was also studied. Equations to convert energy values from one system into another were derived and their precision discussed. With regard to both energy value and production potential, Kellner's starch values deviated most among all systems considered.

Introduction and methods of comparison

Recently several countries have changed to a new feed evaluation system for dairy cows. For many reasons people from dairy farms, feed industries and research institutes wish to compare old and new systems and their own system with those in other countries. In our original study (Steg and van der Honing, 1979) we compared 12 systems, but in this chapter the limited space available allows only a comparison of eight systems: namely Kellner's starch equivalent (STEK), total digestible nutrients (TDN), metabolizable energy in UK according to Technical Bulletin 33 (ME33), metabolizable energy according to the system in Sweden (MESW), the net energy for lactating cows from NRC 1978 (NEL8), the Dutch feed unit system (VEM), the new French feed unit system (UFL) and the system from the GDR, Rostock (EFR). For details about the equations and method of calculation see Steg and van der Honing (1979).

To ensure a good investigation into the differences between systems, a wide range of 32 feedstuffs was examined. Most of these are quantitatively important in dairy feeding but some, like fat, straw, brewer's grains and tapioca, were chosen to emphasize the differences between systems.

Feedstuffs having the same name in different countries may still have differences in chemical composition and digestibility coefficients. Variation in net energy could be due to these differences as well as to the methods of prediction of metabolizable energy (ME) and net energy. The aim of this study was to compare the methods of prediction, excluding variation due to differences in chemical composition and digestibility coefficients. The calculations therefore started with the same input for all systems, and were mainly adopted from the Dutch feeding tables.

The various energy values were not only computed and compared as such but also relative to a standard ration (SR) composed of roughages (20% good quality grass hay, 20% wilted grass silage, 10% high-dry-matter maize silage) and concentrate ingredients (7.5% maize grain, 5% barley, 5% maize gluten feed, 10% beet-pulp, 5% tapioca, 4% cane molasses, 5% coconut expeller, 5% extracted soybean meal and 2.5% rapeseed meal) and 1% of a minerals + vitamins mixture. The calculated composition per kilogram of dry matter (DM) of this SR was 162 g crude protein (XP), 33 g ether extract (XL), 165 g crude fibre (XF) and 543 g N-free extract. The calculated energy values were: STEK= 644, TDN = 0.749 kg, EFR = 616, MESW = 2.733 Mcal (11.43 MJ), ME33 = 11.42 MJ, NEL8 = 1.716 Mcal (7.180 MJ), VEM = 955 and UFL = 0.967.

The productive potential of a certain quantity of a feedstuff in a system depends not only on the net energy value in that system but also on the standards for maintenance and milk production. The amount of SR required for maintenance and to produce 1 kg of milk have, therefore, also been calculated.

Results and discussion

Calculation of ME values showed rather similar averages for all the feedstuffs for EFR, MESW, ME33 and UFL (11.34–11.43 MJ/kg), but VEM was lower (11.05 MJ/kg) and NEL8 was higher (12.09 MJ/kg). Differences in the regression coefficient of digestible crude protein and digestible N-free extract and the effect of a varying ME/DE ratio used in the predictions are at least partly responsible for this.

Table 5.1 shows the energy values of the feedstuffs according to eight systems expressed as a percentage of the energy value of SR in that system. Averages of these relative values for 31 feedstuffs (fat excluded) vary slightly from 93.0 (STEK) to 96.3 (MESW), but the standard deviations are appreciably different: high for STEK (27), low for NEL8, TDN and EFR (16–17) and intermediate for VEM, UFL, ME33 and MESW (18–20). A higher standard deviation in general is caused by lower relative values for poorly digestible roughages and higher values for well digestible concentrate ingredients. Some products such as fodderbeets and molasses have lower values due to a large subtraction for sugar in some systems. In MESW the values of crude protein-rich feedstuffs differ from those in other systems due to a high coefficient for protein in the predicting equation.

There is much similarity between ME33, TDN and NEL8, but there are appreciable differences for some products in other systems, although positive and negative deviations result in almost the same average relative value if fat is not included. Striking differences between VEM and UFL are found for maize silage, brewer's grains, coconut expeller and fodderbeets due to the estimate of gross energy, the prediction of ME or a correction for sugar. The value for fat in the systems is calculated by extrapolation resulting in great differences in relative value: high for EFR, intermediate for VEM and STEK and low for the

Table 5.1 ENERGY VALUES OF DIFFERENT FEED EVALUATION SYSTEMS RELATIVE TO STANDARD RATION = 100

Feedstuff	*STEK*	*TDN*	*EFR*	*MESW*	*ME33*	*NEL8*	*VEM*	*UFL*
Fresh grass, early cut	108.7	104.3	103.7	105.0	102.6	104.5	107.4	105.7
Fresh grass, late cut	89.3	91.1	91.1	88.2	88.2	90.3	87.1	87.4
Grass silage, unwilted	93.5	92.8	95.3	87.1	90.8	92.3	86.8	90.8
Grass silage, unwilted late	78.1	82.0	83.9	75.4	79.2	80.7	74.6	76.8
Wilted grass silage	87.6	93.1	94.5	88.2	91.0	92.5	88.5	90.2
Wilted grass silage, moderate quality	73.0	82.2	83.3	77.2	79.9	81.0	76.3	77.8
Grass hay, good quality	77.8	88.8	88.5	86.4	86.3	87.9	84.8	87.0
Grass hay, moderate quality	64.3	79.4	79.1	75.7	76.7	78.0	74.6	75.5
Fresh alfalfa, early cut	83.2	85.2	82.8	88.2	82.8	84.1	86.2	84.8
Fresh alfalfa, late cut	69.3	76.5	74.5	74.6	73.6	74.8	74.3	73.7
Alfalfa hay, medium quality	56.7	74.9	72.2	72.8	72.0	73.1	72.7	70.8
Maize silage, milky stage	92.2	91.5	92.5	86.4	90.1	90.7	89.8	85.3
Maize silage, dough stage	97.8	95.6	95.9	91.5	95.2	95.2	95.8	91.5
Wheat straw	22.2	54.3	54.5	49.0	50.6	51.1	46.5	45.0
Barley straw	31.5	60.9	61.2	54.9	56.8	58.0	53.0	51.9
Fodderbeets	98.4	113.2	110.1	114.2	116.7	114.0	113.5	99.8
Barley (grain)	126.9	113.5	112.5	117.5	117.1	114.5	117.9	118.7
Maize (grain)	138.2	121.1	123.7	124.4	125.4	122.6	127.9	126.5
Peas	128.1	114.4	108.9	121.5	116.5	115.4	119.1	118.5
Wheat bran	88.0	96.4	98.5	101.0	98.2	96.0	97.2	94.2
Maize gluten feed	120.5	110.3	112.5	116.0	111.6	111.0	114.9	114.4
Beet pulp	95.5	107.1	103.4	105.7	107.3	107.5	107.7	107.0
Beet molasses	85.1	107.2	103.1	110.5	111.2	107.6	108.5	113.3
Cane molasses	79.8	94.0	92.0	95.5	98.1	93.6	91.8	95.4
Brewers grains	89.3	93.2	102.8	95.1	93.1	92.7	94.1	85.8
Citrus pulp	116.0	113.4	114.8	112.0	114.6	114.2	116.0	118.5
Tapioca	124.2	107.9	105.5	109.4	112.3	108.4	111.3	108.6
Coconut expeller	132.8	117.1	129.4	122.6	116.9	118.3	123.5	117.9
Coconut meal solvent-extracted	110.2	100.9	99.4	105.7	100.8	100.9	101.7	99.8
Rapeseed meal solvent-extracted	100.0	93.9	89.0	104.3	94.7	93.4	94.7	94.9
Soybean meal solvent-extracted	124.8	114.3	104.4	128.1	114.6	115.2	120.1	121.6
Fat (vegetable origin)	355.1	285.0	463.5	305.5	284.2	297.9	369.0	303.8
Average (n=31) (fat excluded)	93.0	95.8	95.6	96.3	95.6	95.5	95.4	94.5
s.d. (fat excluded)	27.3	16.0	16.4	19.6	18.2	17.2	19.9	19.7

other systems. Conclusions from these values have to be very restricted because of lack of experimental basis with high-fat diets.

Table 5.2 shows the amount of SR required for maintenance and for production of fat-corrected milk (FCM) according to the standards described in the system; in addition the potential production, maintenance included, from 10, 15 or 20 kg of SR was calculated for a 550 kg cow.

The dry matter (DM) of SR needed for maintenance varied considerably (4.2–5.3 kg), but the per kilogram FCM variation was less, although the extremes still differed by 15%. Compared to VEM and UFL the potential is over-

Table 5.2 REQUIREMENT FOR MAINTENANCE AND MILK PRODUCTION AND POTENTIAL MILK YIELD FROM SR BY A 550 kg COW

	STEK	*TDN*	*EFR*	*MESW*	*ME33*	*NEL8*	*VEM*	*UFL*
SR (kg) required for:								
maintenance (W=50 kg)	4.7	5.3	4.8	5.0	5.1	5.3	4.9	4.7
1 kg of milk (4% fat)	2.34	2.30	2.16*	2.28	2.15	2.32	2.16*	2.25*
Potential milk yield, (kg FCM), including maintenance, from:								
10 kg SR	12.5	10.7	11.3	11.3	10.5	10.9	10.9	11
15 kg SR	24.2	22.2	22.1	22.7	21.3	22.5	21.2	21
20 kg SR	35.9	33.7	32.9	34.1	32.1	34.1	31.1	31

* Without additional corrections for feed level or ration composition

estimated by STEK, but below a production of 25 kg of FCM (about 15 kg of SR) there are only small differences between VEM and UFL, and TDN, EFR, NEL8, ME33 and MESW. Above 25 kg of FCM in VEM the requirements are higher than in TDN, EFR, NEL8 and MESW.

Prediction of the energy value in a system from other systems

By linear regression using 31 or 32 (fat included) feedstuffs, equations were derived to convert the value from one system into that of another. Additional independent variables such as XP, XF and XL were used to improve the accuracy of prediction. In general, one or more of these variables did improve the precision of the conversion equation, although in some cases precision was already good without an additional variable. Prediction of STEK from all other systems was poor and the coefficient of variation (CV) remained at 8–10%, because value numbers and crude fibre corrections have a great influence. As NEL8 is derived from TDN, conversion is easy and accurate. Prediction of TDN and NEL8 from EFR, MESW, ME33 or VEM is accurate (CV $<$ 2.5%) and so is prediction of EFR from TDN, NEL8 and ME33, especially when XL is used as an additional variable. A poor prediction of VEM is obtained from STEK, and from UFL it also showed a rather high CV of 3.7–3.9%. Other systems served as good predictors for VEM with variable additional effects of XP, XL and XF. However, UFL cannot be predicted very precisely from other systems (CV 3.7–4.4%), and the difference in the calculation of ME and the value of fodderbeets is probably the reason for this. Details about the equations and their precision are presented in the report by Steg and van der Honing (1979).

References

STEG, A. and VAN DER HONING, Y. (1979). *Relationships Between Energy Values of Feedstuffs Predicted with Several Feed Evaluation Systems.* Doc. Report No. 49. I.V.V.O.: Lelystad, The Netherlands

6

INFLUENCE OF DIETARY COMPOSITION ON SITE OF ABSORPTION AND EFFICIENCY OF UTILIZATION OF METABOLIZABLE ENERGY IN GROWING PIGS

A. JUST
National Institute of Animal Science, Rolighedsvej 25, 1958 Copenhagen V, Denmark

Summary

The dietary factors influencing the amount of metabolizable energy (ME), site of absorption and efficiency of utilization of metabolizable energy were investigated using diets varying widely in chemical composition. Crude fibre depresses the amount and utilization of ME, excess digestible protein depresses metabolizability and utilization of ME and crude fat improves the utilization of ME. A positive relationship was found between energy absorption in the small intestine and the efficiency of utilization of ME.

Introduction

A proper evaluation of feedstuffs used for pigs is very important as the feed accounts for more than two thirds of the total production cost. It is, therefore, very important that all possible efforts are made to maximize the utilization of the feed.

Several workers have found a negative relationship between digestibility of gross energy (GE) and the concentration of crude fibre in the feedstuffs, whereas increasing concentrations of crude protein and crude fat generally improve the digestibility of GE. In addition Just (1970, 1975) showed that the efficiency of utilization of ME increases with increasing energy concentration.

In order to establish more accurately the quantitative relationship between energy concentration and the utilization of ME and to elucidate the background for that relationship, a number of investigations were planned. In the following, the results obtained so far are summarized, and more detailed descriptions will be published elsewhere.

Materials and methods

The effect of the concentrations of different dietary nutrients on digestibility and metabolizability was studied in four separate balance experiments, and each comprised six litters of seven pigs. Initially one pig from each litter was sacrificed and the littermates distributed – one in each of six groups. The pigs were fed twice daily and three balance experiments were conducted with each pig during the growth period from 20 to 90 kg live weight. Each balance period lasted seven days and faeces and urine were collected quantitatively twice daily.

Deposited energy was determined using the comparative slaughter technique as described by Just (1970, 1975). Net energy (NE) was calculated by adding 78 kcal/kg live weight$^{3/4}$ (326 kJ/kg live weight$^{3/4}$) to deposited energy. The live weights were adjusted for differences in gut fill (Just, Rasmussen and Hansen, 1976).

The absorption in the small and large intestine, respectively, was studied using ileal-caecal re-entrant cannulated pigs weighing 50–70 kg. Digesta were collected through soft plastic tubing attached to the ileal cannula. The tubing was kept immersed in a container filled with ice. As soon as digesta passed through the ileal cannula into the tubing it was squeezed by hand until it was below the ice level in the container.

Four experiments were carried out according to a 3 × 3 Latin Square design. Each test period lasted seven days. Equal amounts of feed were given three times daily at 0700 h, 1500 h and 2300 h. Chromic oxide was used as a marker.

The diets were composed of barley, maize, wheat bran, oat bran, maize starch, ground barley straw, animal fat, soya bean meal, meat and bone meal, skimmilk powder, sugar, lysine, methionine, minerals and vitamins, and they were formulated so that they both varied considerably in chemical composition and fulfilled the requirements of the pigs for essential nutrients. The feedstuff analysis included determination of soluble carbohydrate, and the fat analyses included a hydrochloric acid hydrolysis before ether extraction (Stoldt, 1957).

Results and discussion

The main results of the concluded investigations are summarized in *Tables 6.1, 6.2* and *6.3.* The average results are given only for the two most extreme diets, but these are in accordance with the results of statistical analyses.

The results in *Table 6.1* clearly demonstrate the negative effect of crude fibre originating from cereals on digestibility of GE and on the utilization of ME. The explanation for the negative effect of crude fibre on digestibility is a highly significant negative relationship between crude fibre and the content of soluble carbohydrates (starch) in the nitrogen-free extract substances. The soluble carbohydrates are almost 100% digestible, whereas the remaining part of the nitrogen-free substances (mostly hemicelluloses) has a much lower digestibility.

The depressive effect of crude fibre from barley straw on digestibility is much smaller than that of crude fibre from cereals, whereas the depressive effect on utilization of ME is considerably larger. Crude fat had a positive influence on utilization of ME, but excess protein, as expected, had a negative influence on both the amount and the utilization of ME.

A few of the results obtained with re-entrant cannulated pigs are given in *Table 6.2.* The intention was to feed the cannulated pigs diets composed similarly to those used for the balance experiments, but it became necessary to reduce the crude fibre levels since the high concentrations of crude fibre blocked the cannulas. The pattern of the ileal and faecal digestibility coefficients (DC) for energy is similar to that in *Table 6.2*, but the ileal data are generally vitiated

Table 6.1 THE INFLUENCE OF CRUDE FIBRE, CRUDE FAT AND CRUDE PROTEIN ON DIGESTIBILITY OF GROSS ENERGY AND EFFICIENCY OF UTILIZATION OF ME

	Group 1	*Group* 6	*b*	s_b	*CV*	r^2
% crude fibre[1], cereals	3	16				
DE, % of GE	89	50	−2.87	0.06	2.1	0.99
NE, % of ME	65	59	−0.39	0.07	2.8	0.76
% crude fibre[1], straw	5	15				
DE, % of GE	82	61	−2.09	0.08	2.2	0.96
NE, % of ME	66	59	−0.62	0.13	4.4	0.63
% crude fat[1], animal	4	24				
DE, % of GE	81	80	−0.07	0.04	1.7	0.39
NE, % of ME	63	69	0.33[2]	0.07	3.4	0.65
% crude protein[1]	13	29				
DE, % of GE	79	83	0.20	0.05	3.4	0.80
ME, % of DE	97	94	−0.27[2]	0.01	0.3	0.97
NE, % of ME	62	57	−0.42[2]	0.01	4.4	0.57

[1] In diet dry matter [2] Digestible nutrients
b = Change per percent of nutrient in diet dry matter

Table 6.2 THE INFLUENCE OF DIET COMPOSITION ON ILEAL AND FAECAL DIGESTIBILITY

	Group 1		*Group 3*		*F value for diets*		
	Ileum	*Overall*	*Ileum*	*Overall*	*Ileum*	*Overall*	*Difference*
% crude fibre[1], cereals	4		8				
DC protein	78	89	74	82	10	34	7
DC fibre	3	63	3	56	0	0	1
DC carbohydrates[2]	97	100	96	100	1	16	1
DC energy	74	89	58	81	7	27	2
% crude fibre[1], straw	5		9				
DC protein	80	85	75	75	3	32	2
DC fibre	19	39	13	33	0	2	1
DC carbohydrates[2]	97	100	96	100	7	5	1
DC energy	74	84	63	75	0	28	1
% crude fat[1], animal	5		27				
DC protein	74	82	76	85	2	15	4
DC fat	57	69	86	87	115	216	8
DC carbohydrates[2]	95	100	93	99	4	467	6
DC energy	65	82	69	82	2	0	0
% crude protein[1]	16		33				
DC protein	67	78	79	87	108	28	0
DC fibre	10	21	1	55	1	49	68
DC carbohydrates[2]	97	100	94	100	20	8	11
DC energy	68	79	71	86	1	119	11

[1] In diet dry matter [2] Soluble carbohydrates

by the larger errors which are probably due to sampling and marker problems.

The results also show the significance of the hind gut microflora for fibre digestion because the crude fibre mainly disappears in that region. Soluble carbohydrates are almost 100% digestible and are mainly absorbed in the small intestine. High concentration in the diet of any nutrient seems to increase the amount of soluble carbohydrates and energy passing to the hind gut.

The influence of dietary composition on the site of absorption and on the efficiency of utilization of ME is illustrated in *Table 6.3.* The percentage of energy being absorbed from the hind gut increases with increasing concentrations of crude fibre and crude protein and decreases with increasing concentration of crude fat in the diets. The NE values given in *Table 6.3* are estimated from the data shown in *Table 6.1.* The results indicate a positive relationship between the rate of energy absorption in the small intestine and the efficiency of utilization of ME.

Table 6.3 THE INFLUENCE OF DIETARY COMPOSITION ON THE SITE OF ABSORPTION AND EFFICIENCY OF UTILIZATION OF ME

In diet dry matter	*Ileal DC*	*Faecal DC*	*% of energy absorbed in the small intestine*	*Net energy in % of ME*
4% crude fibre, cereals	74	89	83	64
8% crude fibre, cereals	59	81	73	62
5% crude fibre, straw	74	84	88	65
9% crude fibre, straw	63	75	84	63
5% crude fat, animal	65	82	79	63
27% crude fat, animal	69	82	84	70
16% crude protein	68	79	86	61
33% crude protein	71	86	83	54

The physiological background for that relationship is probably that the nutrients absorbed from the small intestine and the hind gut, respectively, have different chemical compositions. In the hind gut a large proportion of carbohydrates is fermented by the microflora to volatile fatty acids (acetic, propionic and butyric acids) and the protein or amino acids are probably decomposed to ammonia and/or amines since the nitrogen from the hind gut is almost completely excreted in the urine (Just, Jørgensen and Fernández, 1979). This indicates that the carbohydrates and amino acids that disappear in the hind gut are utilized less efficiently than those absorbed from the small intestine.

References

JUST (NIELSEN), A. (1970). 381. beretn. forsøgslab. Copenhagen. 212 pp.

JUST (NIELSEN), A. (1975). *Wld Rev. Anim. Prod.*, **11** (1), 18–30

JUST, A., RASMUSSEN, O.K. and HANSEN, H.L. (1976) In *Energy Metabolism of Farm Animals*, pp. 289–292. Ed. M. Vermorel. EAAP Publication No. 19. Clermont Ferrand: G. de Bussac

JUST, A., JØRGENSEN, H. and FERNANDEZ, J. (1979). *Proceedings of the 30th Annual Meeting of the EAAP, Harrogate*, p. 5

STOLDT, W. (1957). *Landw. Forsch.*, **10**, 273–277

7

THERMOGENIC EFFECT IN PIGS OF DIETS CONTAINING WHEAT OR WHEATFEED

A.J. TAYLOR and C. FISHER*
Unilever Research, Colworth House, Sharnbrook, Bedford, England

Summary

Five experimental treatments, consisting of a basal diet and 200 and 400 g/kg substitutions of wheat or wheatfeed, were imposed on 24 pigs of 22 kg initial weight for 32 days. Faeces and urine were collected and heat production determined using open-circuit respiration chambers. Total body analysis was performed on completion of the experiment and nutrient retention was calculated by comparison with 'baseline' data. The net energy/digestible energy (NE/DE) ratios for wheat and wheatfeed were 0.614 and 0.552 respectively by indirect calorimetry and 0.602 and 0.449 by comparative slaughter.

Introduction

DE values, which are commonly used in pig nutrition, will only predict animal performance accurately if the efficiency of utilization of DE from different sources is similar. Ewan (1976) reported widely different values for the utilization for growth of DE from wheat (0.640) and wheatfeed (0.261; *see also* Pals and Ewan, 1978), based on comparative slaughter experiments with young pigs. This comparison has been repeated in view of the implications of Ewan's findings.

Materials and methods

Five treatments consisted of a basal diet and 200 and 400 g/kg substitutions of wheat or wheatfeed. The basal diet contained (g/kg) ground barley 860, herring meal 90, dicalcium phosphate 35, limestone 6, sodium chloride 3, L-lysine HCl 1.6 and vitamin/mineral premix 4.4. Supplements of lysine, methionine, threonine and tryptophan were added to the test diets to bring the levels of these amino acids up to the basal levels (g/kg) of 'available' lysine 9, methionine 2.8, threonine 6.1 and tryptophan 1.9. Dietary crude protein levels (N × 6.25) ranged from 145 to 161 g/kg.

* Present address: ARC Poultry Research Centre, Edinburgh.

Forty-eight crossbred female pigs 8–9 weeks old were used: 24 to provide 'baseline' data for comparative slaughter estimates and 24 for balance experiments arranged in four blocks. Within each block of six pigs the basal diet was represented twice and each test diet once. The pigs were kept in metabolism crates throughout the experiment and the environmental temperature was maintained at 21°C. Daily feed allowances (kg) of 0.1 W $kg^{0.75}$ were given in two equal parts daily as a wet mash. Overall intakes of water and feed were in the approximate ratio of 2:1 (w/w). Adjustments to the feed allowances were made weekly.

Following 10 days' acclimatization on the experimental diets, there were three five-day balance periods (separated by weekends) during which total collections of faeces and urine were made. Each pig spent 24 hours of each balance period in one of two open-circuit respiration chambers which were similar to those described by Thorbek (1969). At the end of the 32-day experiment the pigs were slaughtered and the whole body (minus gut contents and hair) minced and analysed.

Calculations

Heat production was calculated from respiratory exchange data using the equation given by Brouwer (1965). Corrections for maintenance energy requirements assumed net energy for maintenance (NE_m) = 0.285 W $kg^{0.75}$ MJ/d. This was derived from metabolizable energy for maintenance (ME_m) = 0.475 W $kg^{0.75}$ MJ/d (Verstegen *et al.*, 1973) and net efficiency for maintenance (k_m) = 0.60 (Kielanowski, 1972).

Analysis of carcass data for 24 'baseline' pigs weighing 18.5–27.5 kg gave the following equation for the calculation of initial carcass energy values:

Carcass energy (MJ) = 8.005 W (kg) + 1.432
(r = 0.877, $P<0.001$ and residual s.d. = 10.233)

Results and discussion

Nutrient intakes and growth rate (*Table 7.1*) were similar for all treatments except the diet containing 400 g/kg wheatfeed, for which these parameters were depressed. Body composition data (*Table 7.2*) showed a reduced lipid content of pigs receiving wheatfeed leading to significantly reduced values for lipid retention. Protein and energy retention were also depressed at the higher level of wheatfeed.

Energy balance (*Table 7.3*) showed approximately linear changes with the inclusion of test ingredients. The ratio between ME and DE averaged 0.973 and was constant across diets. Nitrogen digestibility and retention were, however, slightly reduced for the 400 g/kg wheatfeed diet. In comparison with the basal, the NE values of diets increased with the inclusion of wheat and decreased with the inclusion of wheatfeed. In general, heat production was apparently slightly underestimated by indirect calorimetry, although the difference between the two estimates of NE was large only at the higher level of wheatfeed.

Table 7.1 NUTRIENT INTAKES AND GROWTH RATE

	Diet					*Average significant difference (P = 0.05)*
	Basal	*Wheat* (200 g/kg)	*Wheat* (400 g/kg)	*Wheatfeed* (200 g/kg)	*Wheatfeed* (400 g/kg)	
Initial live weight (kg)	22.8	21.2	22.7	22.8	21.3	1.79
Final live weight (kg)	34.8	33.2	34.6	34.9	31.5	2.14
Growth rate (g/d)	377	374	371	378	317	28
Total intakes of:						
dry matter (kg)	29.7	28.5	29.7	30.1	27.4	1.60
nitrogen (g)	918	863	889	927	817	54
DE (MJ)	418	417	436	423	382	23.6
ME (MJ)	407	406	424	412	372	23.3

Table 7.2 BODY ANALYSIS

	Baseline	*Diet*					*Average significant difference (P = 0.05)*
		Basal	*Wheat* (200 g/kg)	*Wheat* (400 g/kg)	*Wheat-feed* (200 g/kg)	*Wheat-feed* (400 g/kg)	
Composition:							
dry matter (g/kg)	324.5	346.1	346.7	357.1	339.0	332.7	13.5
lipid (g/kg)	136.2	144.8	148.1	154.9	129.8	129.2	17.0
crude protein (g/kg)	158.7	170.6	167.9	170.8	173.0	171.6	7.6
ash (g/kg)	29.8	34.3	34.8	32.2	34.6	32.2	2.9
Retention:							
dry matter (kg)	–	4.68	4.52	5.03	4.36	3.50	0.43
lipid (kg)	–	1.96	2.03	2.30	1.42	1.22	0.53
crude protein (kg)	–	2.33	2.14	2.33	2.37	1.96	0.27
ash (kg)	–	0.51	0.47	0.44	0.51	0.34	0.09
gross energy (MJ)	–	128.9	126.5	145.6	120.6	96.2	20.0

Table 7.3 ENERGY AND NITROGEN BALANCE

	Diet					*Average significant difference (P = 0.05)*
	Basal	*Wheat* (200 g/kg)	*Wheat* (400 g/kg)	*Wheatfeed* (200 g/kg)	*Wheatfeed* (400 g/kg)	
DE (MJ/kg DM)	14.06	14.64	14.67	14.05	13.94	0.205
ME (MJ/kg DM)	13.68	14.23	14.27	13.65	13.57	0.211
NE (MJ/kg DM):						
Calorimetry	8.56	9.02	8.95	8.39	8.22	0.479
Slaughter	8.11	8.23	8.66	7.76	7.33	0.737
N digestibility	0.73	0.76	0.75	0.72	0.68	0.026
N retention	0.50	0.51	0.51	0.50	0.47	0.032

The composition of the test samples of wheat and wheatfeed (*Table 7.4*) was typical. Calculation of energy values by regression analysis gave DE values (± standard error) of 15.69 ± 0.636 and 13.77 ± 0.117 MJ/kg dry matter (DM) and ME values of 15.26 ± 0.610 and 13.43 ± 0.209 MJ/kg DM for wheat and wheatfeed respectively. These are in agreement with other published estimates. The calculation of NE values for individual feed ingredients requires assumptions which are possibly invalid, such as the independence and additivity of values. However, NE values for wheat and wheatfeed are also presented in *Table 7.4*. Relative to DE, the utilization of energy from wheat was 0.614 and 0.602 by indirect calorimetry and comparative slaughter respectively and from wheatfeed 0.552 and 0.449 respectively.

Table 7.4 COMPOSITION AND FEEDING VALUE OF WHEAT AND WHEATFEED

	Wheat	*Wheatfeed*
DM (g/kg)	852	855
Ether extract (g/kg)	12	32
Crude protein (g/kg)	142	149
Crude fibre (g/kg)	34	63
Ash (g/kg)	23	38
Calcium (g/kg)	1.8	0.8
Phosphorus (g/kg)	3.4	7.7
Available carbohydrate[1] (g/kg)	610	464
ND fibre (g/kg)	164	299
AD fibre (g/kg)	48	94
Hemicellulose (g/kg)	116	205
Cellulose (g/kg)	12	46
Lignin (g/kg)	12	18
DE[2] (MJ/kg DM)	15.68	13.77
ME[2] (MJ/kg DM)	15.25	13.42
NE[2] (MJ/kg DM) : calorimetry	9.62	7.70
: slaughter	9.44	6.17
ME/DE	0.97	0.97
NE/DE : calorimetry	0.61	0.55
: slaughter	0.60	0.44
NE/ME : calorimetry	0.63	0.57
: slaughter	0.61	0.46

[1] Clegg (1956)
[2] Calculated by regression methods

In conclusion, these data do not confirm the very large difference in energy utilization between wheat and wheatfeed as suggested by Ewan (1976). There is, however, a small difference between these two feedstuffs in thermogenic effect. Possible explanations for this, which cannot be elucidated from the results of this experiment, include:

1. The overestimation of true DE values in intact animals, which may be higher on fibrous diets due to hind gut activity.
2. The end products of digestion from the two feedstuffs may have different metabolic values.
3. High fibre diets may increase protein turnover in the digestive tract.

References

BROUWER, E. (1965). In *Energy Metabolism*, pp. 441–443. EAAP Publication No. 11, Ed. K.L. Blaxter. London: Academic Press

CLEGG, K.M. (1956). *J. Sci. Fd Agric.*, **7**, 40–44

EWAN, R.C. (1976). *Distill. Feed Res. C. Conf. Proc.*, **31**, 16–21

KIELANOWSKI, J. (1972). In *Pig Production*, pp. 183–201. Ed. D.J.A. Cole. London: Butterworths

PALS, D.A. and EWAN, R.C. (1978). *J. Anim. Sci.*, **46**, 402–408

THORBEK, G. (1969). *Beretn. Forsøgslab.*, **373**, 46 pp

VERSTEGEN, M.W.A., CLOSE, W.H., START, I.B. and MOUNT, L.E. (1973). *Br. J. Nutr.*, **30**, 21–35

8

THE EFFECTS OF DIETARY PROTEIN, CARBOHYDRATE, FAT AND FIBRE ON THE NET AVAILABILITY OF METABOLIZABLE ENERGY IN THE CHICK

J. GUILLAUME*, C.W. SCHEELE** and R. KUSSAIBATI
Station de Recherches Avicoles – I.N.R.A., Centre de Tours 37380, France

Summary

Four balance experiments were carried out on 768 male chicks of 1–4 weeks of age to measure the net energy (NE) value of diets with different levels and kinds of protein, fat, carbohydrate and fibre. The 18 diets were fed at three levels to birds individually caged in a controlled-environment chamber. The comparative slaughter technique was used. The availability of metabolizable energy (ME) for maintenance was lower for proteins, irrespective of amino-acid balance, and for fats. The availability of ME for growth, almost unaffected by protein, was clearly higher for saturated fats and seemed lower for high pentosane or high fibre diets. Unsaturated fats appeared less efficient. The effect of fibre seems to be due to an interaction of the feeding level with the digestibility of nutrients.

Introduction

Considerable data have been published concerning the net availability of protein, fat and carbohydrate in the growing chick (cf. de Groote, 1974). However, few studies of the effect of the nature of protein, fat or carbohydrates have been made. The effect of the fibre found by Petersen (1973) also remains unexplained. For this reason four experiments were carried out.

Materials and methods

Each experiment involved a total of 192 male chicks of commercial origin (meat type). At six days of age the birds were weighed, divided into four weight blocks and allotted to each experimental group or to a group killed for the determination of the initial body composition. During the experimental period they were caged and individually fed daily in a controlled-environment chamber. Each diet was fed at three levels with a group of 16 birds per level in most experiments.

At the end of the experiment (20 d) the birds were fasted for 22 h and then killed, stored at –15°C and ground with a Robot Coupe® cutter. An aliquot part of the carcass was lyophilized and reground. Crude energy and nitrogen content were measured for the determination of energy and protein retention. During the second and third week of each experiment the ME was also measured over a three-day period using the total collection and the polyethylene indicator

* Present address: CNEXO-COB, BP 337 29273, Brest Cedex, France
** Present address: Spelderholt Institute for Poultry Research, 7361 DA Beekbergen, The Netherlands

method (Roudibush, Anthony and Vohra, 1974) at each feeding level. Digestibility of protein, starch plus sugar and crude fat were also measured using the methods of Terpstra and De Hart (1974), Thivend and Mercier (1977) and Folch, Lees and Sloane-Stanley (1957).

Most of the diets tested were theoretically isoenergetic on the basis of corrected ME (13 MJ/kg). Most contained 220 g/kg of crude protein and were adequate for all known nutrients when fed *ad lib.* (at least 12.2 g/kg lysine; 9.3 g/kg methionine + cystine; 10 g/kg Ca; 4.4 g/kg available P; *see Table 8.1*).

Birds at each feeding level were fed the same amount of food daily. However, in experiment three the basal diet was diluted with a source of fibre to obtain diets J, K and L (*Table 8.1*) that were fed in amounts that permitted the birds to receive the same quantity of basal diet. In experiment four a high fat diet (Q) was fed at a lower level to provide the same theoretical amount of ME.

Table 8.1 MAIN CHARACTERISTICS OF DIETS USED IN EXPERIMENTS 1–4

Protein level amino acid balance	*Carbohydrate*	*Fibre added*	*Fat level and nature*
A : Control 22% balanced protein	E : Control = A	I : Control = A	N : Control = A
B : 26% balanced protein	F : Same characteristics as A; 25% maize starch	J : I + long fibre cellulose (1 + 0.15)	O : Same characteristics as A; 17% tallow, 8.05% cellulose
C : 35% balanced protein	G : Same characteristics as A; 10% sucrose + 10% glucose	K : I + chromatography cellulose (1 + 0.15)	P : Same characteristics as A; 13% maize oil, 11.55% cellulose
D : 35% unbalanced protein-lysine as in A	H : Same characteristics as A; 10% of pentosane from wheat middling	L : A + ground sunflower hull (1 + 0.15)	Q : Same balance of nutrients as O. No cellulose added.
		M : Same characteristics as I, K and L except high fibre contents from natural ingredients instead of cellulose	R : Same characteristics as A except for lipids (1.5%)
48 birds/diet	48 birds/diet	40 birds/diet except in diets I and K (36)	40 birds/diet except in diets N and R (36)

The net availability of ME was obtained by plotting crude energy retained against ME intake per unit metabolic weight ($W^{0.75}$). Statistical calculations were carried out using variance or covariance analysis.

Results

Classic ME was found to decrease from the highest feeding level (nearly *ad lib.*) to the lowest one, and the digestibility of crude fat and organic matter followed a similar trend. However, the most important differences observed were between

Table 8.2 DIGESTIBLE PROTEIN, FAT, AVAILABLE CARBOHYDRATES AND CLASSIC ME CONTENTS OF DIETS

Experiments	*1*				*2*			
Diets (per kg dry matter)	*A*	*B*	*C*	*D*	*E*	*F*	*G*	*H*
Digestible protein (g)	192	225	313	311	191	193	190	195
Digestible ether extract (g)	63	67	73	73	61	70	62	69
Digestible available carbohydrates (g)	343	292	242	257	349	353	342	310
Digestible organic matter (g)	676	638	628	607	677	652	674	584
Classic ME (MJ)	129	137	136	138	131	134	135	125

Experiments	*3*					*4*				
Diets	*I*	*J*	*K*	*L*	*M*	*N*	*O*	*P*	*Q*	*R*
Digestible protein (g)	202	171	170	172	157	198	204	194	212	196
Digestible ether extract (g)	67	60	58	63	56	72	153	139	164	13.0
Digestible available carbohydrates (g)	352	296	277	279	295	–	–	–	–	–
Digestible organic matter (g)	649	579	569	562	575	650	603	584	634	681
Classic ME (MJ)	133	115	115	116	113	136	142	138	159	130

diets. The mean content in digestible nutrients and the mean classic ME of each diet are given in *Table 8.2.*

In diet H, which contained 10% pentosanes from wheat middling, the digestibility of nitrogen-free extract and ME were very low. In diet M, which had a high fibre content derived from natural ingredients (wheat bran, lucerne and sunflower), the digestibility of protein was also low.

The efficiency of ME of these diets for both production and maintenance requirements are given in *Table 8.3.* The maintenance requirement was higher in experiments one and two when compared with experiments three and four where the temperature of the chamber was held constant and not decreased from 38 to 34°C during the course of the experiment. Other effects on maintenance were as follows:

1. High protein diets (313 or 311 g digestible protein) increased maintenance by 7%
2. The nature of carbohydrates had no influence
3. The addition of 'fibre' caused a slight, but non-significant, increase in maintenance (+ 9% and + 6% for chromatography cellulose and natural fibre respectively)
4. In the high tallow fat, when compared with the low fat diet, maintenance was significantly increased by 7%

With regard to the net availability for growth, the results obtained with a given formula, but with different origins of ingredients (diets A, E, I and N), were clearly different. However, marked differences were not always found: a small non-significant effect on k_f was found in diets containing very high levels of protein (k_f 5% lower).

Table 8.3 ESTIMATIONS OF MAINTENANCE REQUIREMENT (m), NET EFFICIENCY OF ME FOR GROWTH (k_f) AND COEFFICIENT OF CORRELATION BETWEEN RETENTION AND INTAKE (r)

Experiment 1				*Experiment 2*			
Diet	m (MJ/d)	k_f (%)	r	*Diet*	m (MJ/d)	k_f (%)	r
A	0.61	61.2	0.968	E	0.66	71.5	0.979
B	0.61	57.6	0.972	F	0.66	68.5	0.978
C	0.65	61.3	0.967	G	0.64	67.6	0.980
D	0.65	58.3	0.980	H	0.66	64.5	0.969
Experiment 3				*Experiment 4*			
Diet	m (MJ/d)	k_f (%)	r	*Diet*	m (MJ/d)	k_f (%)	r
I	0.54	66.5	0.984	N	0.52	69.1	0.984
J	0.54	63.1	0.986	O	0.48	69.7	0.979
K	0.59	71.0	0.988	P	0.50	65.6	0.977
L	0.52	59.0	0.987	Q	0.53	77.1	0.982
M	0.57	63.1	0.987	R	0.50	63.2	0.979

In experiment four the k_f was significantly greater ($P < 0.01$) in the case of the high tallow diet (Q) compared with the low fat diet (R), and the control (N) was intermediate. The net availability of the ME of oil was significantly lower. Pure cellulose had an apparently depressing effect on k_f (experiments three and four).

Discussion

The variation of k_f for a given formula probably results from the chicks themselves (genetic origin and size), from the actual composition of the diet and from experimental variation. However, the results confirm that in balanced diets the net availability of ME for growth in the chick ranks in the following order, protein, carbohydrate and fat. However, in these experiments protein excess (45% of ME from protein instead of 27%) had little (if any) influence on k_f while excess or deficiency of fat (44% or 4% of ME from fat respectively) clearly influenced it. Dietary fat was probably sufficient or almost sufficient to be converted into body fat (except diet R), and energy arising from protein degradation was used for maintenance. With the low fat diet most of the body fat was synthesized from either carbohydrates or amino acids. With the high fat diet the increase in k_f is also attributable to the fact that birds become more fatty and the amount of energy arising from fat (synthesized with high efficiency) increased.

The relative efficiency of protein energy for maintenance is 74% of that of carbohydrate (Scheele *et al.*, 1976), with no difference found between fat and carbohydrate. The origin of 'available carbohydrates' does not have any noticeable effect on NE availability; only pentosanes seem to be less efficient. Also, amino-acid balance seems to have an effect on the net availability of energy, but this is smaller than in the results of Guillaume *et al.* (1967). The two most surprising results concern the effect of fibre and the difference in efficiency between tallow and oil.

Birds fed the fibre diets J and L that consumed the same amount of nutrients as controls (lot I) had, for the same actual ME intake, a depressed energy retention (–4%). This depression, at least in diet L, appears not to be due to an increase in maintenance (which would have been comprehensible) but to a decrease of k_f. Careful examination of the results indicates that at the low feeding level added cellulose tends to increase the digestibility of fat and to decrease that of protein, whereas at the highest level the opposite effect appears. Therefore the effect of fibre is probably an artefact. Accurate measurement of the digestibility of each nutrient is needed to explain this.

An effect of the nature of the fat on net availability of ME was rarely found. Fuller and Rendon (1977) and Farrell (1979) found similar but less pronounced results.

In conclusion, when ME from a single nutrient ranged from approximately 20 to 50% and 'fibre' content reached 13%, the variations of net availability of ME for growth and k_f remained limited.

Acknowledgement

We thank Liliane Derouet and Joëlle Gomez for technical collaboration.

References

DE GROOTE, G. (1974). In *Energy Requirements of Poultry*, pp. 1–24. Eds. T.R. Morris and B.M. Freeman. Edinburgh: British Poultry Science Ltd.

FARRELL, D.J. (1979). *Br. Poult. Sci.*, **19**, 105–109

FOLCH, J., LEES, M. and SLOANE-STANLEY, G.H. (1957). *J. biol. Chem.*, **226**, 497–509

FULLER, H.L. and RENDON, M. (1977). *Poult. Sci.*, **56**, 549–557

GUILLAUME, J., CALET, C., METAIS, R.F. and FENDRY, M. (1967). In *Energy Metabolism of Farm Animals*, pp. 377–386. Eds. K.L. Blaxter, D. Thorbek and J. Kielanowski. EAAP Publication No. 12. Newcastle: Oriel Press

PETERSEN, B. (1973). In *Energy Metabolism of Farm Animals*, pp. 285–288. Eds. K.H. Minke, H.J. Lantzsch, J.R. Reichl. EAAP Publication No. 14. University of Hoheim

SCHEELE, C.W., JANSSEN, W.M.M.A. and VERSTEGEN, M.W.A. (1976). In *Energy Metabolism of Farm Animals*. Ed. M. Vermorel, pp. 277–281. EAAP Publication No. 19. Clermont-Ferrand: G. de Bussac

ROUDIBUSH, T., ANTHONY, D.L. and VOHRA, P. (1974). *Poult. Sci.*, **53**, 1894–1896

TERPSTRA, K. and DE HART, N. (1974). *Z. Tierphysiol. Tierernähr. Futtermittelk.*, **32**, 306–320

THIVEND, P. and MERCIER, Ch. (1977). Personal communication

9

INFLUENCE OF SPECIALLY EXTRACTED SOYA MEAL ON NITROGEN AND ENERGY METABOLISM IN THE PRERUMINANT CALF

L. NEERGAARD
National Institute of Animal Science, Rolighedsvej 25, 1958 Copenhagen V, Denmark

Summary

With various concentrations of milk protein and soya meal (Danpro®) in four liquid diets fed to 16 calves of 1–6 weeks of age, 96 nitrogen and energy balances were accomplished. Apparent digestibility of nitrogen and energy increased with age for the three diets containing Danpro® and was lower ($P < 0.001$) than in milk. The quotients for nitrogen retention regressed on digested nitrogen were different ($P < 0.5$) for the four diets. Estimated heat loss on the third day of fasting was 153 kJ/kg live weight at 58.9 kg live weight. Maintenance requirement for diets with Danpro was 177 kJ/kg live weight and the partial efficiency of energy utilization for maintenance (k_m) was 0.86. For all Danpro diets the partial efficiency of energy utilization for growth (k_g) was 0.70 ± 0.03. When 17 days old the milk-replacer-fed animals showed relative dehydration.

Introduction

In Denmark the use of milk replacers without milk protein for calves from the fourth day of birth has been extended, since a soya meal concentrate (Danpro®), from which the antitrypsin and soluble sugars have been extracted by special treatment, has been manufactured. The influence on the metabolism of the young calf, because of the decreased transit time of digesta in the abomasum as shown by Sissons and Smith (1976), when Danpro is given, is not known. This chapter describes the energy metabolism of the young calf when fed a milk replacer containing Danpro + whey or this milk replacer with 20% skim-milk or 20% whole milk powder as compared to whole milk powder as the sole source of nutrient.

Materials and methods

Dairy cross-bred male calves, four days old, were transferred to the experimental diets from a salt glucose solution (Richardson, 1967). The animals were kept in metabolism cages at 20°C and fed twice a day by rubber teat. Faeces were collected in plastic bags (Noller *et al.*, 1956) and urine in plastic cans each day in six subsequent periods of seven days each. The animals were weighed on the fourth day in the middle of each period and transferred to the open respiration equipment before feeding (Thorbek and Neergaard, 1970). Fasting metabolism was determined on the 12 calves fed milk replacer. They were all fed diet 2 in the sixth period and subsequently restricted to maintenance level for three days. When transferred to the respiration chambers they were fasted for three days at 24°C.

In two calves on each milk replacer diet, tritiated water space was measured at the end of the second period and twice at 14-day intervals as described by Fayet (1971).

The composition of the experimental diets can be seen in *Table 9.1.* Diet 1 consisted of whole milk powder for human consumption. The milk replacer, diet 2, was composed of 30% Danpro, 33.8% whey powder, 12.5% decarbohydrated whey powder + added vitamins, 0.5% DL-Methionine, 1.2% maize starch, 18.6% emulsified lard, 1.4% emulsifier and 2.0% mineral mixture; diet 3 was made up of 80% of diet 2 mixed with 20% skim-milk powder and diet 4 comprised 80% of diet 2 and 20% whole milk powder.

Table 9.1 COMPOSITION OF EXPERIMENTAL DIETS (PER 100 g DM)

Component	*Diet 1*	*Diet 2*	*Diet 3*	*Diet 4*
Protein	30.1	28.8	31.4	29.5
Fat	26.4	20.5	15.9	21.1
RHC	37.3	22.6	26.1	24.3
RC	–	18.7	17.2	16.6
Ash	6.2	9.4	9.3	8.5
Gross energy (kJ)	2369	2163	2069	2204

RHC = Readily hydrolysable carbohydrates
RC = Residual carbohydrates

All diets were fed each to four calves. Diets 1 and 2 were given for six periods. Diets 3 and 4 were administered for the first five periods, the eight animals thereafter being transferred to diet 2.

Reconstituted diets contained 11–15% dry matter (DM) and were fed daily in constantly increased amounts inside each period.

Results and discussion

The apparent digestibility quotients of DM, nitrogen and energy in relation to age can be seen in *Table 9.2.* Only the milk-fed animals showed no significant differences between periods, whereas in the first period digestibility quotients for all diets containing milk replacer were significantly different ($P < 0.05$) from those in the succeeding periods. For the calves at the age of 16 days on diets 3 and 4, the digestibility was significantly lower ($P < 0.05$) than at the age of 23 days, although some milk protein was present. Mean values are presented in order to compare performances. When diet 2 was fed to all 12 animals on milk replacer, when 45 days of age, no significant differences were obtained between groups, with the percentages of digestibility for DM, nitrogen and energy being 89.2, 86.5 and 88.4 respectively.

Table 9.2 DIGESTIBILITY OF DM, NITROGEN AND ENERGY IN CALVES AGED 1–7 WEEKS. WHOLE MILK (DIET 1) COMPARED TO MILK REPLACERS CONSISTING OF SPECIALLY EXTRACTED SOYA MEAL WITH WHEY (DIET 2) OR DIET 2 + 20% SKIM-MILK (DIET 3) OR DIET 2 + 20% WHOLE MILK (DIET 4), AS %

Diet 1, age d	15.3	22.3	29.3	36.3	43.3	50.3	Mean
DM	97.9	98.1	98.2	98.4	98.4	98.1	98.2[a]
Nitrogen	95.0	96.3	96.9	97.2	97.0	96.7	96.5[a]
Energy	97.7	98.0	98.1	98.4	98.3	98.1	98.1[a]
Diet 2, age d	14.3	21.3	28.3	35.3	42.3	49.3	Mean
DM	80.9	88.2	87.5	88.4	88.8	89.6	87.2
Nitrogen	71.0	83.2	83.1	84.1	84.7	86.4	82.1[d]
Energy	78.4	87.2	86.4	87.6	88.2	88.9	85.7
Diet 3, age d	9.8	16.8	23.8	30.8	37.8	–	Mean
DM	85.7	87.8	87.5	89.3	89.9	–	88.0[bd]
Nitrogen	77.0	82.2	83.0	86.9	88.3	–	83.6[d]
Energy	84.4	87.1	86.9	89.0	89.5	–	87.4[bd]
Diet 4, age d	9.0	16.0	23.0	30.0	37.0	–	Mean
DM	82.9	83.7	87.0	89.3	89.4	–	86.5
Nitrogen	72.3	76.2	83.6	86.4	86.8	–	81.0[c]
Energy	81.2	82.6	86.1	88.9	89.1	–	85.6

[a] Difference from diets 2, 3 and 4 ($P<0.02$)
[b] Difference from diet 2 ($P<0.05$)
[c] Difference from diet 3 ($P<0.05$)
[d] Difference from diet 4 ($P<0.05$)

In order to compare nitrogen retention of whole milk from former experiments with diets 2, 3 and 4, retained nitrogen (Y; mg/kg live weight) was regressed on digested nigrogen (X) and the following equations were obtained:

$$\text{Diet 1: } Y = 1.03\,X - 236;\ CV = 10\%;\ s_b = 0.06;\ r = 0.93;\ n = 46$$
$$\text{Diet 2: } Y = 0.91\,X - 215;\ CV = 6\%;\ s_b = 0.03;\ r = 0.99;\ n = 20$$
$$\text{Diet 3: } Y = 0.76\,X - 169;\ CV = 6\%;\ s_b = 0.03;\ r = 0.99;\ n = 19$$
$$\text{Diet 4: } Y = 0.83\,X - 194;\ CV = 10\%;\ s_b = 0.04;\ r = 0.98;\ n = 20$$

All were significantly different ($P<0.05$).

The regression of log heat loss (HL) on the third day of fasting, for 12 calves weighing 58.9 kg, on log live weight (kg) indicated that the power of metabolic live weight (kg) was close to one for these animals, and HL was 153.1 kJ/kg live weight, CV = 4.6% and r = 0.93.

From *Table 9.3* the average live weights of the experimental animals can be seen together with the average amounts of gross energy (GE) intake per kilogram of live weight for all positive energy balances. The variation of feeding level inside treatments was greater when the milk replacer was fed, because, during the initial periods, the amounts recommended by the factory were applied.

The metabolizability of whole milk was significantly different from all milk replacers, and inclusion of skim-milk decreased the variation but not the percentage metabolizable energy (ME) from GE.

The overall efficiency of utilization of ME was significantly better for the group on whole milk. Because no significant differences were found between diets 2, 3 and 4, or between groups of animals when fed the same diet (2),

Table 9.3 ENERGY METABOLISM IN CALVES AGED 1–7 WEEKS FED FOUR LIQUID DIETS. WHOLE MILK (DIET 1) COMPARED TO MILK REPLACERS CONSISTING OF SPECIALLY EXTRACTED SOYA MEAL WITH WHEY (DIET 2), OR DIET 2 + 20% SKIM-MILK (DIET 3), OR DIET 2 + 20% WHOLE MILK (DIET 4)

Diet	*1*		*2*		*3*		*4*	
No. of observations	24	s.d.	20	s.d.	14	s.d.	16	s.d.
Live weight (kg)	51.7	13.2	49.4	9.0	50.6	7.3	44.5	7.7
Live-weight gain (g/kg live weight)	12.9	–	14.1	–	13.0	–	13.7	–
GE (kJ/kg live weight)	331.1	10.3	419.9	72.2	387.1	60.1	423.1	73.4
ME (% of GE)	95.1[a]	0.9	84.4	3.0	84.5	1.8	84.5	3.2
HL (% of ME)	62.5	2.7	63.3	8.9	66.2	7.2	66.7	8.5
RE (% of ME)	37.6	2.7	36.7	8.9	32.8	7.2	33.3	8.5
RPROT (% of RE)	53.7	4.2	52.3	13.2	59.6	12.5	53.9	13.6
RFAT (% of RE)	46.3	4.2	47.7	13.2	40.4	12.5	46.1	13.6

[a] Difference from diets 2, 3 and 4 ($P<0.001$)

retained energy (RE) from all 50 energy balances was regressed on ME (Y on X). The resulting equation was:

$$Y = 0.70\,X - 123.9;\; CV = 9.9\%;\; s_b = 0.03;\; r = 0.96$$

The maintenance requirement of ME, which is 177 kJ/kg live weight and is not significantly different from that suggested by Jacobsen (1969) of 172 kJ/kg live weight, was calculated on the basis of fasting HL to be utilized with a k_m of 0.86. As can be seen from the equation, the efficiency of ME for k_g was 0.70 as compared with that found for whole milk of 0.85 (Neergaard, 1976).

The faeces from the calves fed milk replacer contained 18.2% DM, but a relative dehydration was observed when spaces of extracellular and intracellular water were compared by means of tritiated water as described by Fayet (1971). When 17 days old the total body water was 74.7%, extracellular fluid was 30.5% and intracellular water was 44.2% of body weight. The respective values obtained 14 days later were 74.5%, 44.7% and 29.8%, and the last two values were significantly different ($P<0.05$) from the corresponding values when first measured, which indicated an improved distribution of body water spaces.

References

FAYET, J.C. (1971). *Br. vet. J.*, **127**, 37

JACOBSEN, N.L. (1969). *J. Dairy Sci.*, **52**, 1316

NEERGAARD, L. (1976). In *Energy Metabolism of Farm Animals*, p. 205. Ed. M. Vermorel. EAAP Publication No. 19. Clermont Ferrand: G. de Bussac

NOLLER, C.H., WARD, G.M., McGILLARD, A.D. HUFFMAN, C.F. and DUNCAN, C.W. (1956). *J. Dairy Sci.*, **39**, 1288

RICHARDSON, A. (1976). *Vet. Rec.*, **80**, 318

SISSONS, J.W. and SMITH, R.H. (1976). *Br. J. Nutr.*, **36**, 421

THORBEK, G. and NEERGAARD, L. (1970). In *Energy Metabolism of Farm Animals*, p. 243. Eds. A. Schürch and C. Wenk. EAAP Publication No. 13. Zurich: Juris Druck and Verlag

10

ENERGY UTILIZATION BY GROWING CALVES: EFFECTS OF AGE, MILK INTAKE AND FEEDING LEVEL

M. VERMOREL, J.C. BOUVIER and Y. GEAY
I.N.R.A. – C.R.Z.V. de Theix, 63110 Beaumont, France

Summary

Four groups of five male calves were fed on two feeding levels and weaned either at 10 weeks of age or fed 6 or 4.5 ℓ of milk in addition to the solid feed. The latter was also given to six mature sheep. Energy balances were determined by indirect calorimetry in sheep and in five, 10, 23 and 34-week-old calves.

No associative effect was observed between milk and solid feed. The solid feed energy digestibility (d_E), nitrogen digestibility (d_N) and energy metabolizability (q) were similar in calves and sheep. The efficiency of metabolizable energy (ME) utilization was not altered by milk feeding or feeding level. It decreased as the calves got older and as the proportion of milk energy dropped, and was similar in 23 and 34-week-old calves and in mature sheep.

Introduction

In suckling calves the weight gain and the feed efficiency increase with the milk intake (Le Neindre, Petit and Muller, 1976). At the age of nine months and for the same live-weight gain, dairy suckling calves are also much leaner than early weaned ones (Geay, 1970). The growth improvement could be explained by a higher protein gain and a better energy utilization of end products of digestion of milk than of those of solid feeds, as shown by Black and Tribe (1973). Therefore, an experiment was designed to study the effects of age, milk intake and feeding level on energy utilization by growing calves. The results will be reported in detail elsewhere (Vermorel, Bouvier and Geay, 1980).

Materials and methods

Four groups of five Friesian male calves were fed a milk substitute diet (18% fat and 22% crude protein on dry matter basis) and a cobbed diet (40% grass hay and 60% concentrate) at two feeding levels: *ad lib.* (H) and 75% of *ad lib.* (L). Two groups (early weaned; EW) were weaned at the age of 73 days. The two other groups (milk fed; MF) were given 6 kg or 4.5 kg of milk per day in addition to the solid feed. The latter was also fed to six mature sheep at two feeding levels (1.08 and 1.76 times maintenance) successively.

Energy and nitrogen balances were determined when the calves were five, 10, 23 and 34 weeks old. Faeces and urine were collected over periods of six, six, 13 and 13 days respectively at these ages and over nine-day periods in sheep. Gas exchanges were measured for four consecutive days, after a three-day adaptation period, using two open-circuit respiration chambers. The calves were kept individually and the sheep in pairs.

Results and discussion

For the whole experimental period, the mean live-weight gains amounted to 1220, 1052, 1062 and 840 g in the (H) MF, (L) MF, (H) EW and (L) EW calves respectively. The solid feed intake of (H) MF calves was 8% lower than that of (H) EW calves, but the metabolizable energy (ME) intake was 20% higher. As the solid feed consumption increased the energy supplied by the milk dropped from 90% to 15% of total energy intake in the MF calves (*Table 10.1*). The lower feeding level of calves corresponded to the higher feeding level of sheep.

Table 10.1 VARIATIONS OF DIGESTIBILITY AND METABOLIZABILITY OF THE DIETS IN CALVES AND SHEEP (MEANS ± STANDARD DEVIATIONS; MEAN VALUES FOR BOTH FEEDING LEVELS IN 10–23 AND 34-WEEK-OLD CALVES)

Animals (age)	*Group*	*Milk energy* (% IE)	*100* d_E	*100* d_N	C_E % I_E	U_E % I_E	*ME* % I_E
Calves	H	90.6	90.6 ± 1.8	88.9 ± 1.7	–	2.0 ± 0.3	88.6 ± 1.9
(5 weeks)	L	81.9	88.0 ± 1.5	87.6 ± 1.1	–	2.4 ± 0.2	85.6 ± 1.4
Calves	MF	40.0	79.9 ± 3.0	81.0 ± 2.3	2.4 ± 0.5	3.2 ± 0.3	74.3 ± 3.0
(10 weeks)	EW	22.0	74.8 ± 2.7	75.5 ± 2.6	2.7 ± 0.4	2.6 ± 0.3	69.7 ± 2.9
Calves	MF	17.2	73.9 ± 2.9	76.4 ± 2.0	3.4 ± 0.6	3.3 ± 0.3	67.3 ± 2.6
(23 weeks)	EW	0	69.8 ± 1.4	71.7 ± 2.1	4.3 ± 0.6	3.7 ± 0.1	61.7 ± 1.1
Calves	MF	14.7	73.4 ± 1.4	74.6 ± 1.7	4.1 ± 0.5	3.7 ± 0.6	65.7 ± 1.8
(34 weeks)	EW	0	71.7 ± 1.3	70.7 ± 2.1	5.6 ± 1.3	3.7 ± 0.4	62.4 ± 1.7
Sheep	H	0	72.0 ± 1.0	72.2 ± 0.6	6.0 ± 1.0	3.6 ± 0.3	62.5 ± 1.2
(2 years)	L	0	77.6 ± 0.5	77.8 ± 0.7	7.7 ± 0.4	4.1 ± 0.3	66.2 ± 0.9

DIGESTIBILITY AND METABOLIZABILITY OF THE DIETS (*Table 10.1*)

The energy and nitrogen digestibility of the rations (d_E and d_N) were higher in MF calves than in EW calves and depended on the proportion of energy supplied by milk (Z) and on the feeding level (L):

$$d_E = 0.2123\ Z - 2.68\ L + 75.21 \qquad r^2 = 0.902 \qquad s.d. = 0.31$$
$$d_N = 0.1946\ Z - 2.98\ L + 76.84 \qquad r^2 = 0.888 \qquad s.d. = 0.25$$

The milk substitutes d_E and d_N were similar to the values obtained previously for the same diet in preruminant calves (Vermorel, Bouvier and Geay, 1976).

The solid feed d_E and d_N were not significantly different in EW calves and in adult sheep for the same feeding level. Furthermore, they were not altered by milk consumption.

Methane and urinary energy losses (% of energy intake) increased with age and were similar in the 34-week-old calves and in the mature sheep for the same feeding level. Methane energy losses were smaller in MF calves than in EW calves when related to total energy intake, but they were not significantly different when related to the solid dietary energy intake. Consequently, the ratio ME/DE decreased from 0.97 to 0.85 as the calves got older and the proportion of energy supplied as milk (Z) dropped. The energy metabolizability (q) of the rations depended on Z, feeding level (L) and age (A, days) :

$$q = 0.263\,Z - 1.92\,L - 0.01125\,A + 67.83 \qquad r^2 = 0.940$$

Energy metabolizability of the solid feed was not significantly different between growing calves (MF and EW) and mature sheep for the same feeding level. These results are in agreement with those of Jentsch, Wittenburg and Schiemann (1976).

UTILIZATION OF ME FOR GROWTH (*Table 10.2*)

In growing calves, four energy balances on the average were valid for each experimental group at each growth stage. The nitrogen balances were satisfactory only in five- and 10-week-old calves. The efficiency of ME utilization for growth (k_g) by calves was obtained by regressing energy retention over ME (RE* over ME*, in kJ/kg $W^{0.75}$). The maintenance requirement was calculated for RE* = 0. The results are given in *Table 10.2.*

Table 10.2 EFFICIENCY OF ME UTILIZATION AND MAINTENANCE REQUIREMENT OF CALVES

Age (weeks)	$k_g \pm s_b$	*Maintenance requirement* (kJ/kg $W^{0.75}$)				*r*
		MF Calves		*EW Calves*		
		H	*L*	*H*	*L*	
5	0.638 ± 0.100	452	452	–	–	0.932
10	0.507 ± 0.033	414	346	414	356	0.969
	(0.57)*	(460)	(402)	(460)	(402)	–
23	0.491 ± 0.036	473	397	473	473	0.966
34	0.545 ± 0.074	460	390	460	422	0.912

* Using 460 and 402 kJ/kg $W^{0.75}$ as maintenance requirements

For the mean ME intake (ME* = 904), RE* decreased with increasing age in the *ad lib.* fed calves (288, 249 and 212, in the 5–10 and 23-week-old calves respectively) but increased in the 34-week-old calves (242). On the other hand, RE* was not significantly different in MF and EW calves at any growth stage. For the same ME* intakes also, RE* was significantly higher in MF calves on the low feeding level than in *ad lib.* fed ones: 0, +15%, +15% and + 21% in five, 10, 23 and 34-week-old calves respectively. In early-weaned calves the 15% increase of

RE* observed in week 10 disappeared in week 23 and amounted to 8% ($P<0.15$) in week 34.

The efficiency of ME utilization was not significantly different in MF calves and in EW calves at any growth stage. This was probably due to a smaller difference in the proportion of energy supplied by milk than in the study of Black and Tribe (1973). Furthermore, it was not significantly altered by the reduction of food intake. However, k_g decreased from 0.638 to 0.491 as the calves got older and as the proportion of energy supplied by the milk was reduced from 86% to 8%.

In the 5-week-old calves the milk supplied 90% or 82% of energy intake: k_g (0.638) was slightly lower but not significantly different from the value obtained by van Es (1970) in preruminant calves. In the 10-week-old calves, milk substitutes supplied 40% or 22% of energy intake, and energy retained as protein (55% of RE) was similar to that of the 5-week-old calves. The low k_g (0.507) was associated with a low maintenance requirement which was not significantly different from that of the five, 23 or 34-week-old calves. If the maintenance requirement was the same at these four growth stages then the k_g would amount to 0.57 in the 10-week-old calves and it would meet the expected value for that proportion of energy supplied by milk.

The rations were almost the same and the milk supplied 16% or 0% of energy intake in the 23 and 34-week-old calves. The k_g amounted to 0.491 and 0.545 respectively. This improvement could be due to an increase in lipogenesis. However, these ME efficiencies were not significantly different from the k_f (0.510 ± 0.008) obtained in mature sheep.

The maintenance requirement (kJ/kg $W^{0.75}$) was not altered by milk supply at any growth stage. It did not decrease significantly between five and 34 weeks of age and was similar to the values obtained by van Es (1970) in preruminant calves and by Hoffmann, Jentsch and Schiemann (1977) in growing bulls. Furthermore, a 25% reduction of food intake was followed by a 15% decrease in maintenance requirement of MF calves at three growth stages and of EW calves at 10 weeks of age (*Table 10.2*), and this was in agreement with the results of Thorbek and Henckel (1976).

Feeding milk substitutes in addition to a solid feed caused increases in ME intake and weight gain, but did not alter either d_E, d_N and q of the solid feed, or k_g and maintenance requirement of calves. On the other hand, a reduction of food intake increased d_E, d_N and q slightly, did not change k_g and reduced the maintenance requirement, at least in milk-fed calves.

References

BLACK, J.L. and TRIBE, D.E. (1973). *Aust. J. agric. Res.*, **24**, 763–773

ES, A.J.H. VAN (1970). *Energy Metabolism of Farm Animals*, pp. 97–100. Eds. A. Schürch and C. Wenk. EAAP Publication No. 13. Zürich: Juris Druck and Verlag

GEAY, Y. (1970). *Journées d'Etude sur la Production de Viande par les Jeunes Bovines*, pp. 95–109. Versailles: SEI Ed. INRA, Route de St-Cyr

HOFFMAN, L., JENTSCH, W. and SCHIEMANN, R. (1977). *Arch. Tierernähr.*, **27**, 545–562

JENTSCH, W., WITTENBURG, H. and SCHIEMANN, R. (1976). *Arch. Tierernähr.*, **26**, 575–585

LE NEINDRE, P., PETIT, M. and MULLER, A. (1976). *Annls Zootech.*, **25**, 521–531

THORBEK, G. and HENCKEL, S. (1976). *Energy Metabolism of Farm Animals*, pp. 117–120. Ed. M. Vermorel. EAAP Publication No. 19. Clermont-Ferrand: G. de Bussac

VERMOREL, M., BOUVIER, J.C. and GEAY, Y. (1976). *Energy Metabolism of Farm Animals*, pp. 217–220. Ed. M. Vermorel. EAAP Publication No. 19. Clermont-Ferrand: G. de Bussac

VERMOREL, M., BOUVIER, J.C. and GEAY, Y. (1980). *Annls Zootech.*, **29**, 67–88

11

THE PREDICTED AND OBSERVED METABOLIZABLE ENERGY VALUES OF MIXTURES OF MAIZE SILAGE AND BARLEY FED TO CATTLE

F.W. WAINMAN, J.S. SMITH and P.J.S. DEWEY
The Rowett Research Institute, Bucksburn, Aberdeen AB2 9SB, Scotland

Summary

Three Hereford × Friesian steers with live weights about 450 kg were fed at about 1.5× maintenance on three mixtures of maize silage and barley. The ratios of maize silage:barley were 77:23, 53:47 and 28:72 in terms of ME. The results of complete energy balance trials showed that the ME values of the dietary components were additive with no evidence of large associative effects.

Introduction

Corn (maize) silage is one of the most important ruminant feedstuffs in the United States where very large acreages are grown. In the United Kingdom maize has become a major forage crop in the south and with new varieties and improved harvesting methods it has proved possible to conserve consistently high yields of dry matter (DM) as silage. The area grown for feeding green or as silage increased spectacularly from 1000 ha in 1970 to 29 000 ha in 1976.

In the United States corn and corn silage are fed together in mixed rations for fattening steers, and Byers, Johnson and Matsushima (1976) fed four ratios of whole corn:corn silage each to three steers. They reported that the metabolizable energy (ME) values of both corn and corn silage when fed in combination were depressed when compared with either fed alone, which indicated large negative associative effects.

The cereal most commonly fed with silage in fattening rations for steers in the United Kingdom is barley. The present experiment was planned to investigate the possible associative effects between maize silage and barley.

Materials and methods

Three Hereford × Friesian steers of approximately 450 kg live weight were fed diets which provided an ME of about 80 MJ/d. The three diets were prepared from maize silage and rolled barley, the chemical compositions of which are shown in *Table 11.1*.

Table 11.1 THE CHEMICAL COMPOSITIONS OF THE MAIZE SILAGE AND BARLEY

	Maize silage (g/kg)	*Barley* (g/kg)
DM as fed (1) Toluene	311	
(2) Oven	303	867
	On a DM basis	
Ash	45	24
Nitrogen	15.5	20.2
Ether extract	37	25
Crude fibre	188	48
Neutral detergent fibre	414	191
Acid detergent fibre	236	60
Acid detergent lignin	20	15
Modified acid detergent fibre	221	58
Starch	305	664
Total VFA	24	
Lactic acid	56	
NH_3 N/Total N	12.4%	
Mean pH	3.55	
Gross energy (MJ/kg DM)	18.54	18.63

DM = dry matter

The maize silage was made at the Grassland Research Institute, Hurley, near Maidenhead, Berkshire; it was blast-frozen and transported by refrigerated lorry to the –20°C cold store at the Rowett Research Institute. Each week enough silage for the diets was removed from store, allowed to thaw, mixed and weighed into daily rations in plastic tubs which were then stored at –5°C until 24 h before feeding when they were removed to room temperature to thaw. The chemical composition shown is the result of analyses of a combined sample from 12 weekly aliquots.

It was planned that the two dietary components should be combined to provide the ratios 75:25, 50:50 and 25:75 of maize silage:barley in terms of ME content. The ME value of the maize silage was determined in a previous experiment with cattle as 11.6 MJ/kg DM, and the barley was assigned the value 12.9 MJ/kg DM (the value determined for barley in the Feedingstuffs Evaluation Unit, Rowett Research Institute (1976); 1st Report 1975). Unfortunately, because there were changes in DM content and small feed refusals, the planned ratios were not exactly achieved and the actual combinations and amounts of DM consumed are shown in *Table 11.2*.

Table 11.2 DIETARY RATIOS, DRY MATTER INTAKES AND DIGESTIBILITIES OF THE DIETS

Dietary ratio silage: barley	*DM intakes* (kg/d)		*Digestibility* (%)	
	Silage	*Barley*	*DM*	*Energy*
77:23	5.380	1.471	76.0	74.5
53:47	3.656	2.885	79.7	78.4
28:72	1.815	4.214	80.4	79.4

Urea was added to the maize silage to raise the crude protein (CP) content (by 15 g/kg DM) to an effective 112 g/kg DM; the barley contained 126 g/kg DM. Mineral supplements were given to all animals at the rate of 112 g/d.

Each diet was fed to each steer for three weeks with excreta being collected over the last six days. For the final three days of each period measurements of gaseous exchange were made in an 'open and shut' respiration chamber (Blaxter, Brockway and Boyne, 1972). The heat production of the animals was calculated using Brouwer's equation (1965).

Results and discussion

The apparent digestibilities of the diets are shown in *Table 11.2.* The DM digestibility increased with the inclusions of barley. The energy data are shown in *Table 11.3* and the metabolizability (q) of the diets increased from 63.8% with 22% barley to 67.8% with 47% barley, but it remained at 67.9% with the highest proportion of barley (72%).

Table 11.3 ENERGY INTAKES, ENERGY LOSSES, ME VALUES, ME/GE%, ME/DE% FOR THE THREE DIETS

Ratios of silage:barley		*77:23*	*53:47*	*28:72*
ME intakes (MJ/d)	Predicted*	81.39	79.63	75.41
	Actual	81.00	82.32	76.33
Energy loss as % GE	Urine	3.3	2.5	3.4
	Methane	7.5	8.0	8.1
ME of diets (MJ/kg DM)	Predicted*	11.9	12.2	12.5
	Actual	11.8 ± 0.2	12.6 ± 0.2	12.7 ± 0.1
ME/GE (%) (q)		63.8 ± 0.72	67.8 ± 1.29	67.9 ± 0.59
ME/DE (%)		85	87	86

* Predicted from a linear model, i.e. assuming no associative effects
GE = gross energy
DE = digestible energy
ME = metabolizable energy

Table 11.3 compares the ME values obtained with the steers with those predicted from a linear model, i.e. assuming no associative effects. There is little evidence for any major associative effects with these two feedstuffs.

Because there were several differences between the experiments, it is difficult to be certain why the results do not accord with those of the American workers. The maize silage used here was of a higher quality: digestible energy and ME values determined in a previous experiment were 13.3 and 11.6 MJ/kg DM in comparison with 12.4 and 10.6 MJ/kg DM for the silage fed by Byers *et al.* (1976), and the starch content of the United Kingdom silage (305 g/kg DM) suggested a grain content higher than the 7% of the American silage. Also, the difference in ME values between the two components of the present diets, 11.6 and 12.9 MJ/kg DM, was less than that between the components of the United States diets, i.e. 10.6 and 13.3 MJ/kg DM respectively.

A more important difference between the two experiments may lie in the physical form in which the grains were fed and in the nature of the starch contents. Results suggest that the sites of digestion of the two grains are to some extent different in ruminants, and that more maize starch escapes ruminal fermentation (Armstrong and Beever, 1969; Ørskov, Fraser and Kay, 1969; Beever, Coehla Da Silva and Armstrong, 1970; Tucker, Mitchell and Little, 1967).

Conclusions

When maize silage and barley were fed together to steers at about 1.5X the maintenance level, there were no major associative effects and thus the ME values of the dietary components could be used additively to calculate the ME of the mixed diets.

References

ARMSTRONG, D.G. and BEEVER, D.E. (1969). *Proc. Nutr. Soc.*, **28**, 121–131

BEEVER, D.E., COEHLA DA SILVA, J.E. and ARMSTRONG, D.G. (1970). *Proc. Nutr. Soc.*, **29**, 43A

BLAXTER, K.L., BROCKWAY, J.M. and BOYNE, A.W. (1972). *Q. J. exp. Physiol.*, **57**, 60–72

BROUWER, E. (1965). *Report of Sub-committee on Constants and Factors. Proceedings of the 3rd Symposium on Energy Metabolism, Troon*, 1954, pp. 441–443. EAAP Publication, No. 11. London: Academic Press

BYERS, F.M., JOHNSON, D.E. and MATSUSHIMA, J.K. (1976). *Proceedings of the 7th Symposium on Energy Metabolism, Vichy*, pp. 253–256. Ed. M. Vermorel. EAAP Publication No. 19. Clermont-Ferrand: G. de Bussac

ØRSKOV, E.R., FRASER, C. and KAY, R.N.B. (1969). *Br. J. Nutr.*, **23**, 217–226

ROWETT RESEARCH INSTITUTE (1976). *Feedingstuffs Evaluation Unit 1st Report 1975.* Edinburgh: DAFS

TUCKER, R.E., MITCHELL, Jr., G.E. and LITTLE, C.O. (1968). *J. Anim. Sci.*, **27**, 824–826

METHANE PRODUCTION IN DAIRY COWS

P.W. MOE and H.F. TYRRELL
Ruminant Nutrition Laboratory, Animal Science Institute, Agricultural Research Science and Education Administration, US Department of Agriculture, Beltsville, Maryland 20705, USA

Summary

The relationship between diet composition, intake level and methane production was investigated: the data were obtained during 404 total energy balance trials with cattle. Methane production in all trials was measured in open-circuit respiration chambers. The most useful predictor of total methane production was the amounts of soluble residue, hemicellulose and cellulose that were apparently digested. The regression was as follows:

CH_4 (MJ) = 1.837 + 1.142 ± 0.063 kg digested soluble residue + 2.142 ± 0.326 kg digested hemicellulose + 5.828 ± 0.406 kg digested cellulose

Introduction

Kriss (1930) found a linear relationship between methane production and dry matter (DM) intake of cattle. Bratzler and Forbes (1940), with cattle, and Swift *et al.* (1948), with sheep, identified a relationship between methane production and the amount of carbohydrate apparently digested. Blaxter and Clapperton (1965) concluded that methane production at maintenance by cattle and sheep could be described by the relationship:

$$CH_4 \text{ (percentage of gross energy consumed)} = 3.67 + 0.062 \text{ digestibility (D; \%)}$$

and that the rate of change in percentage of gross energy lost as methane by an increase in intake equal to one times maintenance could be described by

$$b = 0.050\ D\ (\%) - 2.37$$

for mixed diets. The object of this chapter is to identify the relationship between dietary carbohydrate and methane production in cattle ingesting diets at a substantially greater range of intake than in the studies cited above.

Materials and methods

The data included in the present study are from 404 total energy balance trials in open-circuit respiration chambers with Holstein cows. The experiments were designed to study the following effects on the energy value of dairy cattle diets:

the protein level (Moe and Tyrrell, 1972a); the proportion of concentrate (Tyrrell and Moe, 1972); the relative value of corn grain and barley (Tyrrell and Moe, 1974), oats (Moe and Tyrrell, 1972b), beet pulp (Tyrrell, Moe and Bull, 1973), dried brewers grains, dried distillers grains (Tyrrell and Moe, 1979) or wheat bran (Moe and Tyrrell, 1977b); the physical form of corn grain in the total diet (Moe and Tyrrell, 1977a; Moe, Tyrrell and Hooven, 1973) or the incremental value of corn grain (Moe and Tyrrell, 1979). Methane production was measured during three or four consecutive 24-hour periods during each balance trial. Methane energy was calculated from volume measurement by the factor 39.54 kJ/ℓ (Brouwer, 1965).

The chemical analyses of feed and faeces were based on the fibre methods of Goering and Van Soest (1970). These included acid-detergent fibre (ADF), cell walls (neutral-detergent fibre; NDF), cellulose (ADF–lignin–silica) and hemicellulose (NDF–ADF). An additional fraction (soluble residue) was calculated by subtracting crude protein and ether extract from the neutral-detergent solubles. The soluble residue fraction consists primarily of starch.

Statistical relationships were established by multiple regression analysis with the least squares methods (Searle, 1971).

Results and discussion

The mean and range of carbohydrate components apparently digested (kg/d) were soluble residues 4.64 and 0.86–9.26, hemicellulose 0.86 and 0.01–2.81, and cellulose 0.80 and 0.21–1.94. The maximum amount of total carbohydrate (soluble residue + hemicellulose + cellulose) digested was 11.5 kg/d which is approximately twice that of the study of Bratzler and Forbes (1940).

A regression, in which total methane production was related to intake of crude protein, ether extract, soluble residue, hemicellulose, cellulose and lignin, indicated non-significant terms for crude protein, ether extract and lignin. A second regression related methane production to the remaining terms ($r^2 = 0.67$, $S_{y \cdot x} = 2.59$):

$$\begin{aligned} CH_4\ (\text{MJ/d}) = 3.406 &+ 0.510 \pm 0.075 \text{ soluble residue (kg fed)} \\ &+ 1.736 \pm 0.310 \text{ hemicellulose (kg fed)} \\ &+ 2.648 \pm 0.318 \text{ cellulose (kg fed)} \end{aligned}$$

Total methane production was also related to the amount of each carbohydrate component that was apparently digested as follows ($r^2 = 0.74$, $S_{y \cdot x} = 2.34$):

$$\begin{aligned} CH_4\ (\text{MJ/d}) = 1.837 &+ 1.142 \pm 0.063 \text{ soluble residue (kg digested)} \\ &+ 2.142 \pm 0.326 \text{ hemicellulose (kg digested)} \\ &+ 5.828 \pm 0.406 \text{ cellulose (kg digested)} \end{aligned}$$

Note that the production of methane per gram of cellulose digested is three times that per gram of hemicellulose digested and five times that per gram of soluble residue digested. A part of the latter difference may be due to a substantially lower proportion of soluble residues being fermented in the rumen and a larger proportion being digested and absorbed as glucose in the small intestine. The production of methane from 26 soluble carbohydrates was studied by

Czerkawski and Breckenridge (1969) in 53 experiments with an artificial rumen. They concluded that, with the exception of rhamnose, the amount of methane produced was not very dependent on the type of carbohydrate, but it was related to the amount of carbohydrate fermented. The mean amount of methane produced was 6 kJ for 100 kJ of sugar fermented. If the gross energy value of carbohydrate apparently digested in the present experiment is assumed to be 17.5 kJ/g, then the kJ methane produced/100 kJ of carbohydrate apparently digested was 6.5 for soluble residue, 11.5 for hemicellulose and 33.6 for cellulose. These data indicate major differences in the amount of methane produced during the digestion of various carbohydrate fractions. Experiments of the type described by Czerkawski and Breckenridge (1969) are needed to establish relationships between fermentation of structural carbohydrates and methane production.

The effect of intake level on methane production was studied by deriving separate regressions within four intake level classes. These data are shown in *Table 12.1.* Methane production was not as closely related to digestible carbohydrates at low intake as at higher levels. The differences among the partial regression coefficients for individual digestible carbohydrate fractions were less at intakes of less than 1.5 times maintenance than at higher intake levels. Much of this difference is associated with the digestible cellulose fraction that increased from 2.653 MJ/kg of CH_4 below 1.5 times maintenance to 7.109 MJ/kg of CH_4 above 1.5 times maintenance. The latter rate of methane production is equivalent to 41% of the cellulose energy apparently digested. The high rate of methane production associated with cellulose degradation may indicate a partial shift from a propionate type fermentation to methanogenic fermentation in the rumen. In addition to a shift away from propionate, however, the increase in methane would have to come from sources other than the products of cellulose degradation alone to balance the stoichiometry of the conversion of cellulose to end-products of fatty acids and gases.

Table 12.1 EFFECT OF LEVEL OF INTAKE ON METHANE PRODUCTION FROM DIGESTIBLE CARBOHYDRATE FRACTIONS

Intake level (X Maint.)	*N*	*s.d.*	r^2	*Intercept*	*MJ CH_4 produced per kg carbohydrate digested*		
					Soluble residue	*Hemicellulose*	*Cellulose*
<1.5	103	1.21	0.66	1.93	1.92 ± 0.17	2.25 ± 0.51	2.65 ± 0.45
1.5 – 2.5	119	2.01	0.65	0.95	1.19 ± 0.20	2.64 ± 0.64	6.36 ± 0.68
2.5 – 3.5	152	2.59	0.61	−2.38	1.49 ± 0.20	2.15 ± 0.55	8.19 ± 0.71
>3.5	30	3.39	0.53	−2.10	1.51 ± 0.64	2.55 ± 1.76	5.74 ± 2.43
Intake >1.5	301	2.51	0.65	1.35	1.07 ± 0.10	1.89 ± 0.39	7.11 ± 0.50
All data	404	2.34	0.73	1.84	1.14 ± 0.06	2.14 ± 0.33	5.83 ± 0.41

These data indicate that the production of methane is influenced by the nature of the carbohydrate digested, and that this effect is relatively less important at low levels of feed intake in comparison with high levels of feed intake.

References

BLAXTER, K.L. and CLAPPERTON, J.L. (1965). *Br. J. Nutr.*, **19**, 511–522
BRATZLER, J.W. and FORBES, E.B. (1940). *J. Nutr.*, **19**, 611–613
BROUWER, E. (1965). Proc. 3rd Symposium on Energy Metabolism, Troon, Scotland. *Eur. Ann. Anim. Prod. Publ. No.* **11**, 441–443
CZERKAWSKI, J.W. and BRECKENRIDGE, G. (1969). *Br. J. Nutr.*, **23**, 925
GOERING, H.K. and VAN SOEST, P.J. (1970). *USDA Agr. Hbbk. No. 379.* Washington, DC 20402 : US Govt. Printing Office
KRISS, M. (1930). *J. Agric. Res.*, **40**, 283–295
MOE, P.W. and TYRRELL, H.F. (1972a). *J. Dairy Sci.*, **55**, 318–324
MOE, P.W. and TYRRELL, H.F. (1972b). *J. Dairy Sci.*, **55**, 686
MOE, P.W. and TYRRELL, H.F. (1977a). *J. Dairy Sci.*, **60**, 752–758
MOE, P.W. and TYRRELL, H.F. (1977b). *Northeastern Section ADSA-ASAS Mtgs. (Abstr.)*
MOE, P.W. and TYRRELL, H.F. (1979). *J. Dairy Sci.*, **62**, 447–454
MOE, P.W., TYRRELL, H.F. and HOOVEN, N.W., Jr. (1973). *J. Dairy Sci.*, **56**, 1298–1304
SEARLE, S.R. (1971). *Linear Models.* New York: John Wiley and Sons
SWIFT, R.W., BRATZLER, J.W., JAMES, W.H., TILLMAN, A.D. and MEEK, D.C. (1948). *J. Anim. Sci.*, 7, 475–485
TYRRELL, H.F. and MOE, P.W. (1972). *J. Dairy Sci.*, **55**, 1106–1112
TYRRELL, H.F. and MOE, P.W. (1974). *J. Dairy Sci.*, **57**, 451–458
TYRRELL, H.F. and MOE, P.W. (1979). Unpublished data
TYRRELL, H.F., MOE, P.W. and BULL, L.S. (1973). *J. Dairy Sci.*, **56**, 1384 (Abstr.)

13

A STUDY OF THE NUTRITIONAL ENERGETICS OF SODIUM-HYDROXIDE-TREATED STRAW PELLETS IN RATIONS FED TO GROWING LAMBS

J. ROBB*, P.J. EVANS and C. FISHER**
Unilever Research Laboratory, Colworth House, Sharnbrook, Bedford, England

Summary

A total of 60 growing lambs was used to measure feed conversion and energy retention when fed on rations containing 20% or 40% of untreated straw or sodium-hydroxide-treated straw pellets. Most of the differences in performance and energy retention were due to the increased digestibility and metabolizability resulting from alkali treatment of the straw. The efficiency of conversion of metabolizable energy (ME) to net energy or live-weight gain (LWG) was also higher with the treated straw rations.

Introduction

The utilization of cereal straw in intensive ruminant systems is severely restricted due to the low energy density of straw and limitations on dry matter (DM) intake imposed by physical form. Mechanical processing by fine grinding reduces the residence time of straw particles in the rumen and thus allows greater intakes of straw DM. Grinding and pelleting can reduce the digestibility of straw but the net energy is unaffected due to the increased efficiency with which ME is utilized (van der Honing and van Es, 1973).

More recently, renewed interest in chemical treatment to increase the digestibility of straw has resulted in a range of techniques being developed whereby varying degrees of mechanical processing are carried out simultaneously with chemical treatment (Jackson, 1977).

In the industrial process developed by the Unilever Research Laboratory and briefly described by Wilson and Brigstocke (1977), coarsely ground straw is treated with concentrated sodium hydroxide solution during the pelleting process. The 'dry' industrial process eliminates effluents and the chemical reaction is completed during pelleting.

The nutritional benefits arising from the above process have been attributed primarily to hydrolysis of alkali-labile ester bonds between lignin and the cell-

* Current address: Unitrition International Limited, Basing View, Basingstoke, Hampshire, England
** Current address: Poultry Research Centre, Kings Buildings, West Mains Road, Edinburgh, Scotland

wall carbohydrates, and also to a reduction in the crystallinity of cellulose due to absorption of hydrated sodium hydroxide molecules. The net effect is that both the rate and extent of fermentation of the cell-wall carbohydrates are markedly increased (Evans, 1979).

Materials and methods

A factorial design allowed the following comparisons: straw type (wheat straw and barley straw), straw treatment (ground untreated and nutritionally improved straw; NIS), inclusion level in the ration (0, 20 and 40%) and two levels of intake. A basal ration consisted of 30% chopped hay and 70% of a concentrate mixture composed of 50% barley, 20% soya bean meal, 15% copra meal, 5% groundnut meal, 5% molasses and 5% minerals/vitamins. The untreated straw or NIS was included at the expense of an equal proportion of the total basal ration. The hay was fed in the chopped form whereas all other components of the ration were mixed and pelleted prior to feeding.

A total of 60 North Country Cheviot × Border Leicester lambs were fed one of the nine rations from an initial weight of about 22 kg to a final weight of approximately 42 kg. All animals were individually penned on slatted floors. Complete energy balances were carried out using indirect, gaseous-exchange calorimeters when animals were in the 33–35 kg weight range.

Results and conclusions

The gross chemical composition of the rations concerned with level of straw inclusion and treatment has been summarized in *Table 13.1*. The formulation of the concentrate mixture was intended to supply adequate protein, minerals and vitamins even when straw or NIS replaced up to 40% of the basal ration.

Table 13.1 COMPOSITION OF RATIONS

	Basal ration	*Untreated straw (%)*		*Treated straw (%)*	
		20	40	20	40
Ether extract (g/kg DM)	23	21	19	21	19
Crude protein (g/kg DM)	196	162	130	163	130
Crude fibre (g/kg DM)	143	205	263	192	241
Ash (g/kg DM)	85	82	79	96	106
Gross energy (MJ/kg DM)	18.19	18.21	18.23	17.99	17.79
(MJ/kg OM)	19.87	19.83	19.79	19.88	19.90

DM = Dry matter
OM = Organic matter

Due to a change in environment during the balance trials, intake by animals on the higher level feeding was less than that achieved during the rest of the growth period. Therefore, for simplicity, the energy data have been summarized with respect to straw or NIS inclusion level and alkali treatment only.

As was expected, increasing substitution of the basal ration by untreated straw or NIS resulted in a decrease in energy digestibility and metabolizability

(*Table 13.2*). However, at either level of substitution the coefficients associated with the rations containing treated straw were significantly higher ($P<0.05$) than those observed with the corresponding untreated straw rations. Methane losses, expressed as percentage of digestible energy (DE), were similar for the basal ration and ration containing 20% NIS, but they were significantly higher ($P<0.05$) for

Table 13.2 ENERGY BALANCE RESULTS

		Basal ration	*Untreated straw (%)*		*Treated straw (%)*		*s.e.*
			20	40	20	40	
GE intake	(MJ/d)	20.8	21.5	21.2	22.8	22.2	0.66
	(MJ/kg$^{0.75}$)	1.44	1.48	1.51	1.58	1.54	NA
DE	(%GE)	71.3	62.1	57.0	68.4	62.2	1.26
Methane	(%GE)	7.1	7.6	7.3	6.9	7.8	0.28
	(%DE)	9.9	12.3	12.9	10.0	12.5	0.42
Urine	(%GE)	3.2	2.4	2.3	2.7	2.6	0.18
	(%DE)	4.4	3.9	4.0	4.0	4.2	0.28
ME	(%GE)	61.0	52.0	47.4	58.8	51.8	1.16
	(%DE)	85.7	83.8	83.2	86.0	83.4	0.54
Retained E	(%GE)	10.0	5.2	5.4	11.2	7.7	1.10
	(%DE)	14.0	8.3	9.3	16.3	12.3	1.57
	(%ME)	16.2	9.8	11.1	19.0	14.5	1.83

GE = Gross energy
DE = Digestible energy
ME = Metabolizable energy

all other rations. The ME/DE ratios associated with the basal ration and 20% NIS ration reflected the differences in methane loss and were significantly higher ($P<0.05$) than with all other rations. The overall efficiency with which ME was retained in the body was similar for both rations containing NIS and the basal ration and these were significantly higher ($P<0.05$) than for both rations containing untreated straw.

ESTIMATED MAINTENANCE REQUIREMENT (M_m) AND k_f

As stated earlier, DM intakes during the periods in which energy balances were measured were lower than expected. Therefore, it was not possible to predict precisely by regression analysis the ME required for maintenance for individual treatments. When all of the data were analysed the common intercept was found to be 0.485 MJ/kg$^{0.75}$ of ME per day (s.e.$_a$ = 0.037). However, since differences in the partial efficiency of energy utilization for fattening (k_f) were observed for the main ration types (basal ration, untreated straw and NIS rations) a more accurate estimate of M_m was obtained when the data were allowed to have different slopes but a common intercept. In this case, M_m was found to be 0.410 MJ/kg$^{0.75}$ of ME per day (s.e.$_a$ = 0.031). With this M_m value the k_f values were 32.5% (s.e.$_b$ = 3.3), 23.9% (s.e.$_b$ = 2.0) and 32.9% (s.e.$_b$ = 1.3) for the basal ration, untreated straw rations and NIS rations respectively. The k_f associated with the untreated straw rations was significantly lower ($P<0.05$) than either the basal ration or the NIS rations. The M_m value of 0.410 MJ is 21% higher than that reported recently by Thomson, Fenlon and Cammell (1979), but it is in reasonable agreement with that of Ministry of Agriculture, Fisheries and Food (1975).

PREDICTION OF RATION ME VALUES FROM DIGESTIBLE NUTRIENT CONTENTS

Since it is known that alkali treatment of straw renders a proportion of the hemicellulose fraction soluble (Evans, 1979), conventional crude fibre and nitrogen-free extract (NFE) partition may be misleading when normal equations are used to predict the ME value from digestible nutrient content. Hence, the values predicted from the equation used by Ministry of Agriculture, Fisheries and Food (1975) in the United Kingdom and that proposed by van Es (1978) in the Netherlands have been computed and presented for comparison with the observed ME values in *Table 13.3.* On average, values predicted by Ministry of Agriculture, Fisheries and Food, using Rostock coefficients, were 2.8% lower and those predicted by van Es were 5.9% lower than those actually observed. Also, the ME content of the digestible organic matter (DOM) was not constant, and the values for the basal ration and 20% NIS ration were significantly higher ($P<0.05$) than those associated with all other rations.

Table 13.3 PREDICTION OF RATION ME VALUE FROM DIGESTIBLE NUTRIENT CONTENT

	Basal ration	*Untreated straw (%)*		*Treated straw (%)*		*s.e.*
		20	40	20	40	
Observed: MJ ME/kg DM	11.10	9.47	8.68	10.69	9.23	0.21
Predicted: MAFF, 1975	10.73	9.35	8.55	10.09	9.08	0.18
van Es, 1978	10.37	9.03	8.28	9.76	8.83	0.18
Observed: MJ ME/kg DOM	16.21	15.67	15.50	16.34	15.52	0.13

ME = Metabolizable energy
DM = Dry matter
DOM = Digestible organic matter

LAMB PERFORMANCE AND FEED EFFICIENCY

Although carcass gain was not determined directly, LWG and feed intake were measured during the 20 kg growth trial. The main parameters summarized in *Table 13.4* indicate that although the average daily DM intake (DMI) was similar

Table 13.4 LAMB PERFORMANCE AND FEED EFFICIENCY

	Basal ration	*Untreated straw (%)*		*Treated straw (%)*		*s.e.*
		20	40	20	40	
Duration (d)	99.2	119.0	145.2	99.8	111.4	6.1
Water intake (kg/d)	4.29	3.86	3.68	4.98	5.39	0.29
DMI (kg/d)	1.14	1.18	1.18	1.18	1.22	0.02
LWG (kg/d)	0.221	0.183	0.149	0.213	0.202	0.010
DMI/LWG (kg)	5.31	6.66	8.09	5.68	6.29	0.28
MEI/LWG (MJ/kg)	59.1	62.9	70.3	60.8	58.3	2.9
MEP/LWG (MJ/kg)	31.5	29.8	29.9	32.3	27.1	1.5

DMI = Dry matter intake
LWG = Live-weight gain
MEI = Metabolizable energy intake
MEP = Metabolizable energy intake above maintenance

for the main treatments, marked differences in daily LWG resulted from the lower energy densities and, to a lesser extent, lower k_f values of the untreated straw rations. The overall requirement of ME per kg LWG was significantly higher ($P<0.05$) for the 40% untreated straw ration. However, when the ME required for maintenance (0.410 MJ/kg$^{0.75}$ per day) was deducted from the ME intake (MEI), the ME required per kg LWG was similar with the exception of the 40% NIS ration which was significantly lower ($P<0.05$) than either the 20% NIS ration or basal ration.

Interpretation of the efficiency of utilization of ME above maintenance, either in terms of retained energy as measured using calorimeters or LWG, is dependent on the assumption that animals fed the different rations have a common maintenance requirement. If this assumption is invalid, the partition of energy between maintenance and production could be markedly different. However, the present data indicate that, with the exception of the ration with the high level of untreated straw, differences in performance could be largely attributed to the MEI.

References

EVANS, P.J. (1979). *Proceedings of the Symposium on Straw Decay, Hatfield, April.* In press

JACKSON, M.G. (1977). *Anim. Feed Sci. Tech.*, 2, 105

MINISTRY OF AGRICULTURE, FISHERIES AND FOOD (1975). *Tech. Bull. No. 33.* London: HMSO

THOMSON, D.J., FENLON, J.S. and CAMMELL, S.B. (1979). *Br. J. Nutr.*, **41**, 223

VAN DER HONING, Y. and VAN ES, A.J.H. (1973). *Proceedings of the Sixth Symposium on Energy Metabolism*, p. 209. EAAP Publication No. 14, Universität Hohenheim: Dokumentationsstelle

VAN ES, A.J.H. (1978). *Livestock Prod. Sci.*, **5**, 331

WILSON, P.N. and BRIGSTOCKE, T. (1977). *Process Biochem.*, **12**, 17

14

EFFECTS OF METABOLIZABLE ENERGY CONTENT OF DIET AND FEEDING LEVEL ON THE EFFICIENCY OF ENERGY UTILIZATION BY YOUNG GROWING LAMBS

M. THERIEZ, M. TISSIER and J.P. BRUN
I.N.R.A. – C.R.Z.V. de Theix, 63110 Beaumont, France

Summary

The efficiency of use of metabolizable energy (ME) for fattening (k_{pf}) value was measured by the slaughter technique on five groups of lambs receiving diets composed of lucern wafers and concentrate in different proportions at two feeding levels. The energy concentration had an effect on the k_{pf} value (not significant) but the feeding level did not. The k_{pf} increased when the proportion of energy retained as protein decreased. Body composition was not affected by experimental treatments (covariance analysis).

Introduction

In mature animals, whose body gain is mainly composed of lipids, the efficiency with which ME is used for fattening (k_{pf}) increases with the ME concentration of the diet (q). In the young growing animal, the body gain contains more protein than in the adult, and the protein is deposited with a lower efficiency than fat (0.33 and 0.77 respectively; Ørskov and MacDonald, 1970), so the ME efficiency of gain must be lower than for fattening. Similarly, any increase in energy concentration or feeding level which induces both higher growth rate and percentage of fat in the body gain (Black, 1974) would also improve k_{pf}.

In this experiment the energy and protein gain of young lambs and their ME utilization for gain with different diets were measured by the comparative slaughter technique, and the k_{pf} values were compared with the corresponding k_f (efficiency for fattening alone).

Materials and methods

Thirty intact male lambs (65.5 ± 4.5 days old and 18.0 ± 2.0 kg in body weight (BW) at the beginning of the experiment) were divided into five groups. The animals, fed once daily in individual pens with slatted floors, received a diet composed of dehydrated lucerne wafers (20.7% crude protein) and concentrate (maize, soybean meal and minerals) in different proportions.

The first group received a diet composed of 80% forage (F) and 20% concentrate (C; 80 F/20 C) *ad lib.* Its energy consumption was considered to be at a moderate level. This energy level was offered to two other groups of lambs as diets composed of 50% of each component (50 F/50 C) or 20% F and 80% C. A fourth group of lambs was offered the 50 F/50 C diet *ad lib.* : this was the high feeding level. The last group received the 20 F/80 C diet at this high feeding level.

The digestibility of the five diets was measured with 20 other lambs fed with the same quantity of dry matter (DM) per kilogram of BW as the experimental animals. Thirty-two lambs (three at the beginning of the experiment and 29 at the end; one of the experimental animals was discarded) were slaughtered and analysed for water, ash, proteins and energy content. The difference provided the estimate of lipid content.

Results and discussion

The ME for growth was calculated by estimating the maintenance requirement on 418 kJ/$W^{0.75}$ per day. The k_{pf} values varied from 45.4 to 53.1 between the three groups with the moderate feeding level, and from 46.4 to 50.1 in the high feeding level group (*Table 14.1* and *Figure 14.1*). The differences between feeding levels were not significant and energy concentration had a nearly significant effect in the moderate feeding level group. These values are lower than the corresponding k_f values but the differences were limited and were less than the standard error. This result agrees with Bouvier and Vermorel's (1975) data, but

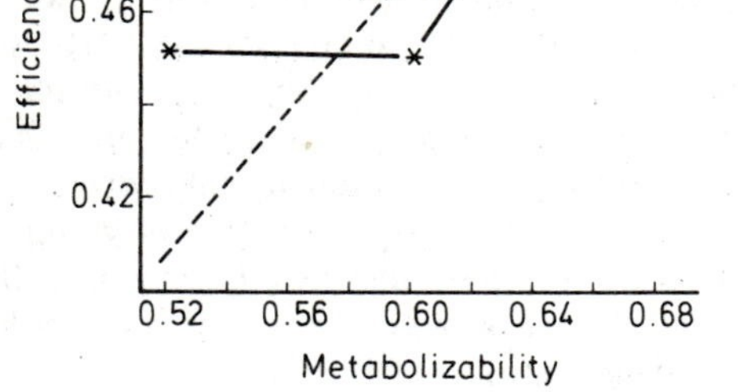

Figure 14.1 Variation of k_{pf} with energy concentration of diet (ME/GE = q).

the k_{pf} of lambs might have been overestimated because the maintenance requirement used in this trial was higher than the value obtained by these authors with animals of the same age and the same breed (363 ± 37 kJ/$W^{0.75}$ per day).

The 80 F/20 C group is quite different from the others, its k_{pf} value being higher than the corresponding k_f value by 4 points. This discrepancy might be explained by the fact that the F contained a high percentage of fine particles. The utilization of ground roughage is known to be better than that of long F. When the

Table 14.1 RETENTION OF ENERGY AND PROTEIN AT DIFFERENT FEEDING LEVELS

Feeding level	*Moderate*			*Significance between levels*	*High*		*RSD*
Diet composition (%)	80 F/20 C	50 F/50 C	20 F/80 C		50 F/50 C	20 F/80 C	
ME/DE	0.85	0.86	0.90		0.87	0.91	
ME concentration (MJ/kg DM)	9.45	11.08	12.83		11.12	13.8	
q = ME/GE	0.52	0.60	0.68		0.61	0.70	
Fattening period							
Length (d)	54.2 ± 7.9	54.2 ± 7.9	52.4 ± 7.5		52.7 ± 6.8	54.2 ± 7.9	
Slaughter weight (kg)	30.8 ± 1.2	31.2 ± 1.5	31.9 ± 2.0		32.5 ± 1.8	33.7 ± 1.9	
Daily gain (g/d)	266^{A}	286^{AB}	326^{B}	XX	317	340	37
ME ingested (MJ)	609.3	614.1	621.0	X	649.7	695.6	82
Proteins retained (g)	1,806	1,915	2,131	X	2,079	2,249	289
Lipids retained (g)	$2,882^{a}$	$3,114^{ab}$	$3,776^{b}$	XXX	$3,527^{a}$	$4,219^{b}$	607
Energy retained (MJ)	153.3^{A}	162.2^{B}	194.2^{C}	XXX	182.7^{A}	214.7^{B}	236
Energy retained as proteins (%)	26.8	26.4	24.7		25.5	23.7	
Utilization of ME: k_{pf}	45.4	45.3	53.1	NS	46.4	50.7	8

At the same level a, b, c : $P<0.10$; A, B, C : $P<0.05$

Between levels X $P<0.10$; XX $P<0.05$; XXX $P<0.01$

ME = Metabolizable energy DE = Digestible energy GE = Gross energy

energy concentration increased, the quantity of fixed proteins and lipids increased (differences were respectively NS and significant at the 10% level) but this had no effect on body composition (after covariance analysis). The increase of feeding level and energy concentration improved the growth rate and fixed energy of lambs (differences highly significant). In the same manner, energy

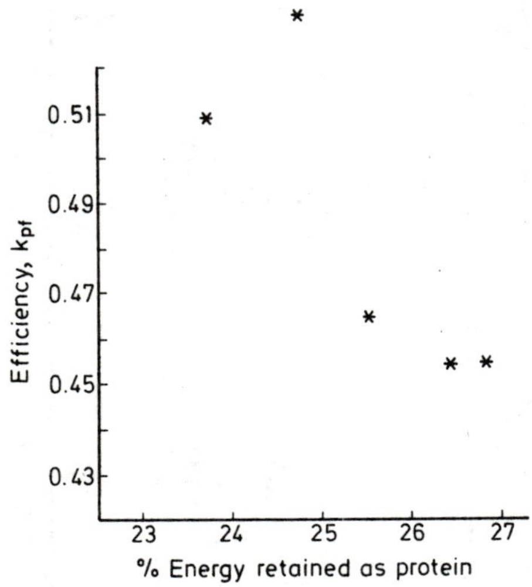

Figure 4.2 Variation of k_{pf} with percentage of energy retained as protein

retained as protein and k_{pf} values decreased (*Figure 14.2*). This is in good agreement with results found by Ørskov and MacDonald (1970) and Bull, Tyrrell and Reid (1976) with sheep, and Geay, Robelin and Vermorel (1979) with cattle.

References

BLACK, J.L. (1974). *Proc. Aust. Soc. Anim. Prod.,* **10**, 211–218

BOUVIER, J.C. and VERMOREL, M. (1975). *Annls Zootechn.,* **24**, 697–710

BULL, L.S., TYRRELL, H.F. and REID, J.T. (1976). *Proceedings 7th Symposium on Energy Metabolism,* pp. 137–140. EAAP Publication No. 19. Clermont-Ferrand: G. de Bussac

GEAY, Y., ROBELIN, J. and VERMOREL, M. (1979). *Proceedings 8th Symposium on Energy Metabolism*, this volume, p. 9

ØRSKOV, E.R. and MacDONALD, I. (1970). In *Energy Metabolism in Farm Animals,* p. 121. Eds. A. Schürch and C. Wenk. EAAP Publication No. 13. Zurich: Juris Druck and Verlag

15

THE INFLUENCE OF DIETARY ENERGY CONCENTRATION ON THE GROWTH EFFICIENCY OF *AD LIB.* FED PIGS

E.H. KEMM
Animal and Dairy Science Research Institute, Irene, South Africa

Summary

The effect of dietary energy density on the metabolizable energy (ME) intake, growth, body composition and ME utilization of *ad lib.* fed pigs was studied. The pigs were slaughtered at two-week intervals at 11–31 weeks of age and chemically analysed. Results show that an increased feed consumption could not compensate fully for the energy intake on low-energy density diets. Hence, dilution of dietary energy content retarded the growth rate and improved the carcass quality. The latter occurred because of a rapid reduction in fat deposition relative to protein gain, which in turn decreased in a pattern very similar to live-mass gain. The statistical model employed proved a splendid tool for quantitating performances at specific mass points.

Introduction

Energy intake greatly determines the amount of fat stored in the pig body (Lewis and Hardy, 1970; Cooke, Lodge and Lewis, 1972; Lodge *et al.*, 1972). Therefore, it is important to quantitate the effects energy intake has on the growth pattern, body composition and the efficiency with which ingested feed nutrients are utilized.

Currently, pigs are slaughtered at a mass of 50 kg for pork and between 85 and 90 kg for bacon, which does not necessarily represent live masses at which optimal efficiency is attained. Consequently, it is desirable to evaluate the effect of energy intake on the performance of the growing pig over its entire period of growth with the aim of establishing the most efficient mass at which the pig must be slaughtered for optimum performance, both biologically and economically. This investigation was thus planned to study the effects of dietary energy concentration on the intake and growth performance of the pig at any specific live mass or live-mass interval.

Materials and methods

Fifty-two castrated male pigs, eight weeks old, were used as experimental animals. Diets were compounded to contain 20% protein (equal in protein quality and origin) but with different energy densities. Dietary energy was diluted by replacement of maize starch with 0, 8, 16, 24 and 32% pine-wood sawdust. Diets 1–5 had dietary ME concentrations of 17.0, 15.6, 14.0, 12.9 and 11.5 MJ/kg dry matter (DM). Pigs were slaughtered at two-week intervals from an age of 11 weeks up to 29–31 weeks. Complete chemical analyses of the whole-pig bodies were then performed.

The Roux mathematical model (Roux and Kemm, 1979) was used to analyse the data. The model allowed treatment effects to be determined at any specific live mass or live-mass interval between 25 kg and about 100 kg (the slaughter limits). For a detailed description of the animals used, the experimental techniques employed and the statistical model and methods used in analysing the data, *see* Roux and Kemm (1979) who obtained, essentially, two sets of estimates of the statistical parameters (slopes and intercepts). The first set consists of the usual estimates for each treatment. The second set is derived from the relationship between some of the parameters and the sawdust proportion as a by-product of an analysis that would allow interpolation between the sawdust levels employed in the experiment. To avoid the possibility of bias from the fit of a possibly inappropriate relationship between sawdust proportion and statistical parameters, the usual estimates for each treatment, as given in *Table 15.1*, were used for the purposes of this chapter.

Table 15.1 STATISTICAL PARAMETERS USED IN CALCULATING THE DATA PRESENTED IN *TABLES 15.2* AND *15.3* (*SEE* ROUX AND KEMM, 1979)

Treatment No.	*1*	*2*	*3*	*4*	*5*
$\bar{\rho}$, ln (per week)	0.9105	0.9090	0.9183	0.9290	0.9292
Common $\hat{\alpha}$, ln (MJ)	8.8716	8.8716	8.8716	8.8716	8.8716
$\bar{\mu}$, ln (MJ)	6.4662	7.3302	6.2356	6.1253	5.9332
b(ME* x mass**)	0.6787	0.6787	0.6787	0.6787	0.6787
a(ME x mass)	−1.0583	−1.0445	−1.0512	−1.0883	−1.1347
b(ME x protein***)	0.5894	0.5894	0.5894	0.5894	0.5894
a(ME x protein)	−2.2984	−2.2984	−2.2984	−2.2984	−2.2984
b(ME x fat****)	1.2141	1.2197	1.3047	1.3476	1.3918
a(ME x fat)	−6.5218	−6.6267	−7.3736	−7.7467	−8.3345

*ME = ln (cumulative ME-intake) as independent variable
**Mass = ln (mass) and dependent variable
*** Protein = ln (protein) as dependent variable
****Fat = ln (fat) as dependent variable

The treatment effects (*Tables 15.2* and *15.3*) were computed for pigs of 45, 70 and 90 kg in live mass, selected to coincide with (a) maximum protein deposition at approximately 45 kg live mass, (b) maximum feed intake at about 70 kg live mass and (c) the maximum mass at which baconers are slaughtered in South Africa, namely 90 kg.

Feed consumption tended to increase as the ME content of the diet decreased. Except for pigs fed the 8% sawdust diet (treatment 2), the other three treatment groups were, however, unable to consume enough feed to attain ME intakes equal to those fed the high energy diet (treatment 1) containing no sawdust. ME intake increased gradually to peak at a live mass of between 73 and 67 kg for treatments 1 and 5 respectively (*Table 15.3*). Maximum ME intake was achieved at a fairly specific live mass, about 70 ± 3 kg, rather than at a specific age, because the pigs in treatment 5 were 36 days older than treatment 1 pigs at maximum feed intake. Pigs fed by Cole *et al.* (1968) consumed about 3.4 kg of feed when fed

Table 15.2 ESTIMATED FEED AND ME INTAKES, GAINS IN LIVE MASS, UTILIZATION EFFICIENCIES OF ME CONSUMED AND BODY PROTEIN AND FAT CONTENTS

Live mass of pigs (g)	45					70					90				
Treatment No.	*1*	*2*	*3*	*4*	*5*	*1*	*2*	*3*	*4*	*5*	*1*	*2*	*3*	*4*	*5*
Diet ME (MJ/kg DM)	17.0	15.6	14.0	12.9	11.5	17.0	15.6	14.0	12.9	11.5	17.0	15.6	14.0	12.9	11.5
Voluntary air-dry feed intake (kg/d)	1.96	2.15	2.15	2.06	2.37	2.32	2.57	2.56	2.40	2.68	2.18	2.43	2.41	2.20	2.35
Voluntary ME intake (MJ/d)	29.6	29.9	26.8	23.7	24.2	35.1	35.6	31.9	27.6	27.5	32.9	33.8	30.1	25.3	24.1
ME utilization (MJ/kg gain in live mass)	42.5	41.6	42.1	44.5	47.5	52.4	51.2	51.9	54.8	58.6	59.0	57.8	58.4	61.7	66.0
Daily live-mass gain (g/d)	697	718	637	533	510	670	695	615	504	469	558	585	515	410	365
Body protein content (kg)	6.87	6.78	6.82	7.05	7.34	10.08	9.96	10.01	10.34	10.76	12.53	12.38	12.46	12.86	13.39
Body fat content (kg)	8.85	8.10	7.13	7.19	6.04	19.52	17.91	16.68	17.28	14.95	30.6	28.14	27.04	28.47	25.03
Ratio of fat: protein in body (kg fat/kg protein)	1.29	1.20	1.05	1.02	0.82	1.94	1.80	1.67	1.67	1.39	2.44	2.27	2.17	2.21	1.87

Table 15.3 MEAN PIG LIVE MASS (kg) AND AGE (d) AT MAXIMUM INTAKE (MJ) OR GAIN (g) OF ME, LIVE MASS, BODY PROTEIN AND BODY FAT

Treatment No.	*1*			*2*			*3*			*4*			*5*		
	Mass	*Age*	*Intake or gain*	*Mass*	*Age*	*Intake or gain*	*Mass*	*Age*	*Intake or gain*	*Mass*	*Age*	*Intake or gain*	*Mass*	*Age*	*Intake or gain*
ME intake/d	73	143	35.1	74	146	35.7	73	157	31.9	70	172	27.6	67	179	27.5
Live-mass gain/d	53	114	705	53	117	727	53	125	645	51	136	537	49	143	511
Protein gain/d	45	103	92.3	46	107	94.0	46	114	83.9	44	123	72.5	42	130	72.3
Fat gain/d	81	156	345	83	160	332	86	179	299	84	202	260	81	210	213

a low nutrient density diet resulting in ME intakes in excess of 40 MJ/d. Furthermore, Cole, Duckworth and Holmes (1967) found pigs to eat less in metabolism crates than when housed in holding pens. The pigs in this experiment were kept individually on perforated metal floor cages, which suggests that their environment may have influenced their absolute intake and also their pattern of intake. Therefore, further evidence is necessary, especially on the finding that feed and energy intake decreases after reaching a peak at a live mass of 70 kg. Although daily ME intake peaked at a live mass of about 70 kg, the live-mass gain increased to reach a maximum (specific for each treatment) at about 50 kg, and thereafter there was a gradual decrease in gain. Maximum daily gain in body protein occurred at a live mass of about 45 kg, while the rate of fat deposition only started to decline after 85 kg.

Body protein content remained fairly constant when calculated at a specific body mass, and increased slightly when treatments 4 and 5 were fed. The relative gains in protein, therefore, decreased in accordance with live-mass gains.

However, body fat content was drastically reduced by the dietary treatments imposed. Relative daily gains were therefore appreciably slower than the gains made in live mass and body protein, thus improving the ratio of body protein to fat (*Table 15.2*).

Conclusions

A dilution in dietary energy retarded the growth rate which resulted in an improvement of carcass quality through a reduction in fat deposition relative to protein deposition. Why protein gain maximized at 45 kg live mass and feed intake at 53 kg, which is contrary to the general trend (Thorbek, 1975), needs further investigation.

The statistical model employed proved a splendid tool for quantitating performance at specific mass points. The model should therefore be used to gather further evidence under different experimental conditions, using animals of different genetic origin and sex.

Acknowledgements

Grateful thanks to Dr C.Z. Roux for his contribution during the statistical analyses of the data. Thanks also to Mr M.N. Ras for technical assistance, to Mr M.P. Olivier, Mrs M. Marias and Miss P. Barnes for chemical analyses.

References

COLE, D.J.A., DUCKWORTH, J.E. and HOLMES, W. (1967). *Anim. Prod.*, **9**, 149–154

COLE, D.J.A., DUCKWORTH, J.E., HOLMES, W. and CUTHBERTSON, A. (1968). *Anim. Prod.*, **10**, 345–357

COOKE, R., LODGE, G.A. and LEWIS, D. (1972). *Anim. Prod.*, **14**, 219–228

LEWIS, D. and HARDY, B. (1970). Retina conference, Majorca, May 1970

LODGE, G.A., CUNDY, M.E., COOKE, R. and LEWIS, D. (1972). *Anim. Prod.*, **14**, 47–55

ROUX, C.Z. and KEMM, E.H. (1979). Submitted to *Animal Production*

THORBEK, G. (1975). Beretning fra Statens Husdyrbrugs forsøg, 424, København, 1975

16

THE INFLUENCE OF MICROBIAL METABOLISM IN THE HIND GUT OF ROOSTERS ON ENERGY UTILIZATION

L.D. CAMPBELL*, A. CHWALIBOG, B.O. EGGUM and G. THORBEK
National Institute of Animal Science, Rolighedsvej 25, 1958 Copenhagen V, Denmark

Summary

Semisynthetic diets without and with antibiotics (0.7% of a 2:1 mixture of bacitracin and neomycin sulfate) were fed to six adult roosters in a total of 36 individual balance trials to determine digestibility and energy metabolism which was measured by open-circuit indirect calorimetry. The addition of antibiotics to the diet produced the following effects:

1. Digestibility of dry matter (DM), organic matter (OM) and readily hydrolysable carbohydrate were significantly reduced ($P<0.001$), indicating a positive influence of microbial metabolism on digestibility of the individual nutrients;
2. Metabolizability of the gross energy (GE) in the diet was significantly reduced ($P<0.001$);
3. Efficiency of utilization of metabolizable energy (ME) was significantly reduced ($P<0.05$).

Introduction

Generally, the influence of microbial metabolism in the hind gut of poultry on nutrient digestion, although not well defined, is considered to be negligible. The extent of microbial fermentation in the hind gut depends largely on the retention time of digesta in the gastrointestinal tract and on the volume of the hind gut reservoir. With respect to the carbohydrate portion of a diet, it has been estimated that microbial digestion followed by the absorption by the host animal of the liberated metabolic products represents approximately 5% and 10% of digestible intake for the rat and pig respectively (Rérat, 1978). Since the rate of passage of the digesta is quite rapid in poultry, a corresponding value for the chicken would be expected to be lower unless the volume of the hind gut reservoir was enlarged, as is the case when fibrous diets are fed (Håkansson, Eriksson and Svensson, 1978).

Materials and methods

Six adult roosters (White Plymouth Rock × White Italian) housed in individual cages (31 × 38 × 48 cm) were used in a series of four balance and respiration trials to determine nutrient digestibility and efficiency of utilization of energy

* Present address: The University of Manitoba, Winnipeg, Manitoba, Canada

Table 16.1 COMPOSITION OF EXPERIMENTAL DIETS[1] (DRY MATTER BASIS)

Component	*Diet 1*	*Diet 2*
Organic matter (%)	94.00	94.04
RHC[2] (%)	62.10	61.20
Nitrogen (%)	2.31	2.33
Crude fibre[3] (%)	4.39	4.96
Fat[4] (%)	3.96	4.02
Gross energy (MJ/kg)	18.13	18.15

1 Antibiotic-containing diets were prepared by adding a 2:1 mixture of bacitracin and neomycin sulfate to each diet at a final concentration of 0.7%. Data presented are an average of each diet without and with antibiotics.
2 Readily hydrolysable carbohydrate.
3 Weende-method.
4 Stoldt-method (HCl-hydrolysis before ether extraction).

from diets both with and without added antibiotics. The mean live weight of the roosters during the experiment was 2233 ± 99 g. The diets (*Table 16.1*) were fed in the following sequence:

Period I, diet 1 – antibiotics;
Period II, diet 2 – antibiotics;
Period III, diet 2 + antibiotics;
Period IV, diet 1 + antibiotics.

During each period the roosters were initially fed the prescribed diet *ad lib.* for 12 days followed by nine days in which the diet was offered twice daily in two 55 g allotments. The last seven days of each period served as the balance and respiration trial and details regarding excreta collection, chemical analysis and conduct of the respiration experiment were as described by Chwalibog, Henckel and Thorbek (1978). The newly-built respiration unit for small animals was used in this experiment (Chwalibog, Lind and Thorbek, 1979). Statistical analyses were as described by Snedecor and Cochran (1967). For technical reasons no measurements of gas exchange were made with roosters No. 1 and 2 in period I, and digestibility data were not collected from rooster No. 5 in period III.

Results and discussion

The moisture content of excreta was higher when antibiotics were added to the diets. For diet 1 the excreta water content was 66.5% without antibiotics increasing to 76.9% with antibiotics, the difference being highly significant ($t = 6.51$). For diet 2 corresponding values were 61.0% and 67.8%, again showing a highly significant difference ($t = 5.09$). The results indicate that the addition of antibiotics to the diets must have an influence on water metabolism either by increasing the intake of water or by reducing water absorption from the intestine or both. In the current investigation no record of water intake was taken.

The digestibility coefficients for DM, OM, readily hydrolysable carbohydrate (RHC) and fat were calculated from the chemical composition of diets and droppings omitting the influence of the negligible amounts in urine of the nutrients in question. No differences in digestibility were evident between diets 1 and 2; hence the results were pooled as shown in *Table 16.2.*

Table 16.2 DIGESTIBILITY BY ADULT ROOSTERS FED DIETS WITHOUT AND WITH ANTIBIOTICS[1] (MEAN VALUES AND STANDARD DEVIATIONS)

	– Antibiotics (n = 12)		*+ Antibiotics* (n = 11)		*Difference between treatments*
	%	s.d.	%	s.d.	
Dry matter	71.6	2.3	67.6	2.3	t = 8.33***
Organic matter	75.3	2.1	70.9	2.1	t = 10.19***
RHC	96.2	0.4	93.9	1.0	t = 14.38***
Fat	69.9	3.6	65.7	2.8	t = 6.29***

[1] Antibiotics: 0.7% of a 2:1 mixture of bacitracin and neomycin sulfate

The data clearly demonstrate that the presence of antibiotics in the diet reduced the digestibility of nutrients to the extent of about 4% (abs.) for DM, OM and fat. A similar difference of approximately 2% was evident for RHC and all mean differences were found to be highly significant using the t-test. Results of previous experiments conducted in this laboratory in which rats were fed a similar level and combination of antibiotics indicated that the microflora were nearly completely eliminated from the hind gut (Eggum *et al.*, 1979). Consequently the differences in digestibility of nutrients observed between diets in the current experiment may be attributed to a reduction in microbial fermentation in the hind gut of roosters fed antibiotics.

The results from measuring the energy metabolism showed no significant difference between diets 1 and 2. Therefore, the data were pooled as shown in *Table 16.3*

Table 16.3 ENERGY UTILIZATION IN ADULT ROOSTERS FED DIETS WITHOUT AND WITH ANTIBIOTICS (MEAN VALUES AND STANDARD DEVIATIONS)

	– Antibiotics (n = 10)	*+ Antibiotics* (n = 11)	*Difference between treatments*
GE (kJ)	1393 ± 163	1321 ± 113	NS
ME (kJ)	1051	950	
ME (% of GE)	75.4 ± 2.3	71.9 ± 1.8	t = 7.87***
HP (kJ)	698	692	
HP (% of ME)	67.4 ± 10.0	72.9 ± 10.9	t = 2.46*
RE (kJ)	353	258	
RE (% of ME)	32.7 ± 10.0	27.1 ± 10.9	t = 2.46*
RPROT (kJ)	82	64	
RPROT (% of RE)	23.2	24.8	NS
RFAT (kJ)	271	194	
RFAT (% of RE)	76.8	75.2	NS

RPROT = retained protein
NS = not significant
HP = heat production
RFAT = retained fat
RE = retained energy

With the intention of feeding the animals *ad lib.* the roosters were offered 110 g feed daily. Feed refusals occurred, which were recorded daily, and the mean intake for the group without antibiotics was 88 g as compared with 82 g for the group with antibiotics. Corresponding GE intakes were 1393 and 1321 kJ respectively. The ME/GE was reduced significantly ($P<0.001$) from 75.4% to 71.9% by the addition of antibiotics to the diet.

Total heat production (HP, CN) was 67% of ME in the group with no addition of antibiotics, which corresponds fairly well with a value of 69% found in an earlier experiment with male chickens weighing an average of 1400 g (Chwalibog, Henckel and Thorbek, 1978). On the addition of antibiotics the heat loss increased to 73% of ME. The overall efficiency of utilization of ME (RE/ME) decreased significantly ($P<0.05$) from 33% to 27% with added antibiotics. The lower accuracy observed in estimating retained energy (RE) and HP compared with that for metabolizability of energy may be explained by the fact that the gas exchange was measured only for 24 hours in the middle of the seven-day-collection period.

Energy retained in fat was about 76% of the total energy retention for both groups while the energy retained in protein was about 24% which indicated that the roosters were probably not fully grown.

From the results it may be concluded that an active microflora in the hind gut of roosters will have a positive influence both on the digestibility of nutrients and on the efficiency of utilization of ME. Presumably volatile fatty acids and lactic acid produced by the fermentation process are involved in these responses, but the rationale for the latter effect requires further clarification.

References

CHWALIBOG, A., HENCKEL, S. and THORBEK, G. (1978). *Z. Tierphysiol., Tierernähr. Futtermittelk.*, **41**, 87–99

CHWALIBOG, A., LIND, J. and THORBEK, G. (1979). *Tierphysiol., Tierernähr. Futtermittelk.*, **41**, 154–162

EGGUM, B.O., FEKADU, M., WOLSTRUP, J., SAUER, W.C. and JUST, A. (1979). *J. Sci. Food Agric.*, **30**, 177–184

HÅKANSSON, J., ERIKSSON, S. and SVENSSON, S.A. (1978). *Report No. 57. Department of Animal Husbandry, Swedish University of Agricultural Sciences, Uppsala*

RÉRAT, A. (1978). *J. Anim. Sci.*, **46**, 1808–1837

SNEDECOR, G.W. and COCHRAN, W.G. (1967). *Statistical Methods, 6th Ed.* Ames, Iowa: Iowa State University Press

17

ENERGY AND PROTEIN EFFICIENCY OF SOME DIETS IN AXENIC AND HOLOXENIC GROWING CHICKENS

G. CHARLET-LERY
Laboratoire de Physiologie de la Nutrition, C.N.R.Z., 78350 Jouy-en-Josas, France
O. SZYLIT and H. BEWA
Laboratoire d'Ecologie Microbienne, I.N.R.A. – C.N.R.Z., 78350 Jouy-en-Josas, France

Summary

The influence of gut microflora on the efficiency of utilization of energy and protein in chickens fed different diets *ad lib.* was studied by the carcass analysis method. Only when the protein concentration of the diet was suboptimal, at 18%, did axenic (germ-free) chickens give better results than holoxenic (conventional) chickens. Could the efficiencies be explained by differences in the level of feed intake or in the pattern and the quantity of lactic acid formed in the crop of holoxenic chickens?

Introduction

Coates *et al.* (1963) reported that axenic (ax) chickens gained more weight than holoxenic (holo) ones. It was explained by Salter (1973) that studies on protein utilization in ax and holo rats gave conflicting results. Salter, Coates and Hewitt (1974) showed that the protein quality of the diet may play a more important role than gut microflora. However, nothing is known about the influence of intestinal microflora on the efficiency of utilization of energy and protein during growth and fattening. The objective of this work was to compare, using the carcass analysis method, the retention of energy and protein in ax and holo chickens fed various diets.

Materials and methods

DIETS

Food and water were freely available. The percentage composition of the semi-purified diets was (g/kg): 26 and 62.9 g or 23 and 65.9 g fish meal and maize starch for 20 and 18% proteins respectively, 4 g maize oil, 2 g cellulose powder (colmacel), 0.7 g salt mixture, 0.5 g vitamin supplement, 0.4 g methionine, 0.5 g celite and 3.0 g maltodextrins with long chains or maltodextrins with short chains.

Tropical tubers were included only in diets with 18% protein. Diets with cassava or yam (*Dioscorea dumetorum*) included tuber, fish meal, peanut oil, cellulose powder, celite, salt mixture, vitamin supplement and methionine in the following proportions: 70.0 and 77.0 g, 23.3 and 15.7 g, 4.0 and 4.0 g, 1.3 and 1.5 g, 0.5 and 0.5 g, 0.7 and 0.7 g, 0.5 and 0.5 g and 0.15 and 0.13 g.

The composition of the salt mixture and of the vitamin supplement was described by Bewa, Charlet-Lery and Szylit (1979). Pelleted diets packed in double bags were sterilized by gamma radiation at 4 Mrad.

CHICKENS AND EXPERIMENTAL PROCEDURE

Chickens of both sexes (Vedette INRA cross) were used; their growth and the chemical composition of their carcass were tested prior to the experiment and were identical up to a body weight of 350 g.

All the chicks born in axenic conditions were fed irradiated maize the first day. Half of them, the ax chickens, were reared in sterilized isolators and kept in metabolism cages (Lecoz, Szylit and Ducluzeau, 1977). The sterility test was carried out on two-day-old chickens and once a week subsequently. The others, the holo chickens, were given water-diluted excreta from conventional cocks and were kept in the same cages and in the same room as the ax chicks.

Body weight and food intake were recorded each week. Material conditions induced some changes as regards the first day of measurement and the duration of the experiment (days 2–5 up to days 19–27). However, for each type of diet the conditions were the same for both ax and holo chicks. In the last three days before slaughter the droppings were collected daily for heat combustion tests.

Chicks were killed by chloroform and the crops and caecal contents were withdrawn. Carcasses were immediately frozen (–25°C), minced, ground and lyophylized. Homogenized samples were analysed for nitrogen and energy content and lipids were calculated by difference.

Results

These are summarized in *Table 17.1.*

The microflora did not affect the level of intake, growth and body retentions in chickens when 20% protein of good biological value (diets 1 and 2) was used, provided the results obtained on one cage (ax diet 2) are not taken into account. Wasting of feeds by the animals made the measurement of metabolizable energy (ME) impossible in some experiments. But the few results obtained with starch maize showed no difference between ax and holo chickens for the two protein concentrations. When the protein supply was reduced to 18% for diets 3, 5 and 6 the level of ingestion was higher in ax chickens than in holo ones by +13, +34, and +19%. Similarly, the growth was better by +26, +52, +40% respectively. Diet 4, with short-chain maltodextrins, did not lead to such improvements.

In all cases the daily amount of retained energy (RE) was larger when the microflora was missing. Also, in the axenic state the efficiencies of utilization for maintenance and growth (RE/IE × 100) and of ME for maintenance and growth (RE/ME × 100) were higher. The decrease in the efficiencies induced by the holoxenic state was small with short-chain dextrins (diet 4): –5.5 and –5.5 respectively, and large with the other three diets: –17.5 and –17.5 for diet 3, –21.5 and –20.0 for diet 5 and –15.8 and –19.0 for diet 6. Variations in protein retention were smaller.

Table 17.1 RESULTS

CP % DM		*20*		*20*		*18*		*18*		*18*		*18*	
Carbohydrate		*maize starch*		*maize starch*		*maize starch*		*maize starch*		*Cassava*		*E. dumetorum*	
Maltodextrins		*long chains*		*short chains*		*long chains*		*short chains*					
		1		*2*		*3*		*4*		*5*		*6*	
		ax	*holo*	*ax*	*holo*	*ax*	*holo*	*ax*	*holo*	*ax*	*holo*	*ax*	*holo*
Cages		2	3	1	3	2	2	2	2	3	3	2	4
IT	(g/d)	23.4	23.5	21.3	24.5	27.0	23.9	25.8	26.3	30.2	22.6	31.7	27.4
W	(g/d)	16.4	16.8	13.7	17.1	15.4	12.2	11.5	11.7	17.5	11.5	19.5	14.8
RE	(kJ/d)	165.9	171.8	150.5	174.6	216.5	157.6	160.5	153.4	211.5	124.6	181.8	134.2
ME/IE	(%)	90.0	90.0	90.0	90.0	87.0	87.0	87.0	87.0	79.6	77.9	74.6	73.6
RE/IE	(%)	37.1	38.0	36.0	37.1	41.6	34.3	32.2	30.4	36.7	28.8	29.8	25.1
RE/ME	(%)	41.3	41.5	40.0	41.2	47.8	39.4	37.0	35.0	46.1	36.9	40.0	32.4
RN/IN	(%)	54.0	56.3	51.0	55.7	57.0	53.0	48.0	44.5	48.0	43.7	50.7	45.1
EXL/RE	(%)	63.5	62.6	66.0	61.5	71.7	66.3	70.4	71.2	67.3	62.0	55.0	51.5

Calculated by difference

DM = dry matter
CP = crude protein
IT = ingested DM
W = weight
RE = retained energy

IE = ingested energy
IN = ingested nitrogen
EXL = deposited lipid energy
ME = metabolizable energy
RN = retained nitrogen

Discussion

Axenic conditions, as compared to conventional ones, induced a higher level of ingestion only when the level of protein was suboptimal. This caused higher energy deposition and to a lesser extent higher protein deposition. This could explain the magnitude of lipid deposition. As less energy was used for maintenance when the birds grew more rapidly, the efficiencies of gross energy and ME were higher and the retention of protein was increased.

Ivorec-Szylit, Raibaud and Schellenberg (1973) observed that micro-organisms modified the pattern of breakdown of starch as they modified the pattern of nitrogen excretion (Salter, 1973). In ax animals dietary starch is hydrolysed only by digestive enzymes, which leads to glucose formation. The enzymes of numerous Lactobacilli, which inhabit the digestive tract of holo birds, modify the final nutrients. They are specially responsible for the production of D+ and L-lactic acid in equal quantities. At the crop level, glucose (Sodermo, Kare and Wasserman, 1961) and lactic acid (Ivorec-Szylit and Sauveur, 1966) are rapidly absorbed. Blood plasma of fowl contains amounts of glucose (250 mg/100 ml) and lactate (30–50 mg/100 ml) that are at least twice those in mammalian blood (Bell and Cuthbert, 1968; Bell, 1971).

In a previous study Szylit (1973) showed that in chickens diets including short-chain maltodextrins or amylase-treated starch induced high production of lactic acid soon after the beginning of the meal and led to very poor growth.

These results suggest that at a low level of protein the higher gross efficiency of utilization of energy and protein could be explained by a higher level of ingestion or by a difference in the nature, the rate and the site of production and absorption of nutrients.

References

BELL, D.J. and CULBERT, J. (1968). *Comp. Biochem. Physiol.*, **25**, 627–637

BELL, D.J. (1971). In *Physiology and Biochemistry of the Domestic Fowl*, Vol. 2, pp. 913–920. Eds. D.J. Bell and B.M. Freeman. New York, London: Academic Press

BEWA, H., CHARLET-LERY, G. and SZYLIT, O. (1979). *Annls Nutr. Aliment*, **33**, 213–231

COATES, M.E., FULLER, R.F., HARRISSON, G.F., LEV, M. and SUFFOLK, S.F. (1963). *Br. J. Nutr.*, **17**, 141–150

IVOREC-SZYLIT, O. and SAUVEUR, M. (1966). *Annls Biol. anim. Biochim. Biophys.* **6**, 517–520

IVOREC-SZYLIT, O., RAIBAUD, P. and SCHELLENBERG, P. (1973). In *Germfree Research*, pp. 225–232. Ed. J.B. Heneghan. New York, London: Academic Press

LECOZ, Y., SZYLIT, O. and DUCLUZEAU, R. (1977). *Sci. Techn. anim. Lab.*, **2**, 27–33

SALTER, D.N. (1973). *Proc. Nutr. Soc.*, **32**, 65–71

SALTER, D.N., COATES, M.E. and HEWITT, D. (1974). *Br. J. Nutr.*, **31**, 307–318

SODERMO, D., KARE, M.A. and WASSERMAN, R.H. (1961). *Poult. Sci.*, **40**, 123–128

SZYLIT, O. (1973). Thesis. Paris VI

18

DISCUSSION PAPER:

FEED EVALUATION, A SURVEY

A.J.H. VAN ES
Research Institute for Livestock Feeding and Nutrition, Lelystad and Department of Animal Physiology, Wageningen, The Netherlands

Introduction

Since 1976 new feed evaluation systems for ruminants have been introduced for application in practical animal husbandry in several countries (Belgium, France, The Netherlands and Switzerland). In other countries preparations for such changes are being made (i.e., West Germany). Several countries still continue to use the old systems although some advantages of the new systems, especially their better physiological basis, are recognized. A similar and more pronounced hesitation to accept modern views can be noted in the field of nitrogen metabolism in ruminants. The new systems proposed in this field mainly appeal because of their superior physiological basis when compared with the old systems based on digestible protein, but in practice the application of the new systems is hampered by their complexity and insufficient quantitative information on some details.

In monogastric animals a few new systems were also proposed some years ago to improve evaluation based on total digestible nutrients (TDN), digestible energy (DE) or metabolizable energy (ME). These were introduced in practice only in East Germany for pigs and poultry and in Denmark and The Netherlands for pigs; elsewhere few changes were made although the need for an improved feed evaluation was voiced at times.

This hesitation to change to another system is due in the first instance to simple conservatism, fear that the new system might hardly be better than the old one or that some aspects of importance for practice might have been overlooked because of the great amount of work needed to introduce new systems. Secondly, it is due to the fact that most new systems are more complicated to apply than the old ones. Finally, with the new systems there are certain cases where they are not well suited, whereas practice prefers a system which is generally applicable.

Practice too little recognizes the fact that an animal's energy metabolism is very complicated, that our knowledge of it still has several gaps and that it is nearly impossible to satisfy all its aspects in one simple system. Therefore, energy systems for practical application cannot be anything more than good compromises.

Desired properties of energy systems for practice

Energy systems of feed evaluation for practice should be precise and generally applicable, and should include unconventional rations and high production levels. Furthermore, they should be simple to use, preferably requiring only two lists of data (one for the feeding value of the feeds and one for animal requirements), and there should be few other rules. Nowadays with the aid of computers a precise but complicated system might also be used with ease. However, such a system has the disadvantage that the user might lose touch with its main aspects and might not know which factors or properties are of primary importance in determining feeding value or feed requirements.

'Simple' systems neglect less important aspects, for example the effect of feeding level or metabolizability (q) and the effect of q on net feed efficiencies (k_m and k_l) in the Ministry of Agriculture, Fisheries and Food system for dairy cows. Another feature of such systems lies in practical approximations, for example, where a correction for the effect of feeding level on the ME content of feedstuffs in the East German and Netherlands systems for dairy cows was included in the energy requirements rather than in the feeding value of the feedstuffs. This permitted the use of one feeding value per feed instead of several. Another very useful procedure in 'simple' systems is to list the feeding values which hold true for the average production level in practical husbandry while applying some corrections, where needed, to the requirements for other levels. This correction was carried out in the new systems used for growing cattle in France, Switzerland and The Netherlands where the feeding values apply to an animal production level (APL) of 1.5.

A more serious problem is that all systems are mainly based on work with rations fed to animals at moderate or low production levels. What is needed is the value of the separate feedstuffs when used in a ration at a high production level as well. Least cost rations, which have the required quality, can only be composed if the values of the separate feedstuffs are known. While deriving the energy systems, the composition of the rations used have been varied. By working with the feedstuffs' content of ME and/or digested nutrients, rather than their composition, the relationships with net energy which hold true for the rations have been said to apply also to the feedstuffs. In fact the main part of the difficulty is simply transferred to another aspect of feed evaluation: the prediction of apparent digestibility and ME content.

The feature of energy systems that is really needed is a rapid and precise method that simulates digestion and predicts the quantities of net absorbed nutrients. Weende analysis does not do so. Van Soest's neutral detergent method might be suited for monogastrics, but it does not inform on endogenous losses and it still gives filtration problems for starchy feeds. For ruminants the detergent methods are lengthy and are not accurate enough because they give insufficient information on the digestibility of the cell walls and on which fermentation products are absorbed. *In-vitro* rumen fluid and cellulase procedures do give precise data on apparent digestibility if they are regularly standardized with samples of known *in-vivo* digestibility, but they do not indicate what is absorbed and moreover they are time consuming. Very rapid instrumental methods are needed which for a monogastric's feeds allow the easy prediction of net absorbed amino acids, fat and monosaccharides, and which for ruminant feeds also inform on the interlinking and crystallinity of the cell-wall fraction, which appear to determine their rate of microbial degradation.

Because there are no such methods at present, we assume that the relationships derived between ME plus q and net energy (NE) or between digested Weende nutrients and NE for rations also holds true for separate feedstuffs. Associative effects between feeds appear small, even in ruminants, with regard to digestibility unless rumen fermentation is seriously changed.

This two-step procedure (first DE or ME and secondly NE) has the advantage that the two processes can be studied separately. It has the disadvantage that for each new feed its DE or ME should be known. For related feedstuffs, for example wheat and its milling byproducts, this is not needed: usually good relationships exist within such groups between NE and one or two simple analytical fractions, and in this example even crude fibre analysis suffices.

The criticism that few energy balances have been measured on high-producing dairy cows and fast-growing animals is true. Comparative slaughter trials with growing animals may help, but precise measurements of ME intake and energy retention require great care, especially in large animals. Feeding trials give little information due to the inability to predict the ME intake and the energy retention precisely.

With respect to animal requirements, energy systems also often lack precision. Due to variation between animals in maintenance requirements or in growth potential, average values may be wrong by 10% or more. Such individual properties can as yet hardly be predicted. In preliminary work clinically healthy sheep showed an increased maintenance requirement due to intestinal worms: there may exist other stress factors whose influence we do not yet know.

Finally, we may conclude that energy systems cannot and will never have the properties of precision, general applicability and simplicity that practice requires. This explains some of the hesitation to adopt new energy systems. Of the latter, however, we can state that most of them are better compromises than the older systems. They are based on far more knowledge and are physiologically more sound; moreover they are more flexible and enable new information to be incorporated easily.

Practical animal husbandry should use energy systems in a realistic way while understanding both their weaknesses and their good points. During the introduction of energy systems they should be evaluated carefully. The level of energy intake should not be linked too closely with the level of production or the rate of weight gain. Energy systems relate energy supply to actual energy needs, and to nothing more. In practice it is often more important to know the energy value of a feed relative to the values of other feeds: energy systems predict relative values more reliably than absolute values.

The present state of energy systems of feed evaluation and problems that need most attention

POULTRY

Feed evaluation is based on ME which is usually determined with adult animals so that the values obtained apply to approximate nitrogen equilibrium. The values are often corrected to zero nitrogen retention to make them more comparable in this respect. This means that most or all protein is supposed to be oxidized.

Sibbald (1975) criticized the conventional procedure used to measure ME which was said to lead to low ME values for unpalatable feeds because of a high proportion of endogenous material in the faeces. Reliable ME values of an unpalatable feed can be obtained without abnormal interference from endogenous secretions when their value is measured when the feed is part of a mixture, but this procedure necessitates very many replications to obtain sufficient precision. Sibbald (1975) proposed a so-called true ME (TME) which is obtained by force feeding fasting roosters 25–40 g of the feed (A), collecting droppings of them (B) and of fasting controls (C) for 24 hours. The TME of the feed is calculated by subtracting the energy in B from the energy in A and adding the energy in C. It is a very rapid method and has a good repeatability if performed correctly. It is questionable as to whether the TME procedure indeed correctly excludes endogenous matter: information is accumulating which shows that endogenous energy excretions with faeces and urine depend upon the amount and kind of feed eaten (Bergner, 1979). Most of these endogenous losses are needed to digest the feed; therefore, for the animal it is (apparent) ME rather than TME that counts. Furthermore, in energy metabolism ME is an easier starting point to study the influence of feed composition on its utilization for maintenance and production than TME. TME values can easily be converted to correct ME values. Thus, in my opinion, for the future of energy metabolism in poultry it is better to continue to work with ME rather than with TME and to use the TME technique as a very valuable tool to produce reliable ME data on feedstuffs.

The main problem in feed evaluation in poultry and also in pigs (and even in man) is whether the ME is always utilized with the same efficiency, for maintenance as well as for production, regardless of its composition. Theoretically protein-ME and fat-ME are considered to contain respectively some 20% and 5% less net energy for maintenance than glucose-ME. For fat production protein-ME is expected to have a 20% lower value than glucose-ME, but fat-ME should surpass glucose-ME. Most of the work with monogastrics supports these theoretical views; however, many of these studies were performed with growing animals in which the protein turnover and physical activity interfere with an easy interpretation of results. It is remarkable that so far few studies have been performed with cocks and laying hens which are well-suited for such work.

When protein-ME, provided it has been corrected toward zero nitrogen retention, is used for protein synthesis an efficiency above 100% can theoretically be expected. The efficiency of the use of protein-ME for protein *deposition* in growing animals, however, has repeatedly been found to be far below 100%. Probably this is due to protein turnover which seems to be related to the rate of protein synthesis. Unfortunately, these rates are not easy to measure with precision. Several contributions to this book deal with this aspect and most of them support the concept that high protein deposition costs in the growing animal are due to turnover of protein, but we need more quantitative information. Reliable information on the rate of protein turnover and on the efficiency of the utilization of protein-ME for protein deposition in eggs and milk is not yet available.

It is clear that more quantitative information from well-planned studies is needed before differentiation of the value of ME according to its composition can be introduced into a practical system. As maintenance in poultry is usually more than half of total metabolism, work with cocks and layers should get more attention. Also, long-term feeding trials are very well suited for the purpose provided that the rations to be compared are so composed that the voluntary intake of ME and essential nutrients is approximately the same, otherwise interpretation of the results will produce more problems. In view of the information on the far-from-negligible amount of maintenance-ME used for physical activity some attention should be paid to animal behaviour in such studies. This seems more advisable when forced-feeding techniques are applied.

Another aspect is the value of the carbohydrate fraction, also very important in the case of pigs. Cell-wall carbohydrates, like celluloses and hemicelluloses, can only be of use to these monogastrics via microbial fermentation and the absorption of the volatile fatty acids. Therefore, theoretically up to 70% of the total energy content of such carbohydrates, if fermented, may be of use to the animals. Several studies with poultry and pigs discussed in this book deal with this subject (*see also* the review on fermentation by Rérat, 1978). Apparently digested cell-wall carbohydrates were found to be of value for the animal's energy supply, although to a lesser degree than carbohydrates like starch and sugar. Measurement of fermentation gases (CH_4, and perhaps H_2), which is not usually done, might indicate the extent of the fermentation which is dependent on the size and the structure of the lower digestive tract and on the ration type.

Weende analysis does not partition correctly between readily hydrolysable carbohydrates and cell-wall carbohydrates. The former are highly digestible while the latter are only partially so via fermentation. This explains why the proposals of de Groote (1974), for poultry, and Henckel (1974), for pigs, make sense, namely that one should work with the fractions, starch plus sugar, and all other apparently digestible carbohydrates, rather than with digestible nitrogen-free extract and digestible crude fibre to predict the energy value. Here again we need more quantitative data before the introduction of this partitioning becomes possible. Priority for work on the extent of fermentation is higher for pigs than for layers, and for the latter it is higher than for chickens.

Fermentation in the crop, usually a lactate fermentation of readily available carbohydrates, is discussed in another chapter. Theoretically such a fermentation decreases the energy of the feed only slightly; however, in this study it also influenced the voluntary intake.

PIGS

At present most energy values of feeds for pigs are based on digestion studies with this species. Either TDN, DE or calculated ME values are used, but in some countries the net energy for fattening is preferred (NEF_s). In view of the greater use in pig nutrition of byproducts, interest in the value of products with higher cellulose contents has increased. The pig's methane production is seldom measured, but for some feeds with a higher cellulose content this might give important information. For studies with sows that are usually fed rations with greater concentrations of cellulose the measurements of methane production are a necessity. In all these cases it would be better to work with ME than with TDN or DE, especially in research.

For the improvement of feed evaluation for slaughter pigs through differentiation of the ME according to its origin, the same holds true as for poultry. As already mentioned with regard to chickens, the unsolved problems regarding the rate of protein turnover and the variation in maintenance requirements (for example, due to physical activity) in growing animals retard progress. A more practical approach might help; for example the performance of a number of series of balance trials at three or four live weights in the range of 20–110 kg with modern pigs at a high rate of gain fed rations either high or low in ratio of structural to readily hydrolysable carbohydrates, in fat or in protein. This will not solve all the problems mentioned, but some of these are probably of more theoretical than practical importance. Even so, such studies require much work and therefore a coordinated approach by several institutes would be useful. Such research could benefit considerably from slaughter trials with pigs and balance and slaughter trials with chickens, already performed or in progress in this field. The results of all these experiments might also be used for a statistical exercise which uses the model:

$$\text{retained energy} = (a \times \text{fat deposited}) + (b \times \text{protein deposited}) + (c \times W^{3/4}).$$

The danger of this model is that too much physiological significance might be given to the 'constants' a, b and c, as pointed out clearly by Pullar and Webster (1977) and by van Es (1978b).

Furthermore, work with nearly mature, non-pregnant sows might lead to more information on the efficiency of the utilization of protein-ME, fat-ME and carbohydrate-ME for maintenance. In slaughter pigs maintenance is responsible for more than one-third of total energy metabolism. Information on the utilization of ME by lactating sows is still hampered by the difficulty of obtaining a reliable estimate of the animal's daily milk energy yield.

DAIRY CATTLE

Van der Honing and Steg (1979) showed that several new energy systems differed little in respect of the value of rations or the amount of feed required for a moderate level of milk production. The difference with the TDN system was also not great. Larger differences were noted for the predicted feeding values of each of the 32 feeding stuffs, although the same digestibilities were used per feed for all systems. To some degree this is due to the use of weakly-based equations to predict ME at the maintenance feeding level (MAFF and West German system, *see* van Es, 1978), and partly it is due to the lack of knowledge about the ME value at higher feeding and production levels. Here more quantitative information is urgently needed, especially for rations containing one-third or more long roughage dry matter. Such rations are preferred in practice for several reasons.

It is the experience at the Lelystad institute that the effect of feeding level on the ME content of rations containing less than one-third long roughage is variable. Some cows can stand such rations much longer than others; moreover, apparently fairly normal rumen fermentations may be upset more easily than in the case of rations with more long roughage.

In the Netherlands energy system ME is corrected simply by a subtraction of 1.8% per multiple of maintenance feeding level for all diets; the extension service in the country stresses the importance of feeding sufficient long roughage to dairy cows to prevent loss of appetite, low milk fat and, probably, less efficient feed utilization. In France, the energy system takes account of the concentrate and feeding level while predicting ME. As said above, both approaches need more supporting data.

The greatest need in feed evaluation for dairy cows is a simple and rapid method to predict a feed's ME content at the maintenance level. Weende and Van Soest analyses suit the purpose only in a few groups of related concentrate feeds. *In-vitro* digestion with rumen fluid or cellulases is a well suited method, but it is laborious and needs continuous checking with samples of known *in-vivo* digestibility. Infrared reflection analysis might be suitable but still requires considerable development. Since interlinking and crystallinity within the plant cell wall affects its digestibility, physicochemical methods that measure such properties might be appropriate.

Prediction of the energy value of concentrate mixtures is a problem which also falls into this category. Such a prediction from a few Weende or other chemical analyses is not sufficiently precise: one should know the percentages of the various feedstuffs that the mixtures contain. In part of West Germany random samples of mixtures on the market were studied in sheep-digestibility trials. Publication of the results narrowed the gap between the claimed and the actual energy values.

The present discussion on feeding the same quantity of concentrates daily throughout the major part of lactation concerns the problem of long-term regulation of milk yield rather than energy utilization. Efficiency calculations in this field should use modern energy systems or TDN rather than systems based on starch value.

Both van der Honing, Steg and van Es (1977) and de Brabander *et al.* (1978), who calculate according to the new Dutch system, compared energy supplied with energy required for maintenance and milk yield in a great number of feeding trials and found a fair agreement when live-weight changes were small. When greater changes occurred the assumed energy content of about 20 MJ/kg live-weight change was considered to be too low.

BEEF CATTLE

Metabolism of growing cattle has received much attention, which indicates that there is still a great need for more knowledge. In energy systems the prediction of the ME at production does not present difficulties as is the case in dairy cows. The feeding level is usually only twice maintenance or less; therefore, feeding level depressions have only a small influence. An exception may be when almost only concentrates are fed and normal rumen fermentation is impaired.

The main difficulties are uncertainty of the following:

1. The relationship between daily live-weight gain (DWG), body weight (W) and energy content of the DWG;
2. The maintenance requirement;
3. The values of k_m and k_g.

All of these difficulties were discussed thoroughly by Webster (1978). With regard to the first aspect, it should be stated that for technical reasons it is extremely difficult to produce precise data on the energy and protein content of DWG in cattle. Similarly, far too little is known of its relationship with DWG, W, breed and genotype (*see also* van Es, 1978a). The main cause of the variation in energy content of the gain is its protein content: deposition of one gram of protein is accompanied by 3–4 g of water, while deposition of one gram of fat (itself containing 1.5 times more energy than protein) is not associated with water deposition. The protein content of the DWG depends on the animal's stage of maturity, breed, type and feeding level.

The efficiency of the use of ME for the deposition of protein energy is far lower than theoretically expected, which is probably due to protein turnover. Maintenance requirements appear to be related to the rate of protein deposition, which is higher for immature and rapidly growing males, and also to the temperament.

The combined research efforts in France, East Germany, West Germany, UK, USA and elsewhere have greatly improved our physiological understanding. What is still needed is accurate quantitative information before a well-based reliable energy system can be offered for practical use. Application of such a system in practice is, however, complicated by group housing and feeding, by feeding roughages or rations *ad lib.* and often by pasturing. Most systems, even new ones, derived their rough precision from empirical information obtained from feeding trials. Their application is most beneficial when circumstances, rations and animal types are the same as those pertaining to the region where the system was developed.

References

BERGNER, U. (1979). *Arch. Tierernähr.*, **29**, 50–51
BRABANDER, D.L. de *et al.* (1978). *Landbouwtijdschrift*, **4**, 621–629
ES, A.J.H. van (1978a). *Livest. Prod. Sci.*, **5**, 331–345
ES, A.J.H. van (1978b). In *Patterns of Growth and Development in Cattle*, pp. 361–382. Eds. H. de Boer and J. Martin. Nijhoff: Den Haag
GROOTE, G. de (1974). *Br. Poult. Sci.*, **15**, 75–95
HONING, Y. van der and STEG, A. (1979). *Documentation Report 49*, 69 pp. Lelystad: Research Institute for Livestock Feeding and Nutrition
HONING, Y. van der, STEG, A. and ES, A.J.H. van (1977). *Livest. Prod. Sci.*, **4**, 57–67
HENCKEL, H. (1974). *Proceedings International Symposium on Energy Management in Mixed Feeds*, pp. 27–33. Brussels: National Renderers Association
PULLAR, J.D. and WEBSTER, A.J.F. (1977). *Br. J. Nutr.*, **37**, 355–363
RÉRAT, A. (1978). *J. Anim. Sci.*, **46**, 1808–1837
SIBBALD, I.R. (1975). *Poult. Sci.*, **54**, 1990–1997
WEBSTER, A.J.F. (1978). *Wld Rev. Nutr. Diet.*, **30**, 189–226

19
GENERAL DISCUSSION

This discussion falls into two parts: one is concerned with ruminants and the other with monograstric animals.

Ruminant animals

The need has been stressed for more fundamental understanding of ruminant digestion and metabolism in order to improve feed evaluation for ruminant farm animals. The partition of total energy requirement between maintenance and productive functions continues to pose serious problems, especially with regard to growing animals. Chemical composition of digestive end-products, breed, rate of growth and the composition of the product formed are all factors that can influence the apparent maintenance requirement. The issue of the interrelation between the value of maintenance and partial efficiency has been raised several times.

With regard to the associative effects of feedstuffs on the energy value of mixed rations, not all contributors agree on the importance of this effect except for specific cases, for example with diets rich in starch. The relevance of metabolizable energy (ME) measured at maintenance intake to ME at higher feeding levels has been questioned and the need to rethink methods of predicting the energy value of mixed rations fed to ruminants emphasized. The latter has led to considerable discussion of the need for the use of dynamic rather than static models to describe nutrient utilization. The necessity to separate factors that describe a particular feedstuff from factors related to the animal has also been stressed. Concern has been expressed that some feed evaluation systems are combining well established and known components of energy metabolism with less well understood aspects with the result that incorporation of new knowledge requires the development of a completely new system of feed evaluation.

In connection with this, the need for a simple system of feed evaluation to predict the performance of a very complex biological model has been discussed, and it has been pointed out that the availability of inexpensive microprocessors enables the relatively easy use of considerably complex mathematical functions. It has been suggested that a breakthrough in the development of a precise, rapid and inexpensive method of predicting ME value and chemical composition of ME for rations fed to productive animals is needed.

Monogastric animals

In experiments on growing pigs and chickens, the estimates of maintenance requirement and ME efficiency for growth can be biassed by several factors such as specific compounds, feeding level, gut fill and environmental temperature. Crude fibre intake may lead to variations in maintenance requirement and to an apparent reduction of ME efficiency in growing chickens. In pigs the fibre intake increases the dry matter and water content of the gut and so affects the live weight and maintenance requirement for unit body weight. It can also alter the composition of body gain and thus ME efficiency.

Methane energy losses are generally not measured; they amount to less than 1% of energy intake but they can reach 3% or 4% of the energy of feeds such as palm corn or coconut meal. If they are neglected the ME intake is overestimated and the ME efficiency is underestimated in studies that use the comparative slaughter technique.

In the evaluation of feed energy systems for pigs and poultry, there are arguments in favour of using digestible energy (DE) instead of ME. These include the extra cost of measuring urine and methane energy in the ME system, the possible non-additivity of ME and the easier measurement of DE of feeds for routine purposes. However, ME has the advantage for feed formulation in that it takes into account urinary energy losses and part of the fermentation losses (methane). Another point is that when DE is used in pigs knowledge of the composition of DE could improve the accuracy of the prediction of the feeding value. For poultry the ME value is preferred, which is either corrected to zero nitrogen-balance or, better, to the average nitrogen-balance occurring during production.

Chemical analysis for feed energy evaluation and carbohydrate analysis for starch and sugar, in addition to the Weende analysis or instead of crude fibre and nitrogen-free extract analysis, is preferred by most researchers, although the evidence put forward does not suggest that the additional determinations greatly improve the prediction except for feeds high in fibre. In fat analysis the incomplete determination of soaps is mentioned, and in nitrogen-balance studies the importance of losses as NH_3 is also considered.

The use of net energy (NE) instead of ME is considered to improve the prediction of the feedstuff's energy value only slightly, but it also assists in a physiological understanding of the energy conversions involved. Corrections for fat and fibre might be applied to the rations instead of to the separate feedstuffs.

For the purpose of practical application it might be useful to study rations containing ME of different origin at production levels found in practice rather than to work only on maintenance requirements and partial efficiencies. Until such information is available, it is considered that DE- or ME-based systems provide the best approach.

References

KIELANOWSKI, J. (1976). In *Protein Metabolism and Nutrition*, pp. 207–215. Eds. D.J.A. Cole, K.N. Boorman, P.J. Buttery, D. Lewis, R.J. Neale and H. Swan. EAAP Publication No. 16. London: Butterworths

KOTARBIŃSKA, M. (1969). Badania nad przemiana energii u rosnących swin. Instytut Zootechniki, Wydawnictwa Własne, Nr 238, Wrocław

KOTARBIŃSKA, A. (1971). *Roczniki Nauk Rolniczych, B-93-1*, pp. 129–135

NEHRING, K., BEYER, M. and HOFFMANN, B. (1970). *Futtermitteltabellenwerk.* Berlin: VEB Deutscher Landwirtschaftsverlag

PART 2

USE OF ENERGY FOR MAINTENANCE AND GROWTH

SOME ASPECTS OF THE ENERGY METABOLISM OF GROWING PIGS FED ON DIETS CONTAINING DIFFERENT CONCENTRATIONS OF PROTEIN

C.W. HOLMES, R. CHRISTENSEN, J.R. CARR and G. PEARSON
Dairy Husbandry Department, Massey University, Palmerston North, New Zealand

Summary

Energy and nitrogen balances were measured with 16 gilts at live weights (LW) of 30, 50 and 90 kg. Two levels of feed intake were offered with four diets which differed in protein concentration. The pooled data provided estimates for partial efficiency for growth, and maintenance requirement, of 0.62 and 0.45 MJ/kg$^{0.75}$ metabolizable energy (ME) daily. Multiple regression analyses showed that when ME was kept constant the energy retained (ER) decreased slightly in association with increases in nitrogen intake (NI). Energy retained as protein in pigs of 50 kg LW was influenced by both ME intake (MEI) and NI; however, MEI had no such effect on pigs of 30 kg LW whereas NI had no such effect for pigs of 90 kg LW.

Introduction

The utilization of dietary energy by young lambs and adult sheep and cattle can be influenced by the concentration of protein in the diet (Walker and Norton, 1971; Blaxter and Boyne, 1978). There appears to be no corresponding information for pigs, although the effects of variations in NI on the nitrogen metabolism of pigs have been studied intensively (Carr, Boorman and Cole, 1977).

Energy and nitrogen metabolism were measured in the present experiment for eight out of 48 treatments used in a larger experiment which was intended to measure the effects of variation in the intake of dietary protein and dietary energy, imposed independently of each other, on the growth of pigs.

Materials and methods

Sixteen gilts (Large White × Landrace) were chosen in groups of four from each of four litters which had been weaned at five weeks of age. Each pig was allocated to a particular treatment when it weighed 20 kg and remained on that treatment until slaughtered at 95 kg LW.

The eight treatments consisted of two levels of feed intake and four levels of dietary protein concentration, and each treatment was replicated once. On the higher level of feeding the daily intake of meal increased from 1.22 kg at 20 kg LW to 2.88 kg at 80 kg LW; the corresponding values for the lower level of feeding were 0.77 kg and 1.82 kg meal respectively. The diets were formulated from barley meal, maize meal, soya bean meal, skimmed milk powder and dried blood meal as main constituents so that the concentrations of threonine, methionine and cystine, isoleucine and tryptophan varied between diets in proportion to the concentration of lysine. The mean value for the gross energy (GE) concentration of the diets was 18.4 MJ/kg dry matter (DM) and some details of the diets are shown in *Table 20.1*

Table 20.1 SOME DETAILS ABOUT THE FOUR DIETS

	Diet					*Significant effects*
	1	*2*	*3*	*4*	*SEM*†	
Crude protein (% DM)	23.8	19.6	15.2	12.1		
Lysine (% DM)	1.46	1.20	0.88	0.59	(± 0.06)	
DE/GE (%)	85.6	84.6	83.7	82.3	(± 0.6)	NS
ME/DE (%)	95.7	95.8	96.6	96.6	(± 0.6)	Diet**

† Pooled standard error of a mean
** $P<0.01$
DE = digestible energy

Measurements of energy and nitrogen balance were undertaken with each pig when it weighed 30, 50 and 90 kg LW. Each balance period was of seven days and generally the methods used were those described previously (Holmes and McLean, 1974).

Results

Linear regression equations that relate ER to MEI are presented in *Table 20.2.*

Table 20.2 LINEAR REGRESSION ANALYSES OF DATA FOR MEI AND ER CALCULATED FOR EACH DIET INDIVIDUALLY AND FOR THE DATA POOLED FOR ALL DIETS (MEASUREMENTS AT 30, 50 AND 90 kg LW INCLUDED IN EACH EQUATION)

Diet	*Regression equation*	*Value for MEI when ER = 0*
1	ER = 0.630 MEI −0.229 (± 0.02)†	0.47
2	ER = 0.622 MEI −0.277 (± 0.03)	0.44
3	ER = 0.608 MEI −0.245 (± 0.02)	0.40
4	ER = 0.627 MEI −0.283 (± 0.04)	0.45
Pooled	ER = 0.621 MEI −0.275 (± 0.03)	0.44

ER = energy retained (MJ ME/$kg^{0.75}$ daily)
MEI = metabolized energy intake (MJ ME/$kg^{0.75}$ daily)
† Standard error of the regression coefficient
All values for $r > 0.98$

Multiple regression analyses were also performed, and two equations calculated for the pooled data are presented in *Table 20.3*: it should be noted that the supposedly independent variables in these equations are correlated (MEI and NI, r = 0.7; energy retained as fat (ER_F) and energy retained as protein (ER_P), r = 0.4; ER_P and $LW^{0.75}$, r = 0.6). The regression coefficient for NI in equation (1) in *Table 20.3* was –0.004 and –0.038 for measurements made at 30 and 90 kg LW respectively.

Table 20.3 MULTIPLE REGRESSION EQUATIONS CALCULATED FOR DATA POOLED FROM MEASUREMENTS AT 30, 50 AND 90 kg LW

Equation	
(1)	$ER = 0.66\ MEI - 0.018\ NI - 0.276$; r = 0.98 (± 0.02) (± 0.008)
(2)	$MEI = 1.44\ ER_F + 1.87\ ER_P + 0.46$; r = 0.99 (± 0.06) (± 0.12)

NI = nitrogen intake (g N/kg $LW^{0.75}$ daily)
ER_P = energy retained as protein (MJ/kg $LW^{0.75}$ daily)
ER_F = energy retained as fat (MJ/kg $LW^{0.75}$ daily)
See Table 20.2
(± standard error of regression coefficient)

The multiple regression equations presented in *Table 20.4* indicate that ER_P increased in association with increases in NI for pigs of 30 and 50 kg LW, but that no such association was evident for pigs of 90 kg LW. In contrast ER_P was not influenced by MEI for pigs of 30 kg LW, but it did increase in association with increases in MEI for pigs of 50 and 90 kg LW.

Table 20.4 MULTIPLE REGRESSION EQUATIONS CALCULATED SEPARATELY FOR DATA MEASURED AT 30, 50 AND 90 kg LW

Weight	*Regression equation*
30 kg LW	$ER_P = 0.075\ NI - 0.005\ MEI - 0.008$; r = 0.95 (± 0.01) (± 0.03)
50 kg LW	$ER_P = 0.033\ NI + 0.078\ MEI - 0.029$; r = 0.90 (± 0.01) (± 0.03)
90 kg LW	$ER_P = 0.002\ NI + 0.088\ MEI - 0.002$; r = 0.79 (± 0.01) (± 0.03)

NI = nitrogen intake (g N/kg $LW^{0.75}$ daily)
ER_P = energy retained as protein (MJ/kg $LW^{0.75}$ daily)
MEI = metabolizable energy intake (MJ ME/$kg^{0.75}$ daily)
(± standard error of regression coefficient)

Daily rates of gain in LW were 0.7 kg and 0.4 kg for the higher and lower levels of feeding respectively, and within each level of feeding it was slowest for pigs fed on diet 4.

Discussion

The values for partial efficiency with which ME was utilized above maintenance (k_g) calculated from *Table 20.2* are slightly lower than those reported recently

(Fuller and Boyne, 1972; Verstegen *et al.*, 1973), whereas the present estimates for ME required for maintenance (ME_M) are similar to those reported by the authors mentioned above. However, since the present calculations of ME_M involve considerable extrapolation below the lower level of MEI they should probably be regarded with caution, a qualification which applies to many estimates of ME_M for the pig.

There were no significant differences between diets with respect to k_g and ME_M (*Table 20.2*), although there was a tendency for both values to decrease to minimum values for diet 3. These results differ from those of Walker and Norton (1971) who reported that k_g increased consistently as dietary protein concentration decreased, whereas ME_M was lowest for a diet of intermediate protein concentration. Nevertheless, equation (1) in *Table 20.3* shows that for a particular value of MEI an increase in NI was associated with a small decrease in ER, an association which was significant for the pooled data ($P<0.05$).

The regression coefficients in equation (2) in *Table 20.3* are similar to those measured for rats and those cited by Pullar and Webster (1977) for pigs. The equations in *Table 20.4* demonstrate an interaction that can occur between energy and nitrogen, but they suggest that this interaction will not occur for pigs of all ages and weights.

The present results demonstrate that dietary NI can exert small influences on energy metabolism independently of dietary energy intake, and that this influence depends on age and live weight or both: the work is being continued with castrated and entire male pigs.

The present use of simple and multiple regression analyses for data of this type is conventional; however, there should be considerable speculation about the validity of any physiological interpretations that might be based on the results of such analyses.

References

BLAXTER, K.L. and BOYNE, A.W. (1978). *J. Agric. Sci., Camb.*, **90**, 47–68
CARR, J.R., BOORMAN, K.N. and COLE, D.J.A. (1977). *Br. J. Nutr.*, **37**, 143–155
FULLER, M.F., and BOYNE, A.W. (1972). *Br. J. Nutr.*, **28**, 373–384
HOLMES, C.W. and McLEAN, A.N. (1974). *Anim. Prod.*, **19**, 1–12
PULLAR, J.D. and WEBSTER, A.J.F. (1977). *Br. J. Nutr.*, **37**, 355–363
VERSTEGEN, M.W.A., CLOSE, W.H., START, I.B., and MOUNT, L.E. (1973). *Br. J. Nutr.*, **30**, 21–35
WALKER, D.M. and NORTON, B.W. (1971). *J. Agric. Sci., Camb.*, **77**, 363–369

21

EFFECT OF BODY WEIGHT ON EFFICIENCY OF UTILIZATION OF ENERGY AND PROTEIN IN PIGS

F. BERSCHAUER, G. GAUS and K.H. MENKE
Institute of Animal Nutrition, University of Hohenheim, 7000 Stuttgart 70, P.O. Box 106-06100, Federal Republic of Germany

Summary

Fifty-six measurements of energy exchange with eight pigs between 40 and 110 kg body weight showed that energy retention was related not only to metabolizable energy (ME) intake but also to body weight. The mean energy maintenance requirement and the partial efficiencies for protein (k_p) and fat (k_f) retention were 477 kJ/$W^{0.75}$, 0.53 and 0.92 respectively. These varied from 497 to 453, 0.57 to 0.60 and 0.91 to 0.82 between the body weight ranges 40–75 and 75–110 kg, respectively.

A 10 kg increase in body weight was associated with a decrease of 0.042 units in protein utilization (k_N). For younger pigs between 3 and 33 kg the equivalent decrease was 0.084 units.

Introduction

It is a well established fact that the efficiency of utilization of ME for protein (k_p) is lower than for fat (k_f) deposition (Thorbek, 1975). For the factorial calculation of energy requirements a prediction equation for protein retention is therefore necessary. Fat retention can then be regarded as a secondary process which depends on ME intake, maintenance energy requirement and protein retention.

Protein retention is a function of exogenous (composition of feed and feed intake) and endogenous (body weight, sex and breed) factors. The present experiments were, therefore, designed to investigate the extent to which the efficiency of utilization of energy and protein changes in relation to the body weight of the animals.

Materials and methods

Eight animals (2 DL, Deutsche Landrasse; 3 DE, Deutsches Edelschwein; 3 PI, Pietrain) were fed a ration containing 9.3 g digestible crude protein (DP) per MJ ME and 14.9 MJ ME/kg dry matter (DM) at feeding levels of 0.75 and 1.0 MJ ME/$W^{0.75}$ (IME^x). The feeding level was changed after each 14-day period (five days preperiod, seven days nitrogen-balance and 48 h respiration measurement). Each animal was subjected to six to eight such experimental periods. In addition, nitrogen-balance experiments with 163 groups of five to seven piglets (DL: 147; DE: 8; PI: 8) were conducted in the body weight range of 3 to 33 kg. The rations contained 140–500 g DP and 10.0–21.8 MJ ME/kg DM. The protein : energy ratios varied between 10 and 25 g DP/MJ ME and the minimum lysine content was 0.90%.

Results of respiration experiments

Neither level of feeding nor body weight influenced the apparent digestibility of protein (dP) and the metabolizability of the ration (ME/GE). Within the weight range of 40 to 75 kg the RPE^x was significantly higher at both the low ($P<0.05$) and the high ($P<0.001$) level of intake when compared with that of the weight range of 75 to 110 kg (*Table 21.1*). On the other hand, the RPE^x was significantly lower ($P<0.001$) when compared with that of the heavier animals.

Table 21.1 PROTEIN AND FAT GAIN OF PIGS IN RELATION TO THEIR BODY WEIGHT AND LEVEL OF FEEDING (MEANS ± s.e.)

	40–75 kg (n = 32)		*75–110 kg* (n = 24)	
IME/W^x (kJ/d)	$761^a \pm 5$	$1041^b \pm 7$	$758^a \pm 2$	$1036^b \pm 7$
dP (%)	$83.2^a \pm 0.5$	$83.2^a \pm 0.4$	$84.2^a \pm 0.4$	$82.5^a \pm 0.3$
ME/GE (%)	$82.9^a \pm 0.5$	$82.7^a \pm 0.4$	$82.7^a \pm 0.4$	$81.9^a \pm 0.2$
RPE/W^x (kJ/d)	$89^a \pm 3$	$127^b \pm 5$	$67^c \pm 7$	$93^a \pm 5$
RFE/W^x (kJ/d)	$108^a \pm 9$	$280^b \pm 9$	$170^c \pm 12$	$351^d \pm 10$

Values with different superscripts are significantly ($P<0.05$) different from one another
RPE = energy retained as protein
RFE = energy retained as fat

For calculation of maintenance energy requirements ($ME_m{}^x$) the intake of energy (IME^x) and the energy retention (RE^x) of each animal measured by respiratory exchange were compared with the mean of the ME intakes and energy retentions of the preceding and the subsequent experiment, which were both at the same feeding level. Maintenance requirement is then calculated by the equation:

$$ME_m{}^x = IME_2{}^x - \left[RE_2{}^x \left(IME_2{}^x - \frac{IME_1{}^x + IME_3{}^x}{2}\right) / \left(RE_2{}^x - \frac{RE_1{}^x + RE_3{}^x}{2}\right)\right] \quad (1)$$

where the subscripts indicate the number of the experiments. In this way six maintenance values could be calculated for an animal with seven balances.

The mean maintenance energy requirement was 474 ± 13 kJ/W^x (mean ± s.e.; n = 39) with higher values for the DL (523 ± 15) and DE (480 ± 13) and lower values for the PI (430 ± 26). Regression of $ME_m{}^x$ and body weight (W) showed a slight but non-significant decrease with increase in body weight:

$$ME_m{}^x = 0.587\ (\pm 0.072) - 0.00159\ (\pm 0.00097)\ W$$
$$(r = 0.25;\ n = 39;\ NS) \quad (2)$$

The efficiency of utilization of ME for energy retention was calculated according to the equation $k_{pf} = RE^x/(IME^x - ME_m{}^x)$, using the means of the 5–7 measurements of $ME_m{}^x$ of each animal. On average k_{pf} was 0.745 ± 0.013, with only slight differences between DL (0.78 ± 0.024), DE (0.758 ± 0.022) and PI (0.708 ± 0.021).

Energy utilization (k_{pf}) increased with W according to the equation:

$$k_{pf} = 0.555\ (\pm 0.051) + 0.0025\ (\pm 0.0007)\ W$$
$$(r = 0.45;\ n = 56;\ P{<}0.01) \quad (3)$$

Energy retention (RE^x) was related to ME intake (IME^x) and W as follows:

$$RE^x = \underset{(\pm 0.0306)}{0.7363\ IME^x} + \underset{(\pm 0.00025)}{0.00119\ W} - \underset{(\pm 0.0288)}{0.4301}$$
$$(r = 0.97;\ n = 56;\ P{<}0.001) \quad (4)$$

If a common model for evaluation of energy-balance experiments is applied, that is, one in which ME intake is related to energy retained in protein (RPE^x) and fat (RFE^x), a rather low regression coefficient for RFE^x is obtained (W = 40–110 kg):

$$IME^x = \underset{(\pm 0.207)}{1.897\ RPE^x} + \underset{(\pm 0.051)}{1.091\ RFE^x} + \underset{(\pm 0.022)}{0.4772}$$
$$(r = 0.96;\ n = 56;\ P{<}0.001) \quad (5)$$

From this equation the partial efficiencies of utilization of ME for protein (k_p) and fat (k_f) deposition as well as $ME_m{}^x$ are:

$k_p = 0.53$ $\qquad$ $k_f = 0.92$ $\qquad$ $ME_m{}^x = 477$ kJ

When the data are divided into two weight classes, the following regression equations are obtained:

$$W = 40\text{–}75\ \text{kg}:\ IME^x = \underset{(\pm 0.41)}{1.74\ RPE^x} + \underset{(\pm 0.11)}{1.10\ RFE^x} + \underset{(\pm 0.036)}{0.497}$$
$$(r = 0.96;\ n = 32;\ P{<}0.001) \quad (6)$$

$$W = 75\text{–}110\ \text{kg}:\ IME^x = \underset{(\pm 0.30)}{1.66\ RPE^x} + \underset{(\pm 0.08)}{1.21\ RFE^x} + \underset{(\pm 0.030)}{0.453}$$
$$(r = 0.97;\ n = 24;\ P{<}0.001) \quad (7)$$

From these equations k_p values of 0.57 and 0.60 and k_f values of 0.91 and 0.82 are derived together with the ME_m requirements of 497 and 453 kJ/W^x within th body weight ranges of 40–75 and 75–110 kg respectively.

Results of nitrogen-balance experiments

Efficiency of utilization of digestible protein (IDN^x, g/W^x) for protein retention (RN^x, g/W^x) was defined as:

$$k_N = (RN^x + B_N)/IDN^x \qquad (8)$$

where B_N is the basal nitrogen metabolism, for which a constant value of 0.23 g/W^x was adopted (Berschauer, 1977).

In the experiment with piglets (3–33 kg W; 163 groups) the efficiency k_N decreased with increasing protein content and W:

$$\begin{aligned} k_N = {} & 1.361 - 0.0325\ DP/ME - 0.0084\ W \\ & (\pm 0.019)\ (\pm 0.0010) \qquad (\pm 0.0006) \\ & (r = 0.94;\ n = 163;\ P < 0.001) \qquad (9) \end{aligned}$$

According to this equation k_N decreased by 0.084 units with each 10 kg increase in body weight.

In growing pigs (40–110 kg W) k_N was influenced by both feeding level (ME intake at constant composition of the ration) and W as follows:

$$\begin{aligned} k_N = {} & 1.059 - 0.105\ IME^x - 0.0042\ W \\ & (\pm 0.085)\quad (\pm 0.089) \qquad (\pm 0.0007) \\ & (r = 0.67;\ n = 56;\ P < 0.001) \qquad (10) \end{aligned}$$

The apparent decrease in protein utilization with each 10 kg increase in W is smaller in older pigs (0.042 units) than in younger animals (0.084 units). Moreover, there seem to be breed differences in the relationship between k_N and W, which can be interpreted as characteristics of early or late maturing animals:

$$\begin{aligned} DL{:}\ k_N = {} & 0.922 \qquad - 0.117\ IME^x - 0.002\ W \\ & (\pm 0.118)\quad (\pm 0.141) \qquad (\pm 0.001) \\ & (r = 0.63;\ n = 14;\ P < 0.05) \qquad (11) \end{aligned}$$

$$\begin{aligned} DE{:}\ k_N = {} & 1.139 \qquad - 0.166\ IME^x - 0.005\ W \\ & (\pm 0.186)\quad (\pm 0.189) \qquad (\pm 0.001) \\ & (r = 0.63;\ n = 21;\ P < 0.01) \qquad (12) \end{aligned}$$

$$\begin{aligned} PI{:}\ k_N = {} & 1.113 \qquad - 0.073\ IME^x - 0.005\ W \\ & (\pm 0.101)\quad (\pm 0.102) \qquad (\pm 0.001) \\ & (r = 0.83;\ n = 21;\ P < 0.01) \qquad (13) \end{aligned}$$

Menke (1979) has proposed a system for the calculation of protein retention capacity from experiments with different feeding levels and from determination of the composition of the animals' gain. In this system k_N decreases with feeding level, depending upon the animals' protein retention capacity (A). Menke applied these equations to the data of Thorbek (1975) and found a negative correlation between maximum energy retention as protein (A, kJ/W^x) and W:

$$A = 1090 - 22.16\ W + 0.1355\ W^2 \qquad (r = 0.91;\ n = 381;\ P < 0.001) \qquad (14)$$

The efficiency of protein utilization k_N is related to protein retention capacity. The change of k_N with feeding level can be calculated. From the data of Thorbek (1975) it can be shown that the efficiency of protein utilization, corrected for 50% decrease of A ($k_{N.5}$), is:

$$k_{N.5} = 0.917 - 0.004\ W \qquad (r = 0.78;\ n = 381;\ P<0.001) \qquad (15)$$

which again shows a similar decrease (0.04 units per 10 kg W) with increasing body weight.

Acknowledgement

This work was supported by Deutsche Forschungsgemeinschaft, SFB 142–B.3.

References

BERSCHAUER, F. (1977). *Hohenheimer Arbeiten, Heft 91.* Stuttgart: Verlag Eugen Ulmer

MENKE, K.H. (1979). *Z. Tierphysiol. Tierernähr. Futtermittelk,* **42**, 173–189

THORBEK, G. (1975). *424 Beretning fra Statens Husdrybrugs forsøg. Rolighedsvej 25, 1958 København V*

22

EFFECT OF DIFFERENT DIETARY LEVELS OF LINOLEIC ACID ON ENERGY METABOLISM AND MITOCHONDRIAL ACTIVITY IN YOUNG PIGS

K. CHRISTENSEN[1], S. HENCKEL[2] and G. THORBEK
National Institute of Animal Science, Rolighedsvej 25, 1958 Copenhagen V, Denmark

Summary

Semisynthetic diets without and with addition of linoleic acid (range 0.04–2.73 energy %) supplied as soya oil were fed to 36 Danish Landrace pigs in a total of 127 individual balance and respiration trials to determine the effect on energy metabolism measured by indirect calorimetry using open-circuit respiration chambers. Mitochondria were isolated from the longissimus dorsi muscle of 16 pigs for determination of activity and oxidative phosphorylation capacity by oxidation of pyruvate plus malate and succinate. The ratio between eicosatrienoic acid (C20: 3 ω 9) and arachidonic acid (C20: 4 ω 6) in plasma lipids indicated that pigs receiving less than one energy % linoleate in their diet were deficient in essential fatty acids. No systematic effect of linoleate levels was found on oxygen consumption, heat production and efficiency of utilization of energy. No effect on ATP formation and coupling integrity of the mitochondria was observed, but the oxidation rates were depressed in the groups receiving less than one energy % linoleate.

Introduction

Previously, Christensen (1974a) observed that oxidative phosphorylation was depressed in mitochondria isolated from skeletal muscle, the heart and the liver of pigs deficient in linoleic acid. A reduction in the synthesis of ATP would lead to an increase in heat production and a decrease in cellular activity. However, no effect on growth rate, feed utilization or deposition of protein and fat in the body was observed (Christensen, 1973, 1974b).

The present results describe the effects of feeding different levels of linoleic acid on energy metabolism measured *in vivo* and *in vitro*.

Materials and methods

Eighteen female and 18 castrated male pigs of Danish Landrace were fed different levels of linoleate and measured individually in 127 balances as shown in *Table 22.1.* The principle for feeding the experimental animals was to supply all of

[1] Present address: The Royal Veterinary and Agricultural University, Copenhagen
[2] Present address: Regnecentralen A/S, Copenhagen

them with the same daily amounts of energy, protein, minerals and vitamins. At least three balance and respiration trials with collection periods of seven days were carried out with each pig in order to determine the efficiency of utilization of energy, as described by Thorbek (1975). In the last period no balance or respiration trials were performed on two pigs from series C because of diarrhoea. In series E one pig died before the fourth collection period because of volvulus. In series G one pig hurt its leg in the cage, and no measurements were carried out for the last two periods.

Table 22.1 ALLOCATION OF PIGS IN THE DIFFERENT SERIES, THEIR LIVE WEIGHTS AND ENERGY INTAKES

Concentration of linoleate energy (%)	*Series no.*	*Number of*		*Mean of live weight*		*Energy intake* (ME kJ/kg$^{0.75}$)	
		pigs	*bal. exp.*	*Start* (kg)	*End* (kg)	*Mean*	*s.e.m.*
0.04	C	4	12	27.8	44.5	1260	68
0.11	E	2	8	17.4	59.4	1282	65
0.15	C	4	10	28.5	44.0	1229	56
0.23	G	4	12	19.0	50.9	1253	108
0.29	D	2	8	13.7	39.3	1248	151
0.77	E	2	10	18.5	57.2	1332	81
1.00	D	2	8	13.3	39.3	1233	155
1.12	G	4	11	18.5	50.0	1258	103
1.49	E	2	10	17.8	60.7	1315	68
2.04	D	2	8	14.8	39.7	1178	145
2.11	G	4	12	18.8	53.8	1253	105
2.23	E	2	10	19.5	62.1	1296	87
2.73	D	2	8	14.4	40.5	1177	170
Total or mean		36	127			1257	110

The pigs from series C and D were killed shortly after the last period. Before killing, the pigs were preanaesthetized intramuscularly with 5 ml Sedaperone® and anaesthetized with 15–20 ml Thiomebumal® intravenously. Then 30 g of the longissimus dorsi muscle was excised from the loin region of the right side as described by Staun and Christensen (1968) and used for isolation of mitochondria essentially according to Cheah (1973).

The oxidation of the substrates was measured polarographically at 25°C with a Clark oxygen electrode in a total volume of 4.0 ml. Protein was determined by a modified biuret method using horse serum albumin (seronorm) as the standard.

The pigs were anaesthetized to a reflexless condition and bled in the normal fashion. Blood was collected for determination of the fatty acid composition of plasma methylesters as described by Christensen (1973). Similar analyses were performed on blood collected from the vena cava of pigs from series E and G after they had finished their last experimental period.

Results and discussion

The ratio between the fatty acids, 20 : 3 ω 9 and 20 : 4 ω 6, of plasma lipids was 2.8, 1.8, 0.8, 0.08, 0.09 and 0.01 for the groups receiving 0.04, 0.15, 0.29, 1.00,

2.04 and 2.73 energy % linoleate respectively. According to Holman (1960), a ratio of 0.4 or more indicates essential fatty acid (EFA) deficiency, and a ratio of 0.2 or less means that the minimum requirement of linoleate has been met. Accordingly, pigs receiving less than one energy % linoleate may be expected to suffer from deficiency symptoms. However, no systematic effect was observed on oxygen consumption, heat production or efficiency of utilization of energy as shown in *Table 22.2.* The present findings correspond with the results obtained in experiments with pigs receiving commercial feed compounds (Thorbek, 1975).

Table 22.2 OXYGEN CONSUMPTION, HEAT PRODUCTION AND EFFICIENCY OF UTILIZATION OF ENERGY IN PIGS RECEIVING DIFFERENT DIETARY CONCENTRATIONS OF LINOLEATE

Linoleate energy (%)	*Series no.*	*Number of bal exp.*	*Oxygen consumption* (ℓ/kg$^{0.75}$)		*Heat production* (kJ/kg$^{0.75}$)		*Energy retained/* $ME_g{}^x$	
			b	s_b	*b*	s_b	*b*	s_b
0.04	C	12	35.8	0.71	737	18.9	0.669	0.021
0.11	E	8	39.1	0.52	809	10.0	0.601	0.021
0.15	C	10	36.8	0.70	760	15.5	0.630	0.021
0.23	G	12	38.9	0.82	779	20.9	0.634	0.028
0.29	D	8	41.6	0.58	852	16.0	0.611	0.025
0.77	E	10	39.6	0.37	832	9.1	0.594	0.014
1.00	D	8	39.0	0.85	791	17.8	0.680	0.020
1.12	G	11	39.0	0.85	764	14.8	0.661	0.016
1.49	E	10	40.7	0.82	832	13.8	0.579	0.018
2.04	D	8	40.1	0.85	811	16.7	0.622	0.023
2.11	G	12	38.6	1.01	760	21.0	0.645	0.021
2.23	E	10	39.0	0.34	796	8.4	0.611	0.013
2.73	D	8	38.9	0.99	786	18.1	0.665	0.021
Total or mean		127	38.7	0.24	788	5.3	0.627	0.006

$^x ME_g = ME - ME_m$ with ME_m (kJ) = 4060 + 210 kg$^{0.75}$

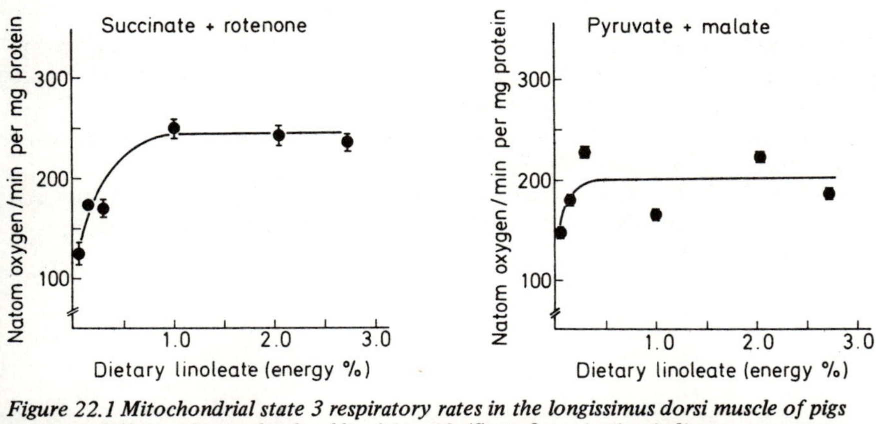

Figure 22.1 Mitochondrial state 3 respiratory rates in the longissimus dorsi muscle of pigs receiving different dietary levels of linoleic acid. (State 3 respiration indicates nanoatoms oxygen used per minute and per mg mitochondrial protein in the presence of ADP)

Table 22.3 RESPIRATORY TRAITS OF MITOCHONDRIA ISOLATED FROM THE LONGISSIMUS DORSI MUSCLE OF PIGS RECEIVING DIFFERENT DIETARY CONCENTRATIONS OF LINOLEIC ACID

Concentration of linoleate energy %	*Number of pigs*	*ADP/O*[1]				*RCI*[2]			
		Pyruvate + malate		*Succinate + rotenone*		*Pyruvate + malate*		*Succinate + rotenone*	
		Mean	*s.e.*	*Mean*	*s.e.*	*Mean*	*s.e.*	*Mean*	*s.e.*
0.04	4	2.77	0.06	1.59	0.08	8.22	1.09	5.82	1.04
0.15	4	2.73	0.06	1.50	0.04	12.06	1.65	5.07	0.49
0.29	2	2.79	0.06	1.28	0.12	8.59	0.89	2.65	0.38
1.00	2	2.73	0.11	1.46	0.13	12.73	2.06	8.70	1.05
2.04	2	2.94	0.11	1.50	0.04	13.12	1.65	5.89	0.42
2.73	2	2.73	0.10	1.26	0.10	9.39	1.40	4.37	0.78

1 Criterion for capacity for ATP synthesis
2 Criterion for coupling integrity of mitochondria

Oxygen consumption and heat production in series C were significantly lower than in the other series, probably owing to digestive disturbances. An increase in heat production, which was observed by Müller (1975) in EFA-deficient rats, could not be confirmed in the present experiments with pigs. The coupling integrity and the physphorylation capacity of skeletal muscle mitochondria were also not affected as shown in *Table 22.3*, but the oxidation rates of succinate and pyruvate plus malate were depressed in the groups receiving less than one energy % linoleate as shown in *Figure 22.1*. Thus, the requirement for linoleate seems to be about 1% of the gross energy as estimated by the energy metabolism criteria employed.

References

CHEAH, K.S. (1973). *J. Sci. Fd Agric.*, **24**, 51–61

CHRISTENSEN, K. (1973). *Z. Tierphysiol. Tierernähr. Futtermittelk.*, **31**, 317–332

CHRISTENSEN, K. (1974a). *Z. Tierphysiol. Tierernähr. Futtermittelk.*, **33**, 10–20

CHRISTENSEN, K. (1974b). *Z. Tierphysiol., Tierernähr. Futtermittelk.*, **33**, 72–80

HOLMAN, R.T. (1960). *J. Nutr.*, **70**, 405–417

MÜLLER, M. (1975). *Untersuchungen über den Bedarf der Ratte an essentiellen Fettsäuren.* Diss. ETH 5636. Zurich

THORBEK, G. (1975). *424. Beretning fra Statens Husdyrbrugsforsøg, Copenhagen*

STAUN, H. and CHRISTENSEN, K. (1968). *Bilag, Landøkonomisk Forsøgslab., efterårsmøde*, 125–132

23

EFFECT OF DIETARY FAT LEVEL AND COMPOSITION ON FAT AND PROTEIN RETENTION AND EFFICIENCY OF ENERGY UTILIZATION BY MALE CASTRATE PIGS BETWEEN 13 AND 40 kg LIVE WEIGHT

J. BOYD and K.J. McCRACKEN

Agricultural and Food Chemistry Research Division, Department of Agriculture, Northern Ireland and The Queen's University of Belfast, Newforge Lane, Belfast, Northern Ireland BT9 5PX

Summary

Large White × Landrace male castrate pigs were given diets of similar energy:protein ratio to provide a daily metabolizable energy (ME) intake of 1.4 MJ/kg$^{0.75}$ from 13 to 40 kg. Three high-fat diets containing either a large amount of stearate, oleate or linoleate were compared with a low-fat-control diet. Fat and protein deposition and energy retention (ER) were determined by slaughter. Protein deposition was significantly higher ($P<0.05$) on the high-fat diets. Fat deposition was similar on the control and high-stearate diets but significantly higher on the oleate ($P<0.01$) and linoleate ($P<0.05$) diets. The ratio of fat energy to protein energy in the gain was highest on the oleate diet and lowest on the stearate diet. Assuming an ME value of 0.48 MJ/W$^{0.75}$ kg for the daily maintenance requirement the efficiency of energy utilization for growth is 0.63.

Introduction

Added dietary fat reduces the rate of lipogenesis from carbohydrate in hepatic and adipose tissue (Leveille and Hanson, 1966; Yeh, Leveille and Wiley, 1970; Steffen *et al.*, 1978). The extent of the reduction varies with the fat source and the species, and between hepatic and adipose sites within the body (Century, 1972; Goodridge, 1972; Raulin *et al.*, 1974; Soorana and Saggerson, 1975; Waterman *et al.*, 1975). Although a reduction in lipogenesis does not necessarily lead to a decrease in net lipid deposition, differences between fat sources might be expected to result in differences in the proportions of fat and protein deposited.

Ehrensvard *et al.* (1976) compared a diet high in linoleic acid with an iso-energetic diet containing a high proportion of medium-chain saturated fatty acids. They found significant improvements in daily gain, feed conversion ratio and nitrogen retention with the diet high in linoleic acid and a reduction in the fat:lean ratio in the carcass gain. Failure to include a low-fat-control group made it impossible to distinguish whether the effects were due to an improvement on the high-linoleate diet or *vice versa* on the high medium-chain saturated fatty acid diet.

To examine further the possibility of a specific fatty acid effect on fat and protein deposition, diets containing high levels of stearate, oleate or linoleate were given to pigs between 13 and 40 kg live weight.

Materials and methods

Two replicate experiments were conducted. In each replicate 25 Large White × Landrace male castrates (five litters × 5 pigs per litter) were allocated to one of five treatments using a Latin-square design. The treatments were: starting control (SC) slaughtered at 13 kg; control (C); stearate (S); oleate (O) and linoleate (L). A typical cereal-based low-fat growers' diet was used as the control and high levels of the three fatty acids were attained by replacing part of the cereal by 70 g/kg added fat as hydrogenated tallow (spray-dried with skim milk), olive oil or sunflower oil. The energy:protein ratios of the diets were kept similar by replacing some barley by soyabean meal in the high-fat diets. The pigs were weighed twice weekly and individually fed to provide a daily ME intake corresponding to 1.4 MJ/kg metabolic bodyweight ($W^{0.75}$). At around 22 kg live weight a seven-day collection of faeces and urine was made to check the ME contents of the diets and the digestible crude protein intake. The pigs were slaughtered at 40 kg live weight and the empty carcasses minced and analysed for crude protein, fat and ash. The ER was calculated from the protein and fat gains using the factors 23.85 and 39.33 MJ/kg respectively. The energy available for production was computed as ME–M where M(MJ) = 0.48 $W_{kg}^{0.75}$ (*see* page 163) and $W^{0.75}$ was integrated on a daily basis over the experimental period. A regression of ER against ME–M was carried out, and a multiple regression of ME–M on protein and fat was also computed assuming zero intercept.

Results

The energy and crude protein digestibilities were lower ($P<0.001$ and $P<0.05$ respectively) in the stearate diet and higher in the oleate diet ($P<0.05$) than in the low-fat control (*Table 23.1*). The energy:protein ratio was 2% lower in the stearate diet than in the control and the difference was significant ($P<0.001$). The mean intakes of ME were close to the levels intended, but were significantly different ($P<0.001$), and the intakes of diets O and L were 2–3% higher as a result of the differences in ME content. Daily protein gain was higher on all three high-fat diets when compared with the control (*Table 23.2*), and the difference was significant ($P<0.05$) for the stearate and oleate treatments. The fat gain was similar on the control and stearate diets but was significantly higher on the oleate and linoleate treatments by 13% and 8% respectively ($P<0.01$ and $P<0.05$). The ratio of energy retained as fat to that retained as protein was highest on the oleate diet (1.91) and lowest on the stearate diet (1.68), but neither value was significantly different from the control (1.80). The proportion of the ME intake retained was higher than the control by 9% and 6% respectively on the oleate and linoleate treatments ($P<0.01$ and $P<0.05$). The efficiency of utilization of energy for production was not significantly different between treatments and averaged 0.630 (s.e. ± 0.005). Multiple regression analysis of the combined data yielded the equation:

$$\text{ME–M (MJ)} = \underset{(\pm 7.9)}{57.9\ P} + \underset{(\pm 7.2)}{44.1\ F} \qquad \text{(protein and fat; kg)}$$

Table 23.1 DIGESTIBILITY RATIO OF ENERGY (DE/GE) AND CRUDE PROTEIN (DCP/CP), ENERGY/PROTEIN RATIO (DE/DCP) AND METABOLIZABLE ENERGY (ME) INTAKE (MJ/kg$^{0.75}$) (10 PIGS/MEAN)

	C	*S*	*O*	*L*	*s.e. of a difference*
DE/GE	0.848	0.820***	0.860***	0.854	0.0051
DCP/CP	0.841	0.826*	0.858	0.852	0.0068
DE/DCP (kJ/g)	91.9	89.7***	91.2	90.2	0.44
ME intake	1.375	1.381	1.423***	1.414***	0.0111

*, ***, significantly different from control ($P<0.05$, $P<0.001$)
DE = digestible energy
GE = gross energy
DCP = digestible crude protein
CP = crude protein

Table 23.2 GAIN OF ENERGY (MJ/d) IN PROTEIN AND FAT AND PROPORTION OF ME INTAKE RETAINED (ER/ME) (10 PIGS/MEAN)

	C	*S*	*O*	*L*	*s.e. of a difference*
Protein gain	2.19	2.33	2.34*	2.31	0.055
Fat gain	3.96	3.91	4.47**	4.27*	0.146
Fat:protein ratio	1.80	1.68	1.91	1.85	0.071
ER/ME	0.395	0.403	0.430**	0.418*	0.0094

*, **, significantly different from control ($P<0.05$, $P<0.01$)
ER = energy retained
ME = metabolizable energy

Discussion

The significant reduction in digestibility of energy in the stearate diet can be directly attributed to the higher level of hydrogenated tallow. Calculation of the digestibility of the hydrogenated tallow by the difference method gave a coefficient of around 0.40, which is lower than the value of 0.55 reported by Bayley and Lewis (1965) for tristearin but similar to those found by Davis and Lewis (1969) for palmitic and stearic acid in tallow. The reduction in protein digestibility is contrary to the trend observed by Hillcoat and Annison (1974) and may be due to partial protection from proteolytic digestion of the milk protein in the spray-dried fatted skim.

The increased gain of protein on the high-fat diets compared to the control may be due to an improvement in protein quality arising from the replacement of barley meal by soyabean meal to maintain similar digestible energy:digestible crude protein ratios. The lysine intake was slightly lower on the control diet, but it did meet the Agricultural Research Council (1967) recommendation. Since the ME intakes were only slightly higher on the oleate and linoleate treatments, it seems unlikely that this factor could have been entirely responsible for the increased fat gain.

The present results do not corroborate the apparent increase in lean-meat production with high linoleic acid reported by Ehrensvard *et al.* (1976). The apparent improvement in the fat:lean ratio on the stearate treatment may be related to the increased inhibitory effect of long-chain saturated acids on adipose tissue lipogenesis reported by Vernon (1977).

The calculated energy of protein deposition is similar to the values reported by Kotarbinska (1969) and Thorbek (1970). The high efficiency of fat deposition (0.89) supports the view that direct incorporation accounted for most of the fat deposition on the high-fat diets.

Acknowledgements

J. Boyd wishes to acknowledge the receipt of a postgraduate scholarship from the Meat and Livestock Commission.

References

AGRICULTURAL RESEARCH COUNCIL (1967). *The Nutrient Requirements of Farm Livestock, No. 3: Pigs.* London: HMSO

BAYLEY, H.S. and LEWIS, D. (1965). *J. agric. Sci., Camb.*, **64**, 373–378

CENTURY, G. (1972). *J. Nutr.*, **102**, 1067–1077

DAVIS, R.H. and LEWIS, D. (1969). *J. agric. Sci., Camb.*, **72**, 217–222

EHRENSVARD, U., BERSCHAUER, F., MENKE, K.H., ROGDAKIS, E. and STURM, G. (1976). In *Energy Metabolism*, pp. 85–88. Ed. M. Vermorel. EAAP Publication No. 19. Clermont Ferrand: G. de Bussac

GOODRIDGE, A.G. (1972). *J. biol. Chem.*, **247**, 6946–6952

HILLCOAT, J.B. and ANNISON, E.F. (1974). In *Energy Metabolism*, pp. 177–180. Eds. K.H. Menke, H.-J. Lantzsch and J.R. Reichl. EAAP Publication No. 14. Stuttgart: Universität Hohenheim, Dokumentationsstelle

KOTARBINSKA, M. (1969). *Wydaw. wlasne Inst. Zootech. Wroclaw, No. 238*

LEVEILLE, G.A. and HANSON, R.W. (1966). *J. Lipid. Res.*, **7**, 46–55

RAULIN, J., LORRETTE, C., LAUNAY, M., LAPOUS, D., GOUREAU-COUNIS, M.G., COUNIS, R. and CARREAU, J.P. (1974). In *The Regulation of the Adipose Tissue Mass*, p. 32. Eds. J. Vague and J. Boyer.

SOORANA, S.A. and SAGGERSON, E.D. (1975). *Biochem, J.*, **150**, 441–451

STEFFEN, O.G., CHAI, E.Y., BROWN, L.J. and MERSMANN, H.J. (1978). *J. Nutr.*, **108**, 911–918

THORBEK, G. (1970). In *Energy Metabolism*, pp. 129–132. Eds. A. Schürch and C. Wenk. EAAP Publication No. 13. Zurich: Juris Druck and Verlag

VERNON, R.G. (1977). *Int. J. Biochem.*, **8**, 517–523

WATERMAN, R.A., ROMSOS, D.R., TSAI, A.C., MILLER, E.R. and LEVEILLE, G.A. (1975). *Proc. Soc. exp. Biol. Med.*, **150**, 347–351

YEH, Y.Y., LEVEILLE, G.A. and WILEY, J.H. (1970). *J. Nutr.*, **100**, 917–924

24

ENERGY METABOLISM IN BOARS, GILTS AND BARROWS, AND BARROWS FED DIETHYLSTILBOESTROL AND METHYLTESTOSTERONE

J.B. LUDVIGSEN
National Institute of Animal Science, Rolighedsvej 25, 1958 Copenhagen V, Denmark

Summary

Energy metabolism and efficiency of utilization of energy was measured in three boars, four gilts and five barrows, three of the last being fed diethylstilboestrol (DES) + methyltestosterone (MT) (Maxymin®) during growth from 23.5 to 80 kg live weight maintained on an isonitrogenous and isocaloric diet daily (16.4% crude protein and from 14.2 to 42.6 MJ gross energy). Heat production as a percentage of metabolizable energy (ME) was found to be significantly lower in barrows than in boars and gilts as well as in the hormone-treated barrows. Energy retained (EP) in protein deposited was significantly lower in barrows when compared to boars, gilts and hormone-treated barrows. The lower heat production as well as the lower utilization of energy in barrows is an effect of the absence of testicular hormones. The suppressing effect of castration in anabolic pathways was abolished through the addition of diethylstilboestrol and methyltestosterone daily to the feed, which readjusted the heat production and the efficiency of utilization of energy of barrows to the levels of boars and gilts.

Introduction

In the United Kingdom DES and MT are approved for commercial use in the pig industry (under the trade mark Maxymin Premix) because it improves feed efficiency and the carcass leaness in finishing pits. Baker *et al.* (1967) and Melliere, Waitt and Jordan (1970) reported that addition of DES + MT (2.2 mg of each per kg of feed) was followed by an increase in nitrogen retention (NR) in barrows as well as in gilts, and this was reflected in an increased loin area and a decrease in backfat thickness as measured by commercial standards.

The data also indicate that DES + MT are more effective in gilts than in barrows when measured by NR and that DES is the actual anabolic agent, as MT alone has a consistent non-effect whereas DES + MT significantly improve NR when compared with the performance of the pigs fed the two compounds individually.

The aim of the present study is to evaluate the effect of DES + MT on the energy metabolism of barrows in a comparative study between boars, gilts and barrows.

Materials and methods

Three boars, four gilts and five barrows (Danish Landrace pigs from genetically closely related litters) were tested individually in five consecutive balance experiments of seven days each. These investigations included a 24-hour-respiration experiment on the middle day of the collection period from an average live weight of 23.5 kg (77 days of age) to 80.0 kg (160 days of age).

The diet was composed of barley, soybean and meat and bone meal, and was kept isonitrogenous (16.4% crude protein) and isocaloric. The gross energy (GE) increased from 14.2 MJ to 42.6 MJ daily.

In the first balance experiment the five barrows were tested without the addition of hormones. From the beginning of the second trial period three of the barrows received 1 g of Maxymin Premix, which is the equivalent of 4.84 mg of DES and MT each per kg of feed, daily for 76 consecutive days.

Results

The experimental data are calculated on the basis of heat production (HP) and energy deposited in protein (EP) as percentages of metabolizable energy (ME).

In *Table 24.1* the average heat production of the four groups of pigs is shown. From this table it can be seen that the average heat production of the barrows fed hormones is on the same level as that of boars and gilts, whereas the heat production of the barrows is lower. Due to the non-significant difference between the three groups, the lower heat production of the barrows was tested statistically against the accumulated results of the three groups omitting the results from the first period, as the barrows were divided into a hormone and a non-hormone treated group. The difference in heat production between the two groups (3.9%) is highly significant: $t = 6.13 > t_{0.001} = 3.55$.

Table 24.1 HEAT PRODUCTION (% OF ME)

	Boars	*Gilts*	*Barrows + Max.*	*Barrows*
n	14	19	11	12
HP (%)	57	56	56	52
n		37		8
HP (%)		56.2		52.3***

n = number

In *Table 24.2* the NR and the EP are shown. From the table it can be seen that NR of the hormone-treated barrows is at the same level as that of the boars and the gilts, whereas NR in the barrows is lower than that of the three other groups followed by a correspondingly lower EP. Statistical evaluation of the observations revealed no difference in EP during all five experimental periods. However, omitting the observations in the second period of the experiment, from the calculations, a statistical difference, although minor, was found between the barrows tested against the three other groups together: $t = 2.61 > t_{0.05} = 2.04$.

Table 24.2 NITROGEN RETENTION AND ENERGY RETAINED IN PROTEIN DEPOSITED (% OF ME)

	Boars	*Gilts*	*Barrows + Max.*	*Barrows*
n	14	19	11	12
Nitrogen retention (g)	23.4	23.6	23.3	22.5
EP (%) Period I–V	14.9	14.8	14.6	14.5
n		28		6
EP (%) Period III–V		13.8		13.0$^{x(x)}$

Discussion

Most studies published on energy metabolism in pigs have been performed with barrows and to a minor extent with gilts, whereas comparative studies on energy metabolism of boars, gilts and barrows from genetically closely related pigs are limited.

Although the material in the present study is limited, significant differences in energy metabolism and the efficiency of energy utilization between boars and gilts on one side and barrows on the other are obvious. On an isonitrogenous diet containing 16.4% crude protein and isocaloric daily feed intake from 14.2 to 42.6 MJ, the heat production of boars as well as gilts is within the range of 56–57% of ME; it is a little higher in the boars than in the gilts. The heat production of barrows is significantly lower, that is 52% as an average. However, no statistical difference between the barrows and boars and gilts could be proved in the first trial period, but the difference increased during the experimental periods.

The same pattern of metabolic events takes place in the metabolism of nitrogen. No significant difference in NR between the three sexes could be found in the first experimental period, but the difference increased with age reflecting itself in an overall reduced NR followed by a significantly lower EP in the barrows. The lower protein deposition and the higher deposition of energy calculated as fat in the barrows account for the lower heat production in castrates versus boars and gilts.

In the present study the pigs were castrated at 49 days of age. In the first experimental period no significant difference in heat production and protein deposition between sexes was detectable at 77 days of age. This may indicate that the effects of absence of testicular hormones in metabolic pathways is a slow and progressive process and, as measured by the influence on energy metabolism and efficiency of energy utilization, it takes more than a month after castration before the effects are measurable.

The daily administration of DES + MT as feed additives to barrows abolishes the effects of castration on heat production, as well as those on EP, and leads to the same levels as those of boars and gilts. The effect of the addition of hormones was not an immediate one in this experiment. The effects on energy metabolism were not detectable in the first experimental period with DES + MT, which seems to indicate that it takes more than three weeks before the effects of the steroids on energy metabolism are measurable in castrates.

The results of the present experiment support earlier investigations in that addition of DES + MT improves protein deposition as measured by commercial standards: improving leanness of the carcass and reducing backfat thickness.

References

BAKER, D.H., JORDAN, C.E., WAITT, W.P. and GOUWENS, D.W. (1967). *J. Anim. Sci.*, **26**, 904

MELLIERE, A.L., WAITT, W.P. and JORDAN, C.E. (1970). *J. Anim. Sci.*, **31**, 1025

25

COMPARISON OF THE ENERGY EXCHANGES OF DUCKLINGS AND CHICKENS

D.J. FARRELL and A.P. SIREGAR
Department of Biochemistry and Nutrition, University of New England, Armidale, New South Wales 2351, Australia

Summary

Calorimetric measurements were made at intervals on groups of ducklings, from 5 to 23 d of age, and chickens, from 11 to 39 d, when starved and when offered high (ME of 14.8 MJ/kg) and low (ME of 11.9 MJ/kg) energy diets. Starvation heat production (kJ/d) was for ducklings 804 $W^{0.70}$ and for chickens 675 $W^{0.74}$. During feeding the ducklings grew much more rapidly, consumed more feed and had a higher energy retention (ER) and efficiency of utilization of metabolizable energy (ME) for gain (0.64 vs 0.50) than chickens. Food conversion differed between diets but not between species. Metabolizability of the diet declined as ducklings aged, particularly on the high-energy diet. Of the dietary energy retained for both species, more was retained as protein than fat, although the ducklings had higher retentions of fat (44%) than chickens (37%). Gaseous ammonia nitrogen was 5% of the nitrogen balance of ducklings and 2.2% of chickens.

Introduction

There is virtually no published work on the energy exchange of ducklings. The much more rapid growth rate and increased fat deposition of ducklings when compared with chickens would suggest that a comparison of their energy exchange and nitrogen metabolism would prove useful. Measurements reported here were made on fed and starved birds.

Materials and methods

Measurements were made on male meat-type White Pekin ducklings and male broiler chickens on four occasions (periods) when ducklings and chickens were of similar live weight (W, kg) but of different ages. Measurements (24 h) of starvation heat production of groups of four birds commenced when ducklings were nine, 16, 23 and 28 d, and when chickens were 15, 21, 33 and 39 d respectively. Corresponding measurements over four days of birds fed *ad lib.* a high (H) or low (L) energy diet commenced at five, 12 and 19 d for ducklings, and at 11, 17 and 29 d for chickens. Both diets contained similar ingredients which are normally used in poultry formulations, but diet H contained more soyabean and in addition vegetable oil (93 g/kg) when compared with diet L which also contained 148 g/kg of bran. The ME (MJ/kg) and crude protein (%) concentrations of diets L and H were 11.9 and 19.7, and 14.8 and 25.2 respectively.

The four respiration chambers used and their principles of operation were as described by Farrell (1972) with modifications by Farrell (1974) and Pym and Farrell (1977). Further modifications were necessary to avoid spillage which is normally associated with the eating and drinking habits of ducklings. Chamber temperature and duration of starvation were appropriate to the age of birds. Chamber air was circulated continuously through a one litre flask containing 1.2–1.5% (W/V) boric acid to trap gaseous ammonia nitrogen; this was subsequently titrated against 0.1 N HCl.

Results

The respiratory quotient was higher ($P<0.01$) for starved ducklings (0.714) than for chickens (0.705). Overall mean daily starvation heat production (M) was higher ($P<0.01$) for ducklings (1064 kJ) than for chickens (951 kJ) and the relationship was calculated, using the method of least squares, and was different ($P<0.05$) for ducklings :

$$M = 804\ W^{0.70},\ n = 16 \qquad (1)$$

and for chickens :

$$M = 675\ W^{0.74},\ n = 16 \qquad (2)$$

Mean composition (g/W per d) of weight loss was also different for ducklings, in that they lost 5.5 g of protein tissue in comparison with 4.5 g by chickens. Gaseous ammonia nitrogen contributed 82.5 mg to the daily nitrogen loss for ducklings and only 28 mg for chickens. Fat (g/W per d), which was calculated by difference, gave values of 19 g for ducklings and 14 g for chickens, but body water loss was not different between the two species. Because heat production (kJ/W per d) decreased as the birds aged, the losses of chemical components (g/W per d) of tissue also declined ($P<0.05$) but the ratio of fat (g) to protein (g) lost remained reasonably constant at 3:1 for both species.

The main results of the feeding experiments are given in *Table 25.1*. Generally, ducklings outperformed chickens with a 60% higher mean daily energy intake, and for both species performance was superior on the H diet. Despite the higher ($P<0.05$) food intake and weight gain of ducklings the food conversion ratio (FCR) was the same. FCR declined ($P<0.05$) as the chickens aged but was not different for ducklings. Although overall metabolizability of the diets was the same, there were some interesting differences ($P<0.05$) observed for ducklings. Metabolizability, uncorrected to zero-nitrogen retention, was for diet L 73.9, 70.6 and 69.6% for periods 1, 2 and 3 respectively; corresponding values for diet H were 79.4, 79.1 and 74.7%. This diet (H) contained 9% vegetable oil.

Nitrogen retention (%) and balance (g/d) were greater ($P<0.05$) for ducklings than for chickens; for both species retention declined ($P<0.05$) as birds aged (*Table 25.1*). However, there was no difference that could be attributed to diet. Gaseous ammonia nitrogen (mg/W per d) excretion was almost four times as

Table 25.1 MEANS (± s.e.) OF PARAMETERS MEASURED DURING CALORIMETRIC STUDIES OVER 12 d IN THREE PERIODS (AGE) EACH OF 4 d ON GROUPS OF THREE DUCKLINGS AND THREE CHICKENS FED A LOW (L) AND HIGH (H) ENERGY DIET (QUANTITIES GIVEN PER d)

		Experimental diet		*s.e. ±*	*Period*[1]			*s.e. ±*	*Overall mean*
		L	*H*		*1*	*2*	*3*		
No. of observations		*6*	*6*		*4*	*4*	*4*		*12*
Parameter									
Weight gain (g/kg W)	D	102^{c}[2]	125^{d}	6.1	131^{d}	123^{d}	87^{c}	2.7	113.6^{b}
	C	65^{a}	76^{b}	2.3	99^{c}	68^{b}	45^{a}	2.8	70.4^{a}
		s.e ± 7.0			s.e. ± 3.4				s.e. ± 3.3
Food intake (g/kg W)	D	210^{c}	200^{c}	8.3	251^{f}	200^{e}	163^{d}	10.1	204.6^{b}
	C	122^{b}	104^{a}	2.7	138^{c}	110^{b}	91^{a}	3.3	112.8^{a}
		s.e. ± 0.08			s.e. ± 7.5				s.e. ± 4.3
Food conversion ratio (g/g)	D	2.10^{b}	1.59^{a}	0.11	1.97^{b}	1.64^{b}	1.91^{b}	0.13	1.84^{a}
	C	1.97^{b}	1.44^{a}	0.05	1.41^{a}	1.66^{b}	2.05^{c}	0.06	s.e. ± 0.06
Metabolizability of diet (%)	D	71.4^{a}	77.7^{b}	0.20	76.6^{c}	74.8^{b}	72.2^{a}	0.25	74.5^{a}
	C	71.6^{a}	78.5^{b}	0.37	75.1^{b}	75.7^{b}	74.4^{b}	0.46	75.1^{a}
		s.e. ± 0.30			s.e. ± 0.37				s.e. ± 0.21
ME intake (kJ/kg W)	D	2477^{c}	2575^{d}	103.5	2790^{b}	2607^{e}	2181^{d}	126.8	2526^{b}
	C	1451^{a}	1600^{b}	23.2	1837^{c}	1540^{b}	1198^{a}	28.5	1525^{a}
		s.e. ± 75.0			s.e. ± 91.9				s.e. ± 53.1
Water intake (g/kg W)	D	912^{b}	831^{b}	66.0	1175^{e}	826^{d}	614^{d}	80.8	870^{b}
	C	266^{a}	251^{a}	6.2	336^{c}	244^{b}	195^{a}	7.6	258^{a}
		s.e. ± 46.5			s.e. ± 57.0				s.e. ± 32.9
Respiratory quotient	D	1.03^{a}	1.03^{a}	0.012	1.03^{a}	1.02^{a}	1.04^{a}	0.009	1.03^{a}
	C	1.02^{a}	1.03^{a}	0.001	1.02^{a}	1.03^{a}	1.02^{a}	0.001	1.02^{a}
		s.e. ± 0.007			s.e. ± 0.009				s.e. ± 0.005
Heat production (kJ/kg W)	D	1444^{c}	1336^{b}	22.2	1529^{d}	1501^{d}	1139^{c}	27.2	1390^{b}
	C	992^{a}	975^{a}	10.6	1170^{c}	969^{b}	811^{a}	13.0	983^{a}
		s.e. ± 17.4			s.e. ± 21.3				s.e. ± 12.3
Nitrogen balance (g/kg W)	D	4.04^{b}	4.37^{b}	0.18	4.84^{d}	4.31^{d}	3.48^{c}	0.23	4.21^{b}
	C	2.09^{a}	2.32^{a}	0.10	2.71^{b}	2.30^{b}	1.60^{a}	0.12	2.21^{a}
		s.e. ± 0.15			s.e. ± 0.18				s.e. ± 0.10
Nitrogen retention (%)	D	61.4^{b}	62.9^{b}	1.02	66.8^{c}	62.1^{b}	57.6^{b}	1.25	62.2^{b}
	C	54.9^{a}	54.7^{a}	0.75	55.6^{ab}	57.3^{b}	51.4^{a}	2.90	54.8^{a}
		s.e. ± 1.56			s.e. ± 1.91				s.e. ± 1.10
Ammonia nitrogen (mg/kg W)	D	216^{b}	161^{b}	30.9	268^{c}	228^{c}	69^{b}	37.8	189^{b}
	C	56.0^{a}	40.5^{a}	5.6	66.6^{b}	40^{a}	39^{a}	6.8	48^{a}
		s.e. ± 22.2			s.e. ± 27.2				s.e. ± 15.7
Energy balance (kJ/kg W)	D	1027^{b}	1190^{b}	89.1	1172^{c}	1116^{c}	1036^{c}	109.2	1108^{b}
	C	447^{a}	612^{a}	33.9	651^{b}	558^{b}	380^{a}	41.5	530^{a}
		s.e. ± 63.0			s.e. ± 77.2				s.e. ± 58.5

Table 25.1 (cont.)

Fat retention (g/kg W)	D	10.2^c	15.0^c	2.30	13.2^c	11.4^c	13.2^c	2.81	12.6^b
	C	3.4^a	6.7^b	0.56	6.2^b	5.4^{ab}	3.6^a	0.69	5.1^a
		s.e. ± 1.09			s.e. ± 1.34				s.e. ± 1.18
Protein retention (g/kg W)	D	26.9^b	28.6^b	1.02	32.2^e	28.7^d	22.4^c	1.30	27.8^b
	C	13.5^a	14.9^a	0.65	17.6^b	14.8^b	10.3^a	1.05	14.2^a
		s.e. ± 0.85			s.e. ± 0.79				s.e. ± 0.60

[1] Age and mean live weight of experimental birds:
D = duck – Period 1 (5–8 d, 377 g); Period 2 (12–15 d, 1014 g); Period 3 (19–22 d, 1933 g)
C = chicken – Period 1 (11–14 d, 618 g); Period 2 (17–20 d, 1211 g); Period 3 (29–32 d, 2469 g)
[2] Means of a particular parameter within rows and columns with different superscripts (a–e) are significantly different ($P<0.05$)

much for ducklings as for chickens and although no differences could be attributed to diet for either species this loss declined as birds aged. Gaseous ammonia nitrogen amounted to 5% of the balance of ducklings and 2.2% of chickens. On average both protein and fat retentions (g/W per d) were twice as much for ducklings as for chickens. Furthermore, fat retention did not change for ducklings as they aged. Overall fat retention was 44% of the energy balance of ducklings and 37% for chickens.

There may be small discrepancies between daily weight gain and the summation of daily nitrogen and fat retentions (*Table 25.1*). The nitrogen balance method used here generally overestimates nitrogen retention, and the calculated amount of protein tissue deposited will depend on the estimate of the water content of that tissue.

Ducklings consumed almost twice as much water as did chickens. The mean ratio of water to food was 4.3:1 for ducklings and 2.3:1 for chickens. Energy balance was much higher ($P<0.05$) for ducklings than for chickens, and although it was higher on diet H than diet L for both species within each species, this difference was not significant. The relationship between ME intake (X, kJ/W per d) and energy balance (Y, kJ/W per d) for ducklings (d) was :

$$Y_d = -472 + 0.64\ (\pm 0.15)\ W \qquad RSD = 168,\ r^2 = 0.80,\ n = 12 \qquad (3)$$

and for chickens (c):

$$Y_c = -228 + 0.50\ (\pm 0.08)\ W \qquad RSD = 72,\ \ r^2 = 90,\ \ n = 12 \qquad (4)$$

Thus the partial efficiency of utilization of ME for gain was 0.64 for ducklings and 0.50 for chickens. It can be calculated from these equations that the maintenance energy requirement (kJ/W per d) was 738 for ducklings and 456 for chickens. If starvation values of energy balance are included in the above two equations, as is frequently done, regression coefficients become 0.79 and 0.77, and maintenance energy requirements become 1073 and 853 for ducklings and chickens respectively. These maintenance values (kJ/W per d) appear to be more realistic in view of the starvation heat production values in equations (1) and (2).

Discussion

The rapid growth rate of ducklings was associated with an increased starvation heat production when compared with chickens, and a higher maintenance energy requirement. The retention of protein (g) to fat (g) was for both species more than 2:1; however, ducklings retained 43% of their energy balance as fat and chickens 38%. The energy efficiency of fat (k_f) deposition is usually higher than that of protein (k_p) deposition (Pym and Farrell, 1977) and is consistent with the higher partial efficiency of ME for gain of 0.64 in comparison with only 0.50 for chickens.

Siregar, Cumming and Farrell (1978) suggested that the protein requirement (% of diet) of young ducklings was considerably less than that of young chickens. Indeed, ducklings on both diets (L and H) retained a significantly higher proportion of dietary nitrogen than did chickens (*Table 25.1*). Ammonia nitrogen contributed substantially to the nitrogen loss of ducklings. Glutamine is probably the major source of urinary ammonia in the fowl, but it is not clear why losses via this route were greater in ducklings than in chickens.

Metabolizability of the diet declined markedly as ducklings aged. This was particularly true of the diet H, which contained added vegetable oil, and to a lesser extent of the low energy diet L. Again, it is difficult to postulate why such a decline might occur as ducklings aged, although ME values were not corrected to zero-nitrogen retention.

For both species weight gain and ME intake were greater on the H diet than on diet L, and although heat production of chickens was the same, ducklings had a lower ($P<0.05$) heat production (kJ/W per d) on diet H than diet L; as a consequence the energy retention of diet H was 46% of ME intake in comparison with 41% for diet L. Therefore, it would appear that a dietary ME concentration of 12 MJ/kg is insufficient for both broiler chickens and ducklings to maximize biological performance. It is recognized that the measurements presented here are for ducklings up to three weeks of age, and for chickens up to 4½ weeks, and, therefore, it is invalid to extrapolate observations beyond these ages.

References

FARRELL, D.J. (1972). *Poult. Sci.*, **51**, 683–688

FARRELL, D.J. (1974). *Br. Poult. Sci.*, **15**, 34–47

PYM, R.A.E. and FARRELL, D.J. (1977). *Br. Poult. Sci.*, **18**, 411–426

SIREGAR, A.P., CUMMING, R.B. and FARRELL, D.J. (1978). In *Recent Advances in Animal Nutrition 1978*, p. 26. Ed. D.J. Farrell. Armidale: University of New England Publishing Unit

26

PROTEIN MASS, PROTEIN SYNTHESIS AND HEAT LOSS IN THE ZUCKER RAT

A.J.F. WEBSTER*, G.E. LOBLEY, P.J. REEDS and J.D. PULLAR
The Rowett Research Institute, Bucksburn, Aberdeen AB2 9SB, Scotland

Summary

Estimates of protein synthesis were compared with measurements of heat loss in fatty and lean Zucker rats at 200 and 350 g body weight and given 240 or 315 kJ of metabolizable energy (ME) per day. The regression equation relating heat loss (H; kJ/d) to protein synthesis (A_S; nitrogen g/d) was $H = H' + 74\ A_S$. The residual term H' was 210 kJ/kg$^{0.75}$ per d for fatty and lean rats, and the correlation coefficient (r) thus obtained was 0.93. Estimates were also made of the contribution made to heat production by protein synthesis and related metabolic processes.

Materials and methods

The requirement of a growing animal for ME has traditionally been calculated by expressing its H as a function of body weight and ME intake. Kielanowski (1976) first attempted to separate by regression analysis the energy costs of maintenance and protein and fat deposition in pigs. Pullar and Webster (1977) used measurements of energy and nitrogen balance in congenitally obese (fatty) and lean Zucker rats to calculate the efficiencies of utilization of ME for protein and fat deposition (0.44 and 0.74 respectively) in a way that was free from *a priori* assumptions about maintenance requirement or the relationship between H and body weight. However, in these animals there were large differences between phenotypes in residual H which were attributed in part to differences in protein synthesis. The present experiment describes attempts to measure protein synthesis in lean and fatty Zucker rats at 200 and 350 g body weight and given 240 or 315 kJ/d of ME. These and other parameters of the energy cost of growth are compared with measurements of H made in a gradient-layer calorimeter.

Rats were infused via a tail vein with either 2,3-^{3}H-L-tyrosine or U-^{14}C-L-tyrosine for 6 h. Measurements of the specific radioactivity (SR) of free and protein-bound tyrosine were as described by Nicholas, Lobley and Harris (1977). The rate of catabolism of leucine was also estimated from $^{14}CO_2$ production during intravenous infusion of 1-^{14}C-L-leucine. In this case an open-circuit respiration apparatus was used which included an ionization chamber for measurement of expired radioactivity. Fractional rates of protein synthesis in muscle, skin,

* Present address: Department of Animal Husbandry, University of Bristol

liver and intestine were calculated from the ratios of SR of protein-bound tyrosine to that of free tyrosine in the plasma ($k_{s,p}$) or in the tissue homogenate ($k_{s,h}$). Neither of these pools truly reflects the precursor from which protein is synthesized (Waterlow, Garlick and Millward, 1978). Whole body protein synthesis was calculated either as the sum of the products of protein mass and fractional synthesis rate in the four major tissue groups ($A_{s,h}$ and $A_{s,p}$) or from leucine flux through the blood ($A_{s,f}$), with flux being the sum of leucine catabolism and leucine incorpoation into protein.

Results and discussion

Table 26.1 summarizes the measurements of $k_{s,h}$ and $k_{s,p}$ in skeletal muscle, liver, intestine and skin of rats at 200 and 350 g body weight. Values for the lean and fatty phenotypes and for the two feed intakes have been pooled because there were no consistent significant differences attributable either to phenotype or to intake. There was a marked difference between $k_{s,h}$ and $k_{s,p}$, especially in the tissues of rapid turnover, the liver and the intestine. The values for $k_{s,h}$ in liver and muscle of rats are similar to those obtained by most other workers using continuous infusion of different labelled amino acids, or from measurements of ^{14}C carbonate decay (Waterlow, Garlick and Millward, 1978, p. 409). Much of the subsequent text of Waterlow, Garlick and Millward (1978) is based on the assumption that $k_{s,h}$ approximately reflects, or slightly overestimates, true protein synthesis.

Table 26.1 POOLED DATA FOR THE FRACTIONAL RATES OF PROTEIN SYNTHESIS IN SKELETAL MUSCLE, LIVER, INTESTINE AND SKIN OF RATS AT 200 g OR 350 g

Group	*Muscle*	*Liver*	*Intestine*	*Skin*
		$k_{s,h}$ (g/kg per d)		
200 g	63.1 ± 2.1 (19)	518 ± 48 (20)	1804 ± 156 (21)	256 ± 26 (21)
350 g	46.0 ± 2.2 (25)**	651 ± 60 (22)	1670 ± 156 (22)	132 ± 8 (23)**
		$k_{s,p}$ (g/kg per d)		
200 g	41.5 ± 2.6 (18)	148 ± 15 (18)	446 ± 67 (18)	100 ± 17 (18)
350 g	32.9 ± 1.8 (25)*	153 ± 16 (25)	459 ± 34 (25)	43 ± 8 (25)*

Statistical differences between body weights in rats of the same phenotype
* $P<0.05$ ** $P<0.01$
Numbers of animals tested in each group are shown in brackets

Table 26.2 presents measurements of heat loss and nitrogen retention and estimates of whole body protein synthesis based on $A_{s,p}$, $A_{s,h}$ and $A_{s,f}$. The agreement between $A_{s,h}$ and $A_{s,f}$ was quite close except for fatty rats on the high ration. If $k_{s,h}$ accurately reflects fractional protein synthesis rate, then $A_{s,h}$ undoubtedly underestimates whole body protein synthesis because the carcass in particular appears to contain proteins which turn over much faster than those in muscle. On the other hand, estimates of fractional protein synthesis in the rat

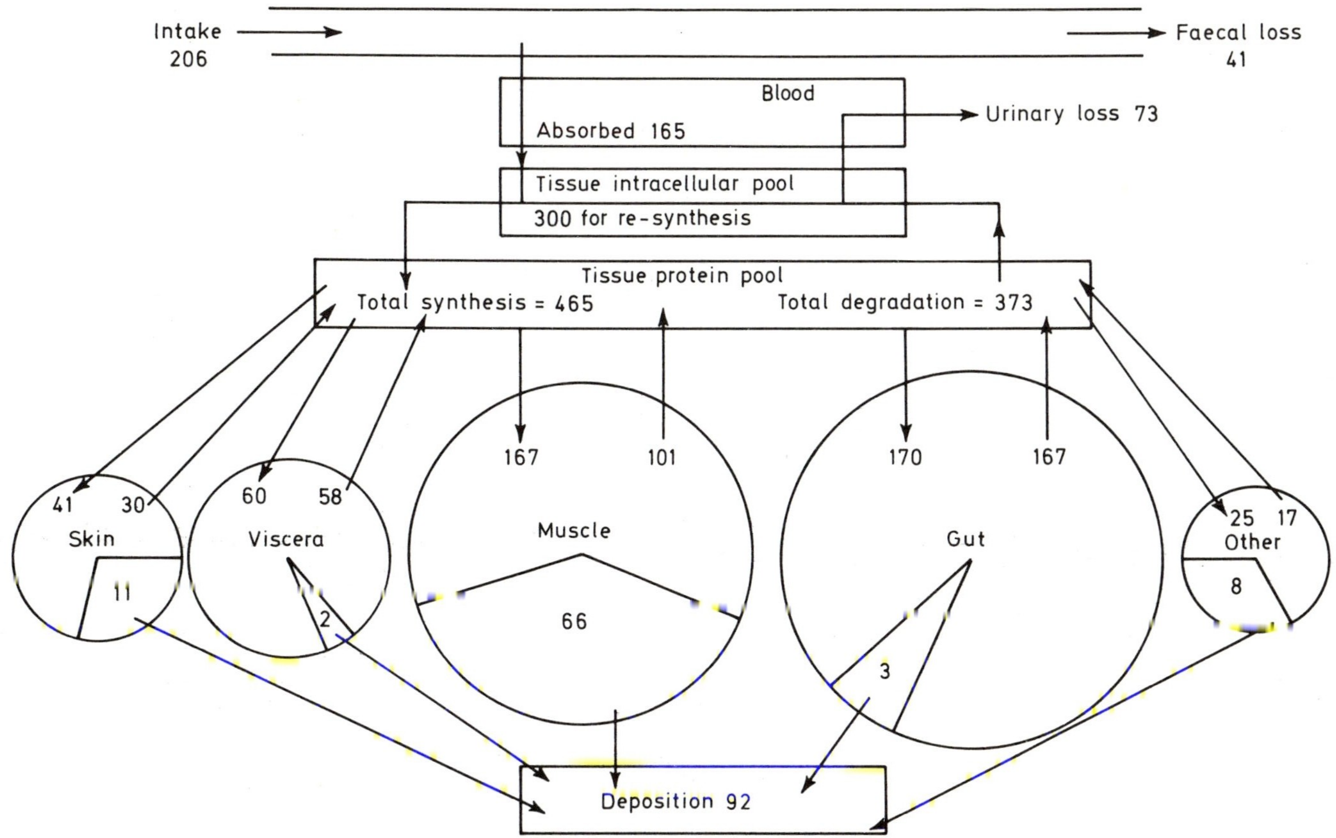

Figure 27.1 A simplified model of protein metabolism in the 25 kg pig. All figures are expressed as grams crude protein/d (nitrogen × 6.25). Circle areas equate to tissue protein synthesis, segment areas equate to protein deposition and the remaining area to protein degradation. See ... e.g. blood, trotters etc., and ... assumed

Discussion

The model illustrates the large contribution that muscle makes to WBPS in the young growing pig. Although having the slowest FSR (0.064/d), the large protein mass of muscle (68% of the carcass protein content) results in 35% of WBPS and 72% of total body protein deposition occurring in this tissue. The viscera and digestive tract contributed extensively to WBPS (approximately 26%) while skin accounted for 12%.

For the 25 kg pig the energy cost of protein deposition (ECPD) calculated from whole animal parameters was 55.8 kJ/g (assuming a maintenance energy requirement of 475 kJ/kg $W^{0.75}$ per d (Verstegen *et al.*, 1973). If it is assumed that the energy cost of peptide bond synthesis is equivalent to 3 kJ/g protein (Buttery and Annison, 1973), then the total body energy consumption associated with peptide synthesis is 1.4 MJ/d. Even assuming that this energy cost of turnover is wholly associated with growth then the calculated value of ECPD was only 39 kJ/g. This would be reduced if a value for maintenance turnover were subtracted. Therefore, the difference normally found between biochemical and whole animal estimates for ECPD (Kielanowski, 1976) cannot be solely attributed to the energy demand of protein synthesis during turnover.

Acknowledgement

B.K.E. acknowledges the receipt of a Science Research Council CASE award.

References

BROUWER, E. (1965). In *Energy Metabolism.* Ed. K.L. Blaxter. EAAP Publication No. 11. London: Academic Press

BUTTERY, P.J. and ANNISON, E.F. (1973). In *The Biological Efficiency of Protein Production.* Ed. J.G.W. Jones. Cambridge University Press

GARLICK, P.J., MILLWARD, D.J. and JAMES, W.P.T. (1973). *Biochem. J.*, **136**, 935–945

KIELANOWSKI, J. (1976). In *Protein Metabolism and Nutrition*, pp. 207–216. Eds. D.J.A. Cole, K.N. Boorman, P.J. Buttery, D. Lewis, R.J. Neale and H. Swan. EAAP Publication No. 16. London: Butterworths

THORBEK, G. (1961). In *Energy Metabolism.* EAAP Publication No. 10. The Netherlands: Wageningen

VERSTEGEN, M.W.A., CLOSE, W.H., START, I.B. and MOUNT, L.E. (1973). *Br. J. Nutr.*, **30**, 21–35

28

ESTIMATES OF METABOLIZABLE ENERGY NEEDS FOR MAINTENANCE AND GAIN IN BEEF STEERS OF FOUR GENOTYPES

F.J. VAN DER MERWE and P. VAN ROOYEN
Department of Animal Science, University of Stellenbosch, South Africa

Summary

Seven young steers of each of the breeds Africander, Drakensberg, Hereford and Simmental were fed at different feeding levels from near maintenance to near *ad lib.* for 120 days. Empty-body-weight (EBW) gains were measured and related to ME intake and indirect estimates of changes in body composition. The equation $ME = aW^{0.75}(1 + b'G)$, fitted to the data, predicted lower maintenance needs for Africander and Simmental but relative weight gains were not significantly different. At 300 kg the live-weight predictions of live-weight gain by the above equation and the metabolizable energy (ME) system were similar.

Introduction

Genotypes in feed lots and performance-testing stations in South Africa range from improved local breeds and strains of predominantly Zebu origin to a variety of exotic breeds and their crosses of which the compact early-maturing British and large-framed late-maturing continentals are typical extremes. For the practical application of the ME system (Blaxter, 1962) a single prediction equation was adopted (Ministry of Agriculture, Fisheries and Food, 1975) for estimating the maintenance allowance of all cattle. Vercoe and Frisch (1974) established a lower requirement for net energy (NE) for maintenance and a higher efficiency of utilization of a low quality diet by Zebu crossbreds when compared with British crossbreds. However, the question of genotypic variance in the 'maintenance component' of ME utilization has not been resolved, and this chapter reports on an experiment carried out with four genotypes in an attempt at validating the use of a single prediction equation under local conditions.

Materials and methods

Twenty-eight male castrates (steers) were used. They were aged from 15 to 18 months of age and represented four different genotypes, namely, Africander, Drakensberg, Hereford and Simmental. After a preliminary 42-day equilibration period of group feeding on the experimental diet at more or less maintenance level, the steers were transferred to individual pens in a closed barn. Each one of the seven members of each breed group was randomly allotted to one of seven levels of daily dry matter (DM) intake which ranged from 65 to 95 g/kg $W^{0.75}$. The trial lasted for 120 days from midsummer to autumn. The steers were weighed weekly and the amounts of DM to be fed were adjusted accordingly. The 70% concentrate and 30% roughage diet contained 145 g crude protein, 124 g crude fibre and 10.5 MJ/kg DM of ME. Digestibilities at all feeding levels were determined with wethers and comparisons were made with values obtained with steers at three levels of intake. There was a slight tendency for digestibility of crude fibre to decrease with the increasing level of feeding, but digestibility of gross energy did not differ significantly and a single ME value was estimated as DE × 0.82. Body weights were recorded as the means of two and three consecutive daily weighings (after 12 h food and water deprivation) at the commencement and the termination of the experiment respectively. Weights of the digestive tract contents were determined, after the 12 h food and water deprivation, at slaughter to estimate empty body weights. To assess changes in body composition the urea space technique, described by Preston and Kock (1973), was applied to all animals at the commencement and termination of the trial. Indirect assessment of carcass composition was obtained from the chemical composition of the 9-10-11th rib, cut, removed and prepared according to the instructions of Hankins and Howe (1946) and analysed according to the procedure suggested by Florence and Mitchell (1972).

Results and discussion

Statistical treatment of the data on ME intakes and corresponding EBW gains was aimed at fitting equations of the form $\text{ME} = aW^p + b\text{G}$ or $\text{ME} = aW^p(1 + b'\text{G})$ by means of a non-linear least-squares method of analysis (Daniel and Wood, 1971). The first equation factorizes ME utilization while the second is intended to predict ME requirements for maintenance and gain; bG is the growth component.

The three variables EWB, empty-body-weight gains (Δ EBW) and ME intakes (MJ/d), are presented in *Table 28.1.* Two direct measures of fat in the carcass and percentage change in urea space, which indirectly indicate the changes in body water content over the experimental period, are also shown in *Table 28.1.*

Mean EBW of Simmental was significantly the highest amomg the genotypes, which is in accordance with observations in performance-testing stations where intact males of 15 months of age and of the four genotypes show a similar live-weight ranking, with Africander, Drakensberg and Hereford not differing significantly.

An analysis of variance of EBW gains showed no significant differences between breed groups. Differences in EBW gains between feeding levels were highly significant ($P \leqslant 0.01$), as was the linear regression of gain on ME intake (r = 0.86** for combined data and 0.89**–0.96** for separate breed groups).

Assuming that a decrease in urea space over the feeding period signifies an increase in body fat *vis-à-vis* body water, the observed changes in urea space are generally in line with the trends observed in kidney and omentum fat (g/kg carcass) and percentage ether extract of 9-10-11th rib cut which are more direct indicators of the degree of fatness. In the present experiment joule value of gain could not be determined directly. These observations, relating to fat in carcass

Table 28.1 MEAN VALUES AND STANDARD ERRORS FOR EMPTY BODY WEIGHTS (EBW), EMPTY-BODY-WEIGHT GAINS (Δ EBW), METABOLIZABLE-ENERGY INTAKES (ME) AND SOME MEASUREMENTS RELATING TO BODY-FAT CONTENT OF BEEF STEERS OF FOUR GENOTYPES FED AT SEVEN DIFFERENT LEVELS FOR 120 DAYS

Genotype and feed level	*EBW* (kg)	Δ *EBW* (g/d)	*ME intake* (MJ/d)	*Kidney and omentum fat* (g/kg *carcass*)	*Ether extraction 9-10-11th rib* (%)	*Decrease in urea space* (%)
Africander	284 ± 7	616 ± 77	57.5 ± 3.8	45 ± 2	17.0 ± 0.9	−11.3 ± 3.3
Drakensberg	299 ± 8	670 ± 96	61.4 ± 3.8	42 ± 3	15.9 ± 1.4	−16.9 ± 2.6
Hereford	317 ± 10	719 ± 102	63.2 ± 4.5	41 ± 3	19.6 ± 1.7	−15.2 ± 2.3
Simmental	393 ± 9	759 ± 82	72.0 ± 4.4	39 ± 2	14.6 ± 1.2	−10.9 ± 2.7
I	304 ± 24	400 ± 13	49.0 ± 2.4	35 ± 2	12.3 ± 0.8	− 8.8 ± 2.5
II	308 ± 22	464 ± 83	54.3 ± 2.2	38 ± 3	12.9 ± 1.4	−10.1 ± 2.3
III	319 ± 27	581 ± 74	58.3 ± 3.7	45 ± 4	17.2 ± 2.4	−17.8 ± 7.0
IV	312 ± 30	723 ± 66	61.6 ± 4.0	44 ± 3	17.5 ± 1.3	−14.5 ± 2.2
V	332 ± 20	800 ± 72	69.1 ± 3.4	42 ± 2	17.7 ± 1.5	−16.0 ± 4.2
VI	345 ± 31	945 ± 66	74.4 ± 4.7	45 ± 2	19.2 ± 0.8	−10.9 ± 2.6
VII	346 ± 28	925 ± 22	77.8 ± 2.5	46 ± 2	20.5 ± 1.2	−15.4 ± 2.8

and body water content, may aid in the interpretation of the differences observed between the genotypes in the utilization of ME.

The equations given in *Table 28.2* are those which showed significant coefficients and exponents. The non-linear least-squares program operates on the iterative principle, and the highly significant coefficients obtained when the exponent p was predetermined at 0.75 might be biased by the fact that the steers were rationed at their different levels according to live weight raised to the power 0.75.

Table 28.2 EQUATIONS RELATING METABOLIZABLE ENERGY INTAKE TO EMPTY BODY WEIGHTS AND GAINS OF GROWING BEEF STEERS

	Equation (1)	*ME (MJ/d)*	$= aW^p$	$+ bG$	S^2
With p open :	Pooled data	ME	= 0.36W$^{0.81}$**	+ 35.8**G	12.1
With p = 0.75 :	Africander	ME	= 0.45**W$^{0.75}$	+ 42.3**G	8.3
	Drakensberg	ME	= 0.57**W$^{0.75}$	+ 29.9*G	12.3
	Hereford	ME	= 0.52**W$^{0.75}$	+ 33.0**G	16.4
	Simmental	ME	= 0.44**W$^{0.75}$	+ 44.0**G	14.7
	Pooled data	ME	= 0.51**W$^{0.75}$	+ 36.1**G	11.7

	Equation (2)	*ME (MJ/d)*	$= aW^p$	$(1 + b'G)$	
With p open :	Pooled data	ME	= 2.24*W$^{0.49}$**	(1 + 0.93**G)	11.5
With p = 0.75 :	Africander	ME	= 0.48**W$^{0.75}$	(1 + 1.16**G)	6.2
	Drakensberg	ME	= 0.60**W$^{0.75}$	(1 + 0.62**G)	12.2
	Hereford	ME	= 0.56**W$^{0.75}$	(1 + 0.69**G)	15.4
	Simmental	ME	= 0.48**W$^{0.75}$	(1 + 0.95**G)	15.4
	Pooled data	ME	= 0.55**W$^{0.75}$	(1 + 0.73**G)	15.5

** Students' t-test shows significance at 1% level
* Students' t-test shows significance at 5% level
W and G in kilograms

With exponent *p* predetermined at 0.75 there is a marked correspondence in the *a* values for a given breed group in the two types of equation, and significant coefficients were obtained in all cases. When the exponent *p* was not predetermined, significant coefficients were obtained with the combined data only. The difference between the exponents 0.81 and 0.49 is difficult to explain but in general it is clear that, in the absence of an error term, W and G in these equations are highly correlated. At a live weight of 300 kg, $0.36\ W^{0.81}$ and $2.24\ W^{0.49}$ converge but at lower or higher live weights they give divergent estimates of maintenance. Although they were not significant and are not presented in *Table 28.2*, the *p* values obtained for the different breed groups were, in order of listing, 0.67, 0.47, 0.64 and 0.52. All of them are smaller than the conventional 0.75 exponent and on average they are of the same order as exponents obtained with young growing pigs (Thorbek, 1974).

The *a* coefficients in *Table 28.2* would indicate noticeable differences in maintenance requirement between breed groups, with Africander and Simmental lower and Drakensberg higher than Hereford. W and G in the equations both relate to EBW. When W in the equations in *Table 28.2* is converted to live weight, uncorrected for gut fill, then all of the *a* coefficients fall within the range of values reported in the literature. As a matter of fact the coefficients of Drakensberg, Hereford and combined data yield maintenance ME requirements which correspond well with the 528 kJ/d per kg $W^{0.75}$ obtained by Webster *et al.* (1976) with Hereford × Friesian steers. However, one would hesitate to ascribe lower maintenance requirements to Africander and Simmental on the basis of these data. One should also credit Drakensberg and Hereford with their higher efficiency of ME utilization for gain as shown by the lower *b* coefficients. Logically these should be related to observed or derived data on compositions of weight gains because with a single diet the ratio of fat:protein joules should be the main determinant of the efficiency of fat deposition. In *Table 28.1* Africander and Hereford showed higher end values for body fat while Drakensberg and Hereford showed the largest decrease in urea space, which signified a higher rate of fat accretion over the experimental period. Contrary to expectation, these two breeds showed lower coefficients for gain with relatively higher maintenance requirements. Significantly, Simmental, which is presumably the latest maturing of these breed groups, had the lowest fat values in carcass and the smallest change in urea space over the growth period under discussion.

References

BLAXTER, K.L. (1962). *The Energy Metabolism of Ruminants*, p. 315. London: Hutchinson Scientific and Technical

DANIEL, C. and WOOD, F.S. (1971). *Fitting Equations to Data*, p. 320. London: Wiley Interscience

FLORENCE, E. and MITCHELL, K.G. (1972). *Proc. Br. Soc. Anim. Prod.*, New Series, 101–107

HANKINS, O.G. and HOWE, P.E. (1946). *U.S.D.A. Techn. Bull.* 926

MINISTRY OF AGRICULTURE, FISHERIES AND FOOD (1975). *Technical Bulletin No.* 33. London: HMSO

PRESTON, R.L. and KOCK, S.W. (1973). *Proc. Soc. exp. Biol. Med.*, **143**, 1057–1062

THORBEK, G. (1974). *Energy Metabolism of Farm Animals*, p. 147. Eds. K.H. Menke, H.-J. Lantsch and J.R. Reichl. EAAP Publication No. 14. Stuttgart: Universität Hohenheim Dokumentationsstelle

VERCOE, J.E. and FRISCH, J.E. (1974). In *Energy Metabolism of Farm Animals*, p. 131. Eds. K.H. Menke, H.-J. Lantsch and J.R. Reichl. EAAP Publication No. 14. Stuttgart: Universität Hohenheim Dokumentationsstelle

WEBSTER, A.J.F., SMITH, J.S., CRABTREE, R.M. and MOLLISON, G.S. (1976). *Anim. Prod.*, **23**, 329–340

29

LEVEL OF ENERGY EFFECTS ON PATTERNS AND ENERGETIC EFFICIENCY OF TISSUE DEPOSITION IN SMALL OR LARGE MATURE SIZE BEEF CATTLE

F.M. BYERS and R.E. ROMPALA
Department of Animal Science, Ohio Agricultural Research and Development Center, Wooster, Ohio 44691, USA

Summary

Composition of empty-body growth was measured at intervals during growth in Charolais and Simmental × Angus steers. Rates of protein deposition were above 90% of physiological maximum limits at 1.0 kg daily gain with little increase with faster rates of gain which resulted in heavier empty-body weight at similar composition for slower gaining steers. Efficiency of energy storage favoured fat, and maintenance requirements reflected protein growth.

Introduction

While the practice of backgrounding and growing beef cattle is widely utilized in the beef cattle industry to allow cattle, especially those of small to average frame size, to develop more protein or 'lean', controversy still exists in the scientific community as to whether or not modification of composition of growth with nutrition is possible. The concept of constancy in composition of growth and in composition at given weights, accepted by some (Reid *et al.*, 1968a,b), dictates that composition of growth cannot be modified. Early (Haecker, 1920; Moulton, Trowbridge and Haigh, 1922) and more recent research (Callow, 1961; Henrickson, Pope and Hendrickson, 1965; Byers, 1977, 1978; Harpster *et al.*, 1978) documents substantial modification in composition of growth with dietary energy levels. Even in the widely cited studies of Winchester and Howe (1955) 'rapidly' growing steers (growing 0.9 kg/d) had more fat at lighter carcass weights than limit-fed cattle. Whether or not cattle of all sizes and with quite divergent genetic potential will respond similarly in composition and efficiency of growth to varying levels of nutrition or whether there might exist optimal 'nutritional niches' which could differ with cattle type is of primary concern.

Approved for publication as Journal Article No. 149–79 of the Ohio Agricultural Research and Department Center, Wooster, Ohio.

Materials and methods

Eight pure-bred Charolais steers and eight Simmental X Angus steers were individually fed either *ad lib.* or at approximately 70% of an *ad lib.* paired steer's intake of whole shelled corn supplemented with soybean meal (908 g/d) and a corn-mineral supplement (454 g/d). Empty-body composition of all steers was measured initially and at approximately 45 kg intervals of weight gain with a D_2O dilution procedure (Byers, 1979a,b). A total of 92 infusion measurements for determination of empty-body weight and components (water, protein and fat) was made. Mature weight of the dam was included in equations predicting tissue growth rates. Feedstuff metabolizable-energy (ME) values used were 14.3, 14.7 and 8.08 kJ/g for shelled corn, soybean meal and corn-mineral supplements.

Data were analysed via least squares (Harvey, 1960) with multiple regression analyses utilized to derive functions predicting protein growth.

Results and discussion

While two levels of nutrition, that is *ad lib.* shelled corn and 70% shelled corn, were provided, an individual steer was paired to and fed 70% of the *ad lib.* consumption of each full-fed steer. Since *ad lib.* consumption varied extensively, consumption of limit-fed steers varied accordingly. Thus, a continuum of energy levels rather than two discrete levels resulted. While this presents a problem in categorization into discrete low or high energy planes, it provides the opportunity to investigate the response to energy level in considerably more depth over an expanded range in growth rates (*Table 29.1*). The energy levels attained modified growth substantially ($P<.01$). As is evident the limit-fed Charolais steers averaging 65.8 kg heavier than *ad lib.* fed steers had accrued nearly similar amounts (111 versus 118 kg) of fat and 14 kg more of protein. Terminal fat averaged 24.6 and 20.4% for limit and *ad lib.*-fed Charolais steers respectively. While limit-fed Angus X steers had 7 kg more empty-body weight (402 versus 395 kg), then had 17.1 kg less fat and 5.8 kg more protein. Final empty-body fat averaged 27.3 and 22.5% for *ad lib.* and limit-fed Angus X steers respectively.

Rates of protein deposition (*Table 29.1*) were similar for Charolais steers gaining above 0.76 kg/d over the total study and decreased with lower rates of empty-body gain. Angus X steers gaining above 0.95 kg/d were similar in protein growth. These responses are in accord with recent research (Byers, 1978; Garrett, 1979), and indicate that protein deposition is essentially maximized at rates of empty-body gain of approximately 0.9 kg/d with little or no additional protein gain with faster rates of animal growth. This concept is illustrated in *Figure 29.1.* Data presented are least-squares regressions relating protein or fat gain to empty-body weight and gain, with mature weight of the dam also included for the Charolais steers. Regressions were based on composition and growth data from the 92 D_2O infusion estimates over time. For the Charolais steers, daily protein gain (kg) = 0.997 – 0.1724 ln empty-body weight + 0.0555 ln daily empty-body gain + 0.00024 (mature weight); $r^2 = 0.84$. Daily protein gain for the Angus X steers = 0.887 – 0.1346 ln empty-body weight + 0.0225 ln daily empty-body gain; $r^2 = 0.59$. As is evident, increasing rate of gain from 1.0 to 1.5 kg/d with either Charolais or Angus X steers resulted in very little additional protein gain. As

Breed: level		Anim. No.	Weight		Days	Protein		Fat		Growth (kg/d)			$\bar{x}$ kg$^{0.75}$	RE/kg$^{0.75}$	ME/kg$^{0.75}$
			Init.	Final		Init.	Final	Init.	Final	E. Body	Protein	Fat			
Charolais:															
Full fed		1	294.6	471.1	231	53.0	81.1	53.8	102.0	0.76	0.122	0.209	86.6	30.37	201
		3	258.9	508.2	231	43.8	78.6	59.9	151.1	1.08	0.151	0.395	86.7	52.37	238
		5	254.0	477.4	311	47.9	78.4	36.3	121.2	0.72	0.098	0.273	83.6	37.15	210
		7	233.5	493.3	311	48.1	87.2	15.0	97.1	0.84	0.126	0.264	83.2	43.97	203
	$\bar{x}$		260.3	478.5	271	48.2	81.3	41.2	117.9	0.85	0.124	0.285	85.0	41.0	213
Limit fed		2	284.7	588.3	493	53.4	92.4	42.0	168.5	0.62	0.079	0.257	95.5	29.81	170
		4	268.5	593.7	610	53.4	106.2	25.9	111.2	0.54	0.087	0.140	94.6	18.94	170
		6	263.9	454.0	473	51.4	84.1	30.3	71.9	0.40	0.069	0.088	82.5	14.64	179
		8	231.7	541.1	610	41.3	98.6	44.1	93.1	0.51	0.094	0.080	87.2	14.62	183
	$\bar{x}$		262.2	544.3	546.5	49.9	95.3	35.6	111.2	0.52	0.082	0.141	90.0	19.50	175.5
Angus X:															
Full fed		11	225.5	405.9	189	37.6	68.3	54.7	95.5	0.95	0.162	0.216	74.9	39.06	231
		13	220.4	411.2	189	42.3	67.5	28.2	104.5	1.01	0.133	0.404	74.9	60.44	260
		15	137.0	363.3	213	26.9	54.6	14.8	115.2	1.06	0.130	0.471	62.9	81.78	300
		17	180.1	397.5	150	34.4	62.6	23.8	115.6	1.42	0.188	0.612	70.2	96.65	307
	$\bar{x}$		190.8	394.5	185.3	35.3	63.2	30.4	107.7	1.11	0.153	0.426	70.7	69.48	274.5
Limit fed		12	204.7	353.7	473	39.3	65.6	26.2	55.6	0.61	0.056	0.062	68.3	12.96	199
		14	221.2	409.8	311	40.6	70.6	36.7	89.1	0.61	0.096	0.168	74.9	28.24	194
		16	163.4	500.0	497	31.3	80.5	21.2	143.2	0.68	0.099	0.245	77.7	36.71	208
		18	174.6	343.5	247	34.7	59.2	17.0	74.6	0.68	0.099	0.233	64.6	42.39	230
	$\bar{x}$		191.0	401.8	382.0	36.5	69.0	25.3	90.6	0.65	0.088	0.177	71.4	30.07	207.8
s.e.			16.6	25.2		2.9	4.2	7.9	16.4	0.07	0.012	0.058	2.9	7.9	10.9

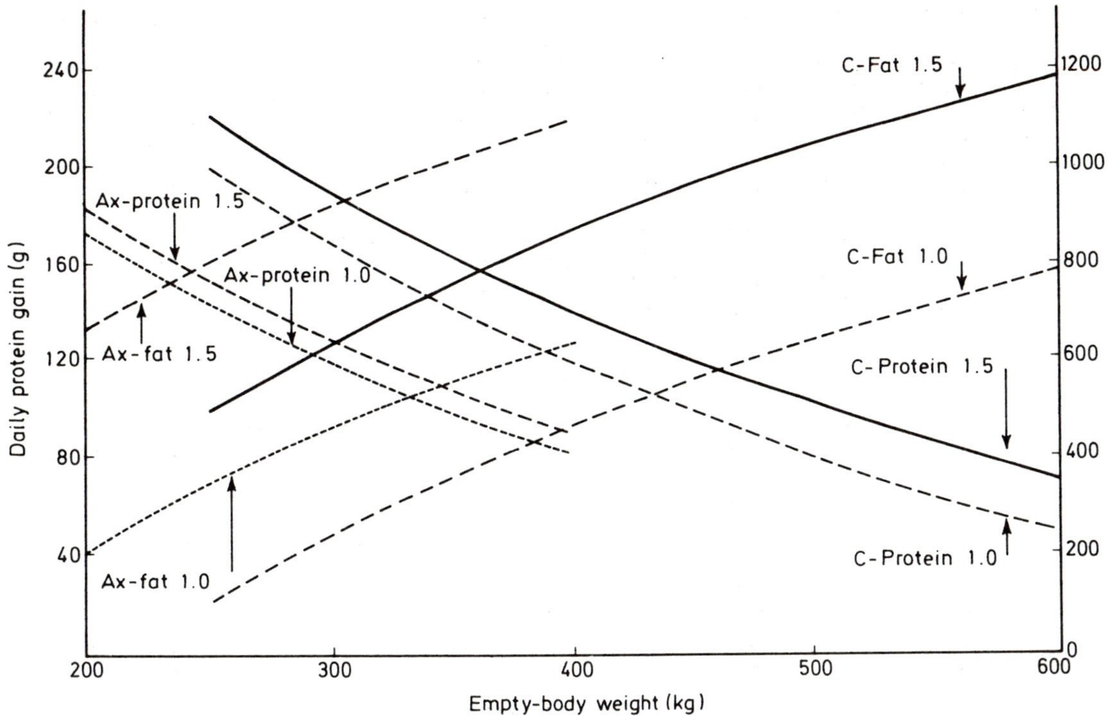

Figure 29.1 Predicted rates of protein and fat deposition for Charolais (C) and Angus × (A×) *steers growing at* 1.0 *or* 1.5 kg/d

anticipated the protein growth decreases with increasing empty-body weight with either cattle type. Rates of fat deposition increase very rapidly with increasing rate of gain from 1.0 to 1.5 kg and with increasing empty-body weight.

These observations concur with other research in rejecting the hypothesis for constancy in composition of growth and provide insight into the regulation and prediction of patterns of tissue growth via nutritional means.

Energy partitioning revealed several responses. Using a linear relationship of energy retention (kJ/kg$^{0.75}$ per d) to metabolizable-energy intake (kJ/kg$^{0.75}$ per d) the equation for the Angus X steers was as follows:

$$RE = -417.6 + 2.6\ (ME)$$

$r^2 = 0.94$ indicating an estimated maintenance requirement of 674 kJ/kg$^{0.75}$ and a 62.0% efficiency of energy use for growth. Similar calculations for the Charolais steers (also in kJ/kg$^{0.75}$ per d) were:

$$RE = -228.2 + 1.87\ (ME)$$

$r^2 = 0.66$ with a maintenance requirement of 510 kJ and a 44.8% efficiency of energy use for growth. The elevated rates of protein growth in the crossbred Angus steers undoubtedly contributed to their higher maintenance requirement, while the increased proportion of protein in the gain of Charolais steers ($\overline{x}$ of 15.29 versus 14.44%) was probably a factor in the lower efficiency of energy use for gain of Charolais when compared with Angus X steers (44.8 versus 62.0%). Within each animal group the efficiency of energy use for growth was positively related to the fraction of empty-body gain as fat.

References

BYERS, F.M. (1977). *National ASAS,* **295**, 116

BYERS, F.M. (1978). *Midwest ASAS,* **106**, 41

BYERS, F.M. (1979a). *Analytical Biochem.* **98**, 208

BYERS, F.M. (1979b). *Measurement of Protein and Fat Accretion in Growing Cattle Through Isotope Dilution Procedures.* OARDC Beef Research Circular, 79–1, 36

CALLOW, E.H. (1961). *J. Agric. Sci., Camb.,* **56**, 265

GARRETT, W.N. (1979). *Feedstuffs,* **51**, 21:20

HAECKER, T.L. (1920). *Minn. Agric. Expt. Sta. Bull.,* 193

HARPSTER, H.W., FOX, D.G., MAGEE, W.T. and BLACK, J.R. (1978). *Proc. Nat. ASAS.* (Abstr.) No. 514, p. 420

HARVEY, W.R. (1960). *USDA. ARS.,* **20**, 8

HENRICKSON, R.L., POPE, L.S. and HENDRICKSON, R.F. (1965). *J. Anim. Sci.,* **24**, 507

MOULTON, C.R., TROWBRIDGE, P.F. and HAIGH, L.D. (1922). *Mo. Agric. Expt. Sta. Bull.,* 55

REID, J.T., BENSADOUN, A., BULL, L.S., BURTON, J.H., GLEESON, P.A., HAN, I.K., JOO, Y.D., JOHNSON, D.E., McMANUS, W.R., PALADINES, O.L., STROUD, J.W., TYRRELL, H.F., VAN NIEKERK, B.D.H., WELLINGTON, G.W. and WOOD, J.D. (1968a). *Proc. Cornell Conf. Feed. Manuf.,* **18**, 37

REID, J.T., BENSADOUN, A., BULL, L.S., BURTON, J.H., GLEESON, P.A., HAN, I.K., JOO, Y.D., JOHNSON, D.E., McMANUS, W.R., PALADINES, O.L., STROUD, J.W., TYRRELL, H.F., VAN NIEKERK, B.D.H. and WELLINGTON, G.W. (1968b). *Nat. Acad. Sci.,* **1598**, 19

WINCHESTER, C.F. and HOWE, P.E. (1955). *USDA Bull.* No. 1108

UTILIZATION OF VOLATILE FATTY ACIDS FOR MAINTENANCE AND FOR ENERGY RETENTION

E.R. ØRSKOV and I. McDONALD
The Rowett Research Institute, Bucksburn, Aberdeen AB2 9SB, Scotland

Summary

Sheep were totally sustained by intragastric infusion of volatile fatty acids (VFA) and protein. Two levels of energy input were used, namely 450 and 900 kJ/$W^{0.75}$, and six mixtures of VFA containing, 35, 45, 55, 65, 75 and 85 molar per cent of acetic acid, and propionic acid decreasing respectively from 55 to 5%. The maintenance requirement for zero energy balance was 450 kJ/$W^{0.75}$ and did not differ between the different VFA mixtures. The efficiencies for energy retention (k_f) for the six mixtures were 0.78, 0.64, 0.57, 0.61, 0.61 and 0.59 respectively. It is concluded that differences in k_f between diets fed to ruminants cannot be explained by differences in the proportion of acetic acid.

Introduction

The importance of the utilization of VFA in ruminants has been recognized and appreciated ever since the Cambridge group of workers (Barcroft, McAnally and Phillipson, 1944) showed convincingly that VFAs were the main source of energy for ruminants and the end-products of anaerobic fermentation in the rumen. It is generally accepted that acetic, propionic and butyric acids constitute the most important VFAs and that of these only propionic acid can give rise to a net synthesis of glucose.

In general, the proportion of acetic acid in the rumen increases and that of propionic acid decreases as the fibre content of the diet increases. This observation led to the proposition that the decrease in efficiency of utilization of metabolizable energy (ME) which was generally observed with increasing fibre in the diet may be causally associated with the increase in the proportion of acetic acid.

In 1957 Armstrong and Blaxter (1957a) showed by infusion of VFA in fasting sheep that there were no substantial differences in the utilization of mixtures of different VFA or differences that could be accounted for by small differences in the stoichiometry of dissimilation of the acids and their yield of ATP. On the other hand, when acetic acid was infused alone there was a high heat increment that could be corrected if about 5 molar per cent was present as propionic acid. In other work using a technique of ruminal infusion of VFA as an addition to a basal diet, Armstrong and Blaxter (1957b) and Armstrong *et al.* (1958) showed that for energy retention the acetic acid was utilized less efficiently (k_f 0.33) than propionic (k_f 0.56) or butyric acid (k_f 0.62).

Since then several groups of workers have studied this problem using various techniques and animals. For instance VFA salts were fed to lambs by Ørskov and Allen (1966) and Howell, Greenhalgh and Wainman (1976). Bull, Reid and Johnson (1970) gave triacetin to lambs and Ørskov *et al.* (1968) used VFA for dairy cows. No clear conclusion or confirmation of earlier results can be made from these experiments except that differences in the proportion of VFA will influence partition of energy in dairy cows and that when VFA salts have been used the efficiency of acetate has varied with the proportion of acetate already present in the rumen from fermentation of the basal diet. The main problem has been the possible interference with the digestion of a basal diet and the inability to supply more than about 15% of ME as VFA. These two problems have made a critical assessment of utilization very difficult.

Recently these problems have been overcome when it was established that ruminants could be nourished at high feeding levels entirely by infusion of VFA and protein (Ørskov *et al.*, 1979a). It became possible then to vary both the level and the proportion of VFA.

In a large-scale experiment six different mixtures of VFA were given to lambs of about 25 kg live weight (W) and about three months old. The compositions of the infusates are given in *Table 30.1* in terms of both molar proportions and energy proportions. Total energy intake per day (X), scaled by metabolic live weight, was either 0.45 or 0.90 $MJ/W^{0.75}$ of which 78% was supplied by the VFA infusion and 22% by casein infusion. For the mixture containing only 50 mmol/mol of propionic acid, metabolic problems did not allow a level of infusion to be reached in excess of 1.5 times maintenance. Daily energy retention (Y; $MJ/W^{0.75}$) was measured in closed-circuit respiration chambers. Linear regressions of Y on X were calculated for each infusate and all six lines could be constrained to intersect at energy balance (Y = 0). The estimated intersection point was $X = 0.45\ MJ/W^{0.75}$ (the maintenance requirement, E_m), and the slopes of the lines (efficiencies of utilization for growth, k_f) are given in *Table 30.1*. This model fitted the data better ($P<0.05$) than the best single line ($k_f = 0.63$), and this was largely because of the results for the infusate with only 350 mmol/mol acetic acid. Over the range of 450–750 mmol/mol, which corresponds to the range of practical dietary conditions, the differences in k_f were far too small to account for the differences that are found between diets in efficiency of utilization of ME.

Table 30.1 THE EFFECT OF DIFFERENT PROPORTIONS OF VOLATILE FATTY ACID (VFA) ON EFFICIENCY (k_f) FOR ENERGY RETENTION USING A COMMON ESTIMATE OF E_m (0.45 $kJ/W^{0.75}$).

VFA proportions (mmol/mol)			*VFA proportions* (kJ/MJ)			k_f
Acetic	*Propionic*	*Butyric*	*Acetic*	*Propionic*	*Butyric*	
350	550	100	200	620	160	0.78
450	450	100	300	530	170	0.64
550	350	100	390	430	180	0.57
650	250	100	480	330	190	0.61
750	150	100	590	210	200	0.61
850	50	100	720	70	210	0.59

As a test of the reliability of this conclusion the k_f values for a series of other values of E_m were calculated. Although k_f values were reduced by lowering E_m, the differences from one infusate to another were almost unaffected (*Figure 30.1*). The changes in E_m from the optimum value (0.45) to 0.375, 0.40 and 0.50 gave a poorer fit to the data, with the residual variance being increased by 62%, 32% and 39% respectively, and there was not the slightest indication that the E_m should have been estimated separately for each infusate.

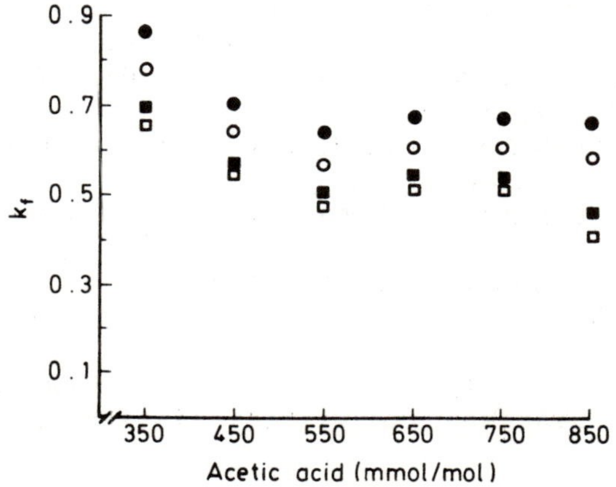

Figure 30.1 The effect of acetic acid proportion on k_f when the maintenance requirement was taken to be 0.500 (●), 0.450 (○), 0.400 (■) or 0.375 (□) MJ/$W^{0.75}$

It was calculated that the daily intake which resulted in energy balance was associated with a positive daily nitrogen balance of about 0.2 g/$W^{0.75}$, and this may have enhanced the energy requirement since the efficiency of utilization of ME for protein deposition is lower than that for maintenance. Taking values of 0.33 and 0.8 respectively for these efficiencies (Ørskov and McDonald, 1970, 1976), it can be estimated that had nitrogen balance been achieved simultaneously with energy balance then the value of E_m might have been reduced by about 0.05 MJ/$W^{0.75}$.

The experiment was not designed to estimate k_f for each VFA separately, but such estimates can be obtained if some assumption is made regarding butyric acid. It was assumed from the similarity in metabolism that k_f is the same for butyric as for acetic acid, and it was found that the best fit to the results in *Table 30.1* is provided by a combined value of 0.55 together with one of 0.78 for propionic acid. Weighting these two values according to the VFA proportions by energy in each infusate yields k_f estimates of 0.69, 0.67, 0.65, 0.62, 0.60 and 0.57 respectively for the six infusates in order of increasing acetic acid proportion. There is no significant difference in closeness of fit to the original data between these six 'graded' estimates of k_f and the individual least-squares estimates given in *Table 30.1*. The revised range in k_f is still only from 0.67 down to 0.60 as acetic acid varies over the practical range of 450–750 mmol/mol.

In practice large and important differences in k_f occur between diets, and the values range from about 0.3 to 0.6. It is clear from the present results that such differences cannot be attributable to differences in the efficiency of utilization between the different VFAs.

References

ARMSTRONG, D.G. and BLAXTER, K.L. (1957a). *Br. J. Nutr.,* **11**, 247

ARMSTRONG, D.G. and BLAXTER, K.L. (1957b). *Br. J. Nutr.,* **11**, 413

ARMSTRONG, D.G., BLAXTER, K.L., WAINMAN, F.W. and GRAHAM, N. McC. (1958). *Br. J. Nutr.,* **12**, 177

BARCROFT, J., McANALLY, R.A. and PHILLIPSON, A.T. (1944). *J. exp. Biol.,* **20**, 120

BULL, L.S., REID, J.T. and JOHNSON, D.E. (1970). *J. Nutr.,* **90**, 47

HOVELL, F.D., GREENHALGH, J.F.D. and WAINMAN, F.W. (1976). *Br. J. Nutr.,* **35**, 353

ØRSKOV, E.R. and ALLEN, D.M. (1966). *Br. J. Nutr.,* **20**, 519

ØRSKOV, E.R., FLATT, W.P., MOE, P.W., MUNSON, A.W., HEMKEN, R.W. and KATZ, I. (1968). *Br. J. Nutr.,* **23**, 443

ØRSKOV, E.R., GRUBB, D.A., SMITH, J.S., WEBSTER, A.J.F. and CORRIGALL, W. (1979a). *Br. J. Nutr.,* **41**, 541

ØRSKOV, E.R., GRUBB, D.A., WENHAM, G. and CORRIGALL, W. (1979b). *Br. J. Nutr.,* **41**, 553

ØRSKOV, E.R. and McDONALD, I. (1970). In *Energy Metabolism in Farm Animals,* p. 121. Eds. A. Schürch and G. Wenk. EAAP Publication No. 13. Zurich: Juris Druck and Verlag

ØRSKOV, E.R. and McDONALD, I. (1976). In *Energy Metabolism,* p. 131. Ed. M. Vermorel. EAAP Publication No. 19. Clermont-Ferrand: G. de Bussac

31

ENERGY REQUIREMENTS FOR THE GROWING PIG

V.R. FOWLER and M.F. FULLER
The Rowett Research Institute, Bucksburn, Aberdeen AB2 9SB, Scotland
W.H. CLOSE
ARC Institute of Animal Physiology, Babraham, Cambridge CB2 4AT, England
C.T. WHITTEMORE
East of Scotland College of Agriculture, West Mains Road, Edinburgh EH9 3JG, Scotland

The authors have been concerned with reviewing the literature that relates to the energy requirements of pigs for the working party set up by the Agricultural Research Council (ARC) in 1974. The contents of this chapter do not represent the final views of the group nor should they be construed as being the official ARC estimates of requirements because the discussions are still at an interim stage. The chapter is primarily aimed at presenting a few of the issues that have been central to the deliberations, at highlighting some of the difficulties of deriving a framework for expressing energy requirement and at obtaining a consensus from the diverse literature.

The classic format for reviewing energy requirements has been to consider the degree of agreement between two approaches – the 'empirical' one which consists of examining dose/response relationships in comparatively simple measures such as growth rate and the 'factorial' approach whereby the energy cost of each major process contributing to an energy requirement is considered separately.

The empirical approach may lack the scientific substance of the factorial one, but it is rooted in realities and can often be translated directly into practical circumstances. Its applicability, however, tends to be restricted to the conditions of the original experiments and it may be difficult to accommodate differences in genotype, sex, live weight or environmental temperature.

The attraction of the factorial approach is that it appears to give the opportunity of synthesizing the 'requirement' for any level of performance specified in terms of the factors.

Choice of a factorial model

Kielanowski (1966) proposed a model whereby metabolizable energy (ME) was regarded as the summation of three factors: maintenance, the total cost of protein accretion and the total cost of fat accretion, i.e.

$$ME = ME_m + \frac{1}{k_p} P + \frac{1}{k_f} F$$

Where ME is metabolizable energy; ME_m is metabolizable energy required for maintenance; P is energy retained as protein; F is energy retained as fat; k_p is efficiency of energy utilization for protein accretion and k_f is efficiency of energy utilization for fat accretion.

The advantages of the model are that many estimates are available for the pig of the value of each of the coefficients of the independent variables P and F. Its limitations are: first that it assumes, in its simplest form, constant efficiencies of tissue accretion, secondly, that the estimates of the constants have often been made from experiments in which the independent variables were themselves correlated and, lastly, the estimate of maintenance that it provides has a very particular meaning (this is discussed later).

Although calculations based on such a model allow a 'requirement' to be calculated for given rates of fat and protein accretion, information on these specific products is rarely available in practice. If the equation could be solved in reverse to predict changes in the rates of protein and fat deposition resulting from given changes in ME, then the practical use of the factorial approach would be greatly extended. To achieve this a considerable amount of additional information is required which relates to the nature and shape of the input/output responses. The strategies for describing and modelling such changes will be dealt with in the final section.

Requirements of metabolizable energy for maintenance

Estimates of the energy requirement for maintenance have been made from a number of types of experiment. The classic approach used by Breirem (1939) involved the establishment of the fasting heat production. This was then scaled first by an efficiency factor for the utilization of ME below maintenance and also by an activity factor. The arbitrary nature of the activity factor, together with sensitivity to previous nutritional history, make the fasting metabolism unsatisfactory as a predictor of the maintenance requirement.

More commonly maintenance is measured as the ME corresponding to zero-energy retention and is usually derived from regression analyses of data from experiments in which several levels of feeding were used. This approach raises the interesting question, what is happening metabolically at zero-energy balance? The data of Close, Mount and Brown (1978) indicate that at energy equilibrium there is substantial lipid mobilization and nitrogen can be accreted at as much as 4–7 g/d in pigs of 35 kg. Similar data are provided by the experiment of Fuller *et al.* (1976) which suggests nitrogen-retention values at energy equilibrium of 3–5 g in pigs of 30 and 60 kg. Retentions of this magnitude represent from a quarter to a third of the potential nitrogen retention and show quite clearly that estimates of the partition of energy above energy equilibrium between lipid and protein accretion could be subject to very serious errors.

The alternative approach, following Kielanowski's model, is to estimate the maintenance requirement as the intercept of the multiple regression equation in which ME is the dependent variable and lipid and protein accretion are the independent variables. This provides an estimate of the ME when neither protein nor lipid are being accreted or decreted, and is the maintenance term appropriate to the factorial model described at the outset. Essentially, it is a statistical concept of maintenance since the growing animal, when fed normal diets, would not be in this physiological state; even at very low levels of intake protein is still accreted at the expense of energy drawn from the lipid reserves.

In *Table 31.1* estimates are given from the literature of maintenance, which was calculated as the ME at energy equilibrium (above the line) and as the intercept of the multiple regression equation of Kielanowski (1966) (below the line).

Table 31.1 ESTIMATES OF THE DAILY ME REQUIREMENTS OF PIGS FOR MAINTENANCE, 'a' IN THE EQUATION $ME_m = aW^{0.75}$, FOR TWO SETS OF EXPERIMENTS. IN THOSE ABOVE THE LINE THE VALUE WAS ESTIMATED AS ME INTAKE AT ZERO ENERGY BALANCE. FOR THOSE BELOW THE LINE THE VALUE WAS ESTIMATED FROM MULTIPLE REGRESSION EQUATIONS INCORPORATING P AND F AS INDEPENDENT VARIABLES. VALUES FOR THE EFFICIENCY OF ENERGY RETENTION ABOVE MAINTENANCE (k_w) AND FOR k_p AND k_f ARE ALSO GIVEN.

Body weight (kg)	*a* (MJ/d)	k_w	k_p	k_f	*Source*
2–6	0.542	0.76–0.88			Jordan andBrown (1971)
2–6	0.524	0.74–0.88			Jordan (1974)
13–30	0.455	0.64			Sharma, Young and Smith (1971)
13–30	0.570	0.70			Sharma, Young and Smith (1971)
20–30	0.522	0.74			McCracken and Gray (1972)
23–42	0.418	0.66			Verstegen *et al.* (1973)
20–55	0.429–0.469	0.76			Verstegen and van der Hel (1974)
20–70	0.354–0.385	0.64–0.68			Holmes (1973)
20–70	0.418	0.71			Holmes (1974)
20–90	0.439	0.67–0.75			Fuller and Boyne (1972)
2.5–8.5	0.573		0.76	0.81	Kielanowski (1965)
5–21	0.601		0.76	0.78	Burlacu *et al.* (1973)
20–90	0.473		0.48	0.77	Thorbek (1975)
8–15	0.682		0.80	0.70	Hoffmann *et al.* (1977)
15–25	0.523		0.56	0.83	Hoffmann *et al.* (1977)
25–35	0.561		0.56	0.91	Hoffmann *et al.* (1977)
35–45	0.536		0.62	0.83	Hoffmann *et al.* (1977)
23–42	0.511		0.58	0.70	Close, Verstegen and Mount (1973)
20–50	0.440		0.63	0.70	Close and Mount (1976a,b)
20–90	0.418		0.35	0.73	Kotarbinska (1969)
20–90	0.429		0.37	0.91	Houseman and McDonald (1973)
30–110	0.397		0.52	0.70	Gädeken, Oslage and Fliegel (1974)

k_p = efficiency of energy utilization for protein accretion
k_f = efficiency of energy utilization for fat accretion
k_w = efficiency of energy retention above maintenance

Before a preferred estimate describing maintenance over a range of live weights can be calculated the published data must be unified in a single equation. For this purpose an equation of the form $ME_m = aW^b$ has been generally found most appropriate. The data of *Table 31.1* were recalculated as coefficients of $W^{0.75}$, but this clearly did not describe the whole range. It is far from elegant to postulate a series of exponents or variable values for 'a' in the equation:

$$ME_m = aW^{0.75}$$

where ME_m is daily requirement of ME for maintenance and W is live weight. While, ideally, only those values from below the line in *Table 31.1* should be considered, the variability was found to be such that all estimates had to be combined to give a stable relationship. Recalculation of the original values in each of the publications given in *Table 31.1* gave an overall pooled relationship

$$ME_m = 0.641\ W^{0.66}.$$

Estimates of k_p and k_f

The estimated values of k_p and k_f in *Table 31.1* are quite variable. It has been suggested by Kielanowski (1976) and others that for young animals k_p is higher than for older ones, and at first sight this is supported by the estimates in *Table 31.1.* However, the error mentioned earlier of ignoring nitrogen retention at energy equilibrium applies to the data of Kielanowski (1965) and Burlacu *et al.* (1973) and may account for their high estimates. In the case of one set of data of Hoffman *et al.* (1977) it is curious that k_f is estimated to be lower than k_p and there is an exceptionally high heat loss ascribed to maintenance. If Hoffman's data sets are treated as one and the experiments of Kielanowski (1965) and of Burlacu *et al.* (1973) are left out, then the mean value for k_p is 0.52. The mean value of k_f for pigs of 20 kg and above is 0.73.

Modelling the responses to changes in daily metabolizable energy

If the nature of the response of the animal in terms of nitrogen retention to changes in feed intake is known, then the above factors allow the partition of retained energy between protein and lipid to be computed. In addition, by the application of appropriate factors, the changes in chemical composition can be related to changes in physical composition.

There are a number of possible ways of modelling the response of nitrogen retention to changes in ME intake. Linear response with plateau (a) was examined by Whittemore and Fawcett (1974,1976), linear with positive nitrogen retention at zero-energy balance (b) by Fowler (1978) and curvilinear (c) by ourselves during the current study. By fixing nitrogen retention at three different levels (12, 16 and 20 g/d) at an ME of 25 MJ/d for a pig of 60 kg live weight it was possible to estimate (assuming linearity between 30 and 90 kg) the average rates of deposition of tissue/d predicted by each of the models. This could then be compared with actual data from empirical experiments. A major set of such empirical data is that from the dissections of carcasses by the Meat and Livestock Commission (MLC). The data were derived from their commercial product evaluation tests in which pigs from different breeding companies are compared at one central testing station. As an example of some of our results a comparison is given in *Figure 31.1* of the lean tissue growth rates (g/d) predicted from the three types of relationship (a, b and c) with the overall means of restricted- and *ad lib.*-fed pigs killed at 91 kg taken from the second, third and fourth tests (Meat and Livestock Commission, 1976, 1977, 1978).

The main assumption in the linear model (b) was that nitrogen retention at energy equilibrium was 4 g/d. For the curvilinear model (c) the data of Fuller and Boyne (1972) were used to generate a quadratic relationship with ME intake (I):

$$N = 0.00459\, I^2 + 0.161\, I - 0.227$$

The quadratic coefficient was scaled to give nitrogen retentions of 12, 16 and 20 g/d at 25 MJ.

The results in *Figure 31.1* show that for the linear model the MLC data correspond most closely with the example of nitrogen retention of 12 g/d at an

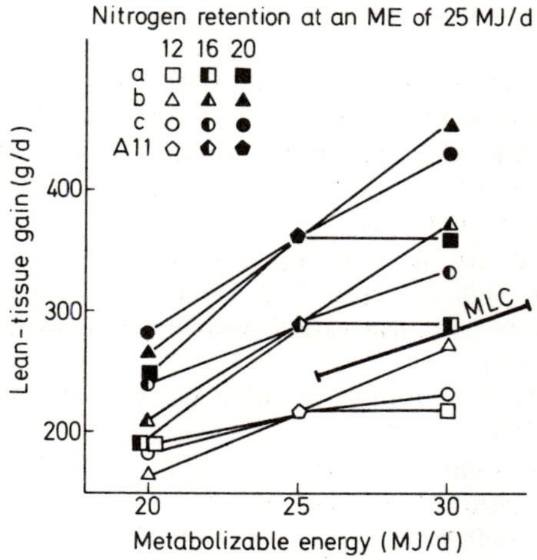

Figure 31.1 Comparison of daily lean-tissue gain in pigs growing from 25 to 90 kg, fed restrictedly and ad lib. (Meat and Livestock Commission, 1976, 1977, 1978) with modelled responses for pigs of 60 kg live weight retaining either 12, 16 or 20 g nitrogen/d at an ME of 25 MJ/d, using linear/plateau (a), linear (b) and curvilinear (c) response curves.

ME of 25 MJ/d, but the slope was less steep for the empirical data. The curvilinear model (c) corresponded to the MLC data at a nitrogen retention of about 12 g/d but the slope tended to be too shallow. In this example the plateau model did not match the empirical data; if higher maxima for nitrogen retention had been taken it would have been similar to the linear model.

Conclusions

The factorial approach to energy requirements has been most productive in improving our understanding of energy metabolism. Its value in incorporating it into models of the response of pig growth to different inputs is limited by the data available for modelling the response of nitrogen retention to feed intake particularly at levels of feeding close to appetite or to energy equilibrium.

References

AGRICULTURAL RESEARCH COUNCIL (1967). *The Nutrient Requirements of Farm Livestock: No. 3 Pigs.* London: Agricultural Research Council

BREIREM, K. (1939). *Tierernährung,* **11**, 487–528

BURLACU, G., BAIA, G., DUMITRA IONILA, DOINA MOISA, TASCENCO, U., VISAN, I. and STOICA, I. (1973). *J. agric. Sci., Camb.,* **81**, 295–302

CLOSE, W.H., MOUNT, L.E. (1976a). *Proc. Nutr. Soc.*, **35**, 60A

CLOSE, W.H. and MOUNT, L.E. (1976b). In *Energy Metabolism*, pp. 173–177. Ed. M. Vermorel. EAAP Publication No. 19. Clermont-Ferrand: G. de Bussac

CLOSE, W.H., MOUNT, L.E. and BROWN, D. (1978). *Br. J. Nutr.*, **40**, 423–431

CLOSE, W.H., VERSTEGEN, M.W.A. and MOUNT, L.E. (1973). *Proc. Nutr. Soc.*, **32**, 72A

FOWLER, V.R. (1978). In *Patterns of Growth and Development in Cattle*, pp. 325–340. Eds. H. de Boer and J. Martin. C.E.C. Seminar. The Hague: Martinus Nijhoff

FULLER, M.F. and BOYNE, A.W. (1972). *Br. J. Nutr.*, **28**, 373–384

FULLER, M.F. and CROFTS, R.M.J. (1977). *Br. J. Nutr.*, **38**, 479–488

FULLER, M.F., WEBSTER, A.J.F., MacPHERSON, R.M. and SMITH, J.S. (1976). In *Energy Metabolism*, pp. 177–180. Ed. M. Vermorel. EAAP Publication No. 19. Clermont-Ferrand: G. de Bussac

GÄDEKEN, D., OSLAGE, H.J. and FLIEGEL, H. (1974). In *Energy Metabolism of Farm Animals*, pp. 169–172. Eds. K.H. Menke, H.-J. Lantsch and J.R. Reichl. EAAP Publication No. 14. Stuttgart: Universität Hohenheim Dokumentationsstelle

HOFFMANN, L., JENTSCH, W., KLEIN, M. and SCHIEMANN, R. (1977). *Arch. Tierernähr.*, **27** (7) 421–438

HOLMES, C.W. (1973). *Anim. Prod.*, **16**, 117–133

HOLMES, C.W. (1974). *Anim. Prod.*, **19**, 211–220

HOUSEMAN, R.A. and McDONALD, I. (1973). *Anim. Prod.*, **17**, 295–304

JORDAN, J.W. (1974). In *Energy Metabolism*, pp. 189–192. Eds. K.H. Menke, H.-J. Lantsch and J.R. Reichl. EAAP Publication No. 14. Stuttgart: Universität Hohenheim Dokumentationsstelle

JORDAN, J.W. and BROWN, W.O. (1970). In *Energy Metabolism*, pp. 161–164. Eds. A. Schürch and C. Wenk. EAAP Publication No. 13. Zurich: Juris Druck and Verlag

KIELANOWSKI, J. (1965). In *Energy Metabolism*, pp. 13–20. Ed. K.L. Blaxter. EAAP Publication No. 11. London: Academic Press

KIELANOWSKI, J. (1966). *Anim. Prod.*, **8**, 121–128

KIELANOWSKI, J. (1976). *Livestock Prod. Sci.*, **3**, 257–269.

KOTARBINSKA, M. (1969). Badania nad Przemiana Energii u Rosnacyck swin. Instytut Zootechniki, Wroclaw

McCRACKEN, K.J. and GRAY, R. (1972). *Proc. Br. Soc. Anim. Prod.*, **1**, 138 (abstr. 31)

MEAT AND LIVESTOCK COMMISSION (1976, 1977, 1978). *Results of Commercial Product Evaluation.* Bletchley: Meat and Livestock Commission

SHARMA, V.D., YOUNG, L.G. and SMITH, G.C. (1971). *Can. J. Anim. Sci.*, **51**, 761–770

THORBEK, G. (1975). *Studies on Energy Metabolism in Growing Pigs*, p. 424. København: Beretning fra Statens Husdyrbrugs forsøg

VERSTEGEN, M.W.A., CLOSE, W.H., START, I.B. and MOUNT, L.E. (1973). *Br. J. Nutr.*, **30**, 21–35

VERSTEGEN, M.W.A. and VAN DER HEL, W. (1974). *Anim. Prod.*, **18**, 1–11

WHITTEMORE, C.T. and FAWCETT, R.H. (1974). *Anim. Prod.*, **19**, 221–231

WHITTEMORE, C.T. and FAWCETT, R.H. (1976). *Anim. Prod.*, **22**, 87–96

32

ENERGY METABOLISM IN GROWING BOARS

M. WAŁACH-JANIAK, M. KOTARBINSKA and J. KIELANOWSKI
Institute of Animal Physiology and Nutrition, 05-110 Jablonna, Poland

Summary

The intake of metabolizable energy (ME) and the deposition of protein and fat in the body are taken as variables in computations aiming at the estimation of the energy costs of maintenance and of protein deposition. These costs, estimated by a linear equation, are shown to be higher in boars than in castrated males. It is demonstrated further by an exponential equation that the unitary cost of protein deposition depends probably on the live weight of the growing animal and on the amount of protein deposited.

Introduction

This chapter is based on the results of experiments with growing Norwegian Landrace boars carried out during extensive studies on breeding methods. Data on the intake of ME and on the deposition of chemical protein (P) and fat (F) in the body allowed statistical investigation of the energy metabolism of the experimental animals.

After weaning at the age of six weeks all boars were placed in separate boxes and fed individually. The experiments began at 13 weeks of age. Six boars, taken at random, were slaughtered and their bodies were analysed chemically to establish the initial contents of P and F. In the first experiment boars in four analogous groups of 24 were fed on a pelleted mixture according to prearranged schedules, and each group was fed on a different plane of nutrition. In the second experiment three groups of 13 boars, fed on another mixture with a higher concentration of energy and protein, were compared. Boars of two groups were fed, as previously, according to schedules and those of the third group were fed to appetite with a record of the daily feed intake.

During the experiment the digestibility of nutrients in the mixtures was determined using Cr_2O_3 as marker. The net energy and ME of the diet were computed from the contents of digestible nutrients by means of the equations proposed by Nehring, Beyer and Hoffmann (1970). In the first experiment (groups 1–4) the pelleted mixture, composed of barley (80%), fish meal (6%), dried skimmilk (8%), dried yeast (2%), lucerne meal (2%), minerals and vitamins, contained an ME of 11.88 MJ/kg and 11.5% digestible protein. The other mixture (groups 5–7), composed of dehulled barley (80%), soya bean oilmeal (6%), fish meal (7%), dried yeast (3%), dried lucern (1.6%), minerals and vitamins, contained an ME of 13.87 MJ/kg and 15.3% digestible protein.

At 182 days of age all experimental boars were slaughtered and their bodies were analysed for the contents of P (nitrogen × 6.25), F (ether extract) and ash as described by Kotarbińska (1971). The mean experimental results for each group are presented in *Table 32.1.*

Table 32.1 RESULTS OF EXPERIMENTS WITH BOARS FED ON DIFFERENT PLANES OF NUTRITION FROM 91 TO 182 DAYS OF AGE (ARITHMETIC MEANS ± STANDARD ERRORS)

Group	*n*	*Daily intake of ME* (MJ)	*Live weight*			*Daily deposition of*	
			Initial (kg)	*Final* (kg)	*Average metabolic* ($kg^{0.75}$)	*Protein* (g)	*Fat* (g)
1	24	26.45	32.0 ± 0.8	92.3 ± 0.8	21.97	110.7 ± 2.2	144.7 ± 3.8
2	24	28.56	32.0 ± 0.8	95.1 ± 0.8	22.33	111.0 ± 2.4	170.3 ± 5.6
3	24	31.06	33.1 ± 1.0	101.3 ± 1.3	23.27	122.6 ± 2.3	194.6 ± 6.2
4	24	31.56	34.9 ± 0.9	104.3 ± 1.6	23.90	122.6 ± 1.9	199.6 ± 7.8
5	13	31.92	30.9 ± 1.0	105.6 ± 1.6	23.51	138.7 ± 4.0	187.0 ± 9.0
6	13	33.15	30.6 ± 0.8	109.4 ± 2.1	23.95	143.3 ± 3.3	200.3 ± 7.3
7	13	41.42 ± 1.64	33.9 ± 1.3	121.2 ± 3.2	25.86	155.8 ± 4.4	254.7 ± 18.1

In order to simplify the investigations it has been assumed that the energy cost of fat deposition (C_F) is constant and corresponds to the net energy (computed from the equation given by Nehring, Beyer and Hoffmann, 1970), equal to the heat of combustion of fat, i.e. to 39.6 MJ/kg. If C_F is to be expressed in units of ME, this value must obviously be multiplied by the ratio of ME to net energy contents in the diet. Since the variability of this ratio was small, the mean value of 57.4 MJ/kg of ME was adopted as C_F for all experimental groups. Energy cost of maintenance (C_M) is expressed usually in units of ME per unit of metabolic live weight ($W^{0.75}$). If, therefore, the intake of ME (IME) and the deposition of F (ΔF) and of P (ΔP) are divided by $W^{0.75}$, the following linear equation will be obtained:

$$(IME - C_F\ \Delta F)/W^{0.75} = C_M + C_P\ \Delta P/W^{0.75} \qquad (1)$$

in which the difference between the intake of ME and the energy cost of total F deposition divided by the metabolic live weight is the dependent variable, and P deposition divided by the metabolic live weight is the independent variable, the regression coefficient C_P is the energy cost of deposition of a mass unit of protein, and the regression constant C_M is the maintenance cost per unit metabolic live weight. Since IME was equal within each group from 1 to 6, the mean results for these groups, shown in *Table 32.1*, were taken as variables for solving equation (1). The following equation was obtained (in this and all further equations kg is adopted as the mass unit and MJ as the energy unit):

$$(\text{IME}-57.4\ \Delta F)/W^{0.75} = 0.466 \pm 0.040 + (73.3 \pm 7.5)\ \Delta P/W^{0.75} \pm 0.007 \quad (1a)$$
$$r = 0.980;\ n = 6 \text{ group arithmetic means (122 animals)}$$

Equation (1a) can be expanded into equation (2) which presents the full energy balance of the experimental boars (the multiple correlation coefficient r was found in this and in the further equations by regressing the actual IME on that computed from the equations):

$$\text{IME} = (0.466 \pm 0.040)W^{0.75} + (73.3 \pm 7.5)\ \Delta P + 57.4\ \Delta F \pm 0.159 \quad (2)$$
$$r = 0.998;\ n = 6 \text{ group arithmetic means (122 animals)}$$

Kotarbinska (1969) in experiments with castrated male pigs obtained the following equation:

$$\text{IME} = (0.425 \pm 0.048)W^{0.75} + (66.8 \pm 8.0)\ \Delta P + 54.2\ \Delta F \pm 0.382 \quad (3)$$
$$r = 0.992;\ n = 12 \text{ group arithmetic means (54 animals)}$$

Although the differences are non-significant, because of the high standard errors, the experiments seem to indicate that both the maintenance cost and the cost of protein deposition are higher in boars than in castrated males.

All of the foregoing equations were built on the assumption that the maintenance cost is a linear function of the metabolic live weight and that the unitary energy cost of protein deposition is independent of both the amount deposited in the body and the size of the animal. However, in a discussion of this question (Kielanowski, 1976) it has been supposed that the deposition of protein in the growing animal involves an increase of the overall energy metabolism, which is probably partly due to an increased protein turnover, and that this increase is a function of the amount of protein deposited. If this is so, then the energy cost of protein deposition, apart from the portion corresponding to the heat of combustion and the energy cost of synthesis of the amount deposited, should include an increment of the maintenance cost, as compared with the maintenance requirement at zero-nitrogen balance. To examine this supposition, using the data from the experiment described, it has been assumed that the increase of the overall metabolism is an exponential function of protein deposition ($\exp K\ \Delta P$), by which the maintenance requirement at zero-nitrogen balance (C_{MO}) should be multiplied. If the energy cost of synthesis plus heat of combustion of a weight unit of protein is designated as C_{PO}, then the following equation presents the energy metabolism of the growing animal:

$$\text{IME} = C_{MO}\ W^{0.75} \exp K\ \Delta P + C_{PO}\ P + C_F\ \Delta F \quad (4)$$

If definite values of C_{PO} and C_F are assumed and IME, $W^{0.75}$, ΔP and ΔF are known, then equation (4) can be transformed into an easily solvable logarithmic equation:

$$\ln\ [(\text{IME} - C_{PO}\ \Delta P - C_F\ \Delta F)/W^{0.75}] = \ln C_{MO} + K\ \Delta P \quad (5)$$

For the solution of this equation the value 57.4 MJ/kg was substituted for C_F and 31.0 MJ/kg for C_{PO} (Kielanowski, 1976). The values for $W^{0.75}$, ΔP and ΔF, given in *Table 32.1* for all seven groups, were taken as variables. The following equation was obtained:

$$IME = (0.422 \pm 0.58)W^{0.75} \exp(4.08 \pm 1.05)\Delta P + 31.0\,\Delta P + 57.4\,\Delta F \pm 0.54 \quad (4a)$$

r = 0.988; n = 7 group arithmetic means (135 animals)

In a similar way an equation was obtained for castrated males from data published by Kotarbińska (1969):

$$IME = (0.428 \pm 0.043)W^{0.75} \exp(3.49 \pm 1.04)\Delta P + 31.0\,\Delta P + 54.2\,\Delta F \pm 0.47 \quad (4b)$$

r = 0.989; n = 12 group arithmetic means (54 animals)

Equations (4a) and (4b) are almost identical which supports the hypothesis on which they are founded. The difference in exponents might mean that protein deposition is connected with a greater increase of metabolism in boars than in castrated males. More detailed investigations are needed in this respect, but the conclusion seems justified that in growing animals the energy costs of maintenance and of protein deposition should be considered jointly rather than separately.

If the maintenance requirement is assumed to be constant (equal to C_{MO}), then the unitary energy cost of protein deposition, computed from equations (4a) and (4b), becomes greater when the amount deposited and/or the live weight of the animal increases, as demonstrated in *Table 32.2.* This corresponds to the finding that estimates of the energy cost of protein deposition obtained with very young animals are usually lower than those obtained with older ones.

Table 32.2 THE UNITARY ENERGY COST OF PROTEIN DEPOSITION (MJ/kg) AS A FUNCTION OF THE LIVE WEIGHT OF THE ANIMAL AND THE AMOUNT OF PROTEIN DEPOSITED DAILY IN THE BODY (UPPER FIGURES PERTAIN TO BOARS, COMPUTED FROM EQUATION 4a, LOWER TO CASTRATED MALES, COMPUTED FROM EQUATION 4b)

Live weight (kg)	*Daily protein deposition* (kg)			
	0.050	*0.075*	*0.100*	*0.125*
25	52.3 49.2	53.5 50.1	54.8 51.0	56.1 51.9
50	66.9 61.7	68.8 63.1	70.9 64.6	73.2 66.2
75	79.6 72.6	82.3 74.5	85.1 76.6	88.2 78.7
100	91.4 82.6	94.6 85.0	98.2 87.5	102.0 90.2

References

KIELANOWSKI, J. (1976). In *Protein Metabolism and Nutrition*, pp. 207–215. Eds. D.J.A. Cole, K.N. Boorman, P.J. Buttery, D. Lewis, R.J. Neale and H. Swan. EAAP Publication No. 16. London: Butterworths

KOTARBINSKA, M.(1969). *Badania nad przemianą energii u rosnących świń.* Wydawnictwa Własne, Nr 238, Wrocław: Instytut Zootechniki
KOTARBIŃSKA, M. (1971). *Roczniki Nauk Rolniczych*, B-93-1, pp. 129–135
NEHRING, K., BEYER, M. and HOFFMANN, B. (1970). *Futtermitteltabellenwerk.* Berlin: VEB Deutscher Landwirtschaftsverlag

33

PLANE OF NUTRITION AND THE MAINTENANCE REQUIREMENT

R. GRAY
Agricultural and Food Chemistry Research Division, Department of Agriculture, Newforge Lane, Belfast, Northern Ireland BT9 5PX
K.J. McCRACKEN
Agricultural and Food Chemistry Research Division, Department of Agriculture and the Queen's University of Belfast, Newforge Lane, Belfast, Northern Ireland BT9 5PX

Summary

The hypothesis that the maintenance energy requirement of growing pigs is responsive to plane of nutrition has been investigated with 25 kg and 10 kg pigs. It is concluded that the best estimate of the maintenance requirement is the heat production (MH, corrected to zero-energy balance) on the second or third day following reduction to an approximate maintenance intake (M).

The mean values of MH with 25 kg pigs previously fed approximately 2, 2.5 or 3 M were 407, 470 and 481 kJ/kg$^{0.75}$ and the uncorrected values of MH with 10 kg pigs previously fed 2 M or 3 M were 579 and 667 kJ/kg$^{0.75}$ ($P<0.01$).

Introduction

There have been numerous reports of a reduction in the fasting metabolic rate of adult animals and humans subjected to submaintenance levels of nutrition (Benedict, 1938; Marston, 1948; Blaxter, 1962; Westerterp, 1977). Kielanowski (1967) suggested that the maintenance energy costs in the growing pig are related to the level of feeding, but van Es (1972) concluded after surveying the literature that 'no reliable indication was found that the maintenance requirement changes with the level of production'. Similarly, Thorbek and Henckel (1976) reported that 'no influence on heat production in relation to feeding level before starvation could be found' in experiments with chickens.

If one accepts the hypothesis that the maintenance requirement is responsive to plane of nutrition it is easy to see why, in the growing animal, such a hypothesis is difficult to prove. The only method that would appear to offer a means of detecting such a relationship would be the measurement of heat production (corrected to zero-energy balance) of animals kept for a suitable length of time on different planes of nutrition and then reduced to the maintenance level. Ideally the animals on the different treatments should be the same age and weight at the time the measurements are made but this is difficult to achieve in practice coincident with different production levels. The other problem that arises is to determine the optimum period between reducing intake and measuring MH. In order to ensure that MH is not overestimated due to the presence of absorbed nutrients arising from the previous feeding level some time must elapse before measurements can be considered valid. Marston (1948) found that the previous level of nutrition of adult sheep affected the time required for the effect of previously ingested nutrients to become minimal. On the other hand, if MH is responsive to plane of nutrition then the fact of reducing intake to the maintenance level will tend to reduce the metabolic rate and minimize any real differences due to previous plane of nutrition. The studies to be described are not considered by the authors to provide any definitive answer but rather to ask two questions: Is the maintenance requirement of a growing animal responsive to plane of nutrition? Is it possible to design satisfactory experiments to test this hypothesis?

Materials and methods

EXPERIMENT 1 – TO ASSESS THE OPTIMUM PERIOD FOR MEASUREMENT OF MAINTENANCE

Singly-caged pigs, 20 kg live weight, were given a standard growers' diet to provide a daily ME intake of 10.5 MJ above the estimated maintenance (M) requirement (i.e. approximately 3 M) for 10 d. After one 23 h period in a closed-circuit respiration chamber on this feeding level each pig was reduced to an approximate maintenance intake and heat production was measured for a further five consecutive days.

EXPERIMENT 2 – EFFECT OF PLANE OF NUTRITION ON THE MAINTENANCE HEAT PRODUCTION OF SINGLY-CAGED 25 kg PIGS

Large White × Landrace male castrate pigs, initial weight 18 kg, were randomized to one of three planes of nutrition which were, respectively, 6.3, 8.4 or 10.5 MJ/d above the estimated maintenance requirement and fed once daily at 09.00 h for 16 d. Heat production was measured for 22 h on day 10, 13 and 16 in order to estimate energy retention by difference. After 16 d the energy intake was reduced to approximately maintenance and heat production was measured on day 3 at 29°C.

EXPERIMENT 3 – EFFECT OF PLANE OF NUTRITION ON THE MAINTENANCE HEAT PRODUCTION OF 10 kg PIGS KEPT IN GROUPS

This experiment is continuing. So far three replicates have been conducted. In each replicate six pigs from one litter, selected to give the least weight range, were weaned at 12 d and fed a pelleted semisynthetic diet (metabolizable energy (ME) content 18.3 MJ/kg dry matter (DM)) to appetite for 14 d. At 26 d of age they were separated into two groups. The three lightest pigs were given feed almost to appetite which provided an energy intake around 3 M. The three heaviest pigs were restricted to an intake corresponding to 2 M. After 10 d the 3 M pigs were reduced to approximately M (200 g/pig per d given in one feed at 09.00 h) and daily heat production was measured for three consecutive days. Immediately afterwards the 2 M pigs were treated similarly.

Results and discussion

EXPERIMENT 1

The daily heat production declined rapidly following reduction to M intake (*Table 33.1*). Although the reduction appeared to continue up to day five the difference between day three and day five was small. The lowest respiratory quotient (RQ; 0.94) was obtained on day two. If this is taken as an indication that the direct effect of previously ingested nutrients had ceased within the first 24 h then the 7 per cent difference in heat production between day two and day three can at least partly be attributed to a decrease in metabolic rate associated with the reduced plane of nutrition.

Table 33.1 EXPERIMENT 1. MEAN HEAT PRODUCTION (CORRECTED TO EMPTY-BODY WEIGHT), RATIO OF HEAT PRODUCTION ON DAYS 1–5 RESPECTIVELY TO THE VALUE ON DAY 0 AND MEAN RQ ON DAYS 0–5. DAY 0, FEEDING LEVEL 3 M, DAYS 1–5, FEEDING LEVEL, M (THREE PIGS PER MEAN).

Day	*0*	*1*	*2*	*3*	*4*	*5*
H (kJ/kg $^{0.56}$)	1355	1088	968	894	873	860
R		0.80	0.71	0.66	0.64	0.63
Mean RQ	1.20	1.00	0.94	0.97	0.97	0.98

H = mean heat production
R = ratio of heat production

The pattern of heat production on days one to three following reduction to M (*Figure 33.1*) would also tend to support the view that the contribution of

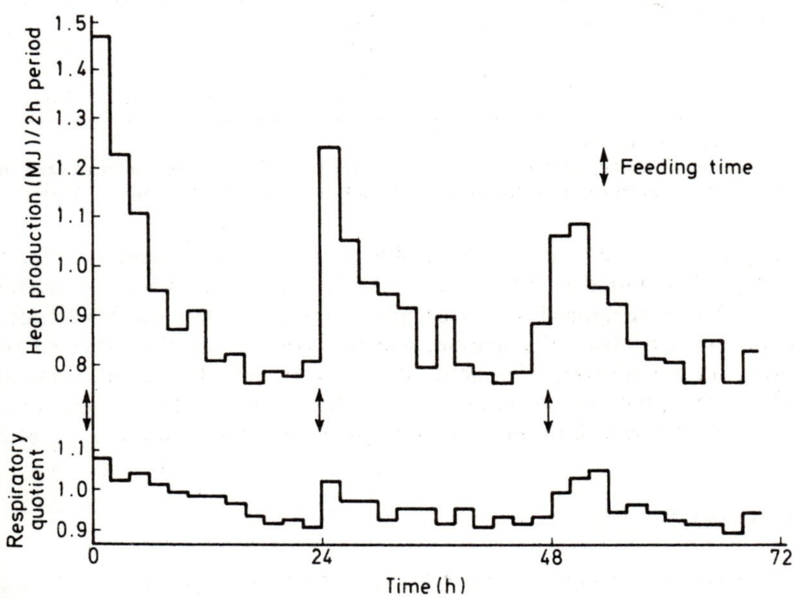

Figure 33.1 Heat production and respiratory quotient of 60 kg pig during the first three days on maintenance intake.

previously ingested nutrients to heat production is minimal on the second day. However, there is sufficient uncertainty about this and about the amount of protein deposition occurring to cast doubt on the validity of estimates based on day two. An acceptable compromise would seem to be the use of the day three value.

EXPERIMENT 2

There was a trend of increasing MH with increasing plane of nutrition (*Table 33.2*). The higher RQ for the animals previously on the low plane reflected the fact that considerable energy retention occurred since the feeding level corresponded to approximately 500 kJ/kg$^{0.75}$. Since the effect of plane of nutrition on MH was not statistically significant, a mean value of 453 kJ/kg $W^{0.75}$ was applied to the energy retention results obtained on the appropriate planes of nutrition in order to calculate the efficiency of energy utilization for production (E_1). However, this method was found to produce a significant ($P<0.05$) treatment effect that was removed when the individual values of MH were applied. It would seem, therefore, that although the differences in MH due to the previous plane of nutrition failed to attain statistical significance they are of biological importance. Furthermore, the fact that, even when the individual values of MH were applied, the efficiency of utilization was still lowest for the pigs on the high plane of nutrition, which is contrary to the expected trend arising from the relative efficiencies of fat and protein deposition, lends some weight to the earlier discussion that heat production on day two may be a better measure of the true maintenance requirement under normal conditions of feeding.

Table 33.2 EXPERIMENT 2. MAINTENANCE HEAT PRODUCTION (CORRECTED TO ZERO-ENERGY BALANCE) AND MEAN RQ ON DAY 3 AFTER INTAKE REDUCTION, CALCULATED EFFICIENCY OF UTILIZATION OF ENERGY FOR PRODUCTION USING THE MEAN VALUE OF MH AND CALCULATED EFFICIENCY OF UTILIZATION FOR PRODUCTION USING INDIVIDUAL VALUES OF MH FOR PIGS GIVEN M + 6.3, M + 8.4 OR M + 10.5 MJ/d.

	M + 6.3	*M* + 8.4	*M* + 10.5	*Significance*	*s.e. of a difference*
MH (kJ/kg$^{0.75}$)	407	470	481	NS	33
RQ	1.01	0.94	0.95	NS	0.034
E_1 (%)	73.1	68.3	64.4	*	1.7
E_2 (%)	64.7	68.9	65.9	NS	1.3

MH = maintenance heat production
E_1 = calculated efficiency of utilization of energy for production using the mean value of MH
E_2 = calculated efficiency of utilization for production using individual values of MH

EXPERIMENT 3

On the basis of the limited results available there was no clear evidence of a difference in heat production or RQ between day two and day three with pigs weighing 10 kg (*Table 33.3*). Analysis of the hourly values of heat production

Table 33.3 EXPERIMENT 3. MEAN WEIGHT, MAINTENANCE HEAT PRODUCTION AND RQ OF 10 kg PIGS ON DAYS TWO AND THREE FOLLOWING REDUCTION TO M FROM 3 M OR 2 M (M TAKEN AS 3.46 MJ METABOLIZABLE ENERGY/PIG PER d).

		Mean weight in chamber (kg)		*Maintenance heat production* ($kJ/kg^{0.75}$)		*RQ*	
	Day	3 M	2 M	3 M	2 M	3 M	2 M
Litter 1	(2)	10.6	10.8	667	596	0.85	0.89
	(3)	10.5	10.8	663	595	0.88	0.90
Litter 2	(2)	–	10.2	–	534	–	0.95
	(3)	10.4	10.3	645	545	0.89	0.95
Litter 3	(2)	10.0	10.3	692	623	0.89	0.89
	(3)	10.0	10.4	666	581	0.89	0.96
Mean		10.3	10.5	667**	579	0.88	0.92

** Significant difference ($P<0.01$)

indicated that even on day one the metabolism of absorbed nutrients was complete within 12 h. It would seem, therefore, that with the younger pig the measurement of heat production on the second day following reduction to M should not overestimate the true energy cost of maintenance. The difference between the values of MH for 3 M and 2 M pigs is large and significant ($P<0.01$). It could be argued, however, that the difference is not due to plane of nutrition but to the deliberately imposed bias in the selection of the groups in order to equalize body weight and age at the time of measurement. This aspect of the problem is still under investigation. The overall mean (623 $kJ/kg^{0.75}$) is close to estimates based on fasting heat production measurements (McCracken and Caldwell, *see* pages 445–448) and to the value of 644 obtained by regression analysis on data covering a range of energy intakes with pigs of 8–32 d (McCracken, Eddie and Stevenson, 1980).

References

BENEDICT, F.G. (1938). *Vital Energetics.* Carnegie Institute Publication No. 503. Washington: Carnegie Institute

BLAXTER, K.L. (1962). *The Energy Metabolism of Ruminants.* London: Hutchinson

KIELANOWSKI, J. (1967). In *Proceedings of the Ninth International Congress on Animal Production*, pp. 212–224. Edinburgh: T and A Constable

MARSTON, H.R. (1948). *Aust. J. Sci. Res.*, **31**, 93–129

McCRACKEN, K.J., EDDIE, S.M. and STEVENSON, W.G. (1980). *Br. J. Nutr.*, **43**, 305–319

THORBEK, G. and HENCKEL, S. (1976). *Eur. Ass. Anim. Prod.* No. **19**, 117–120

VAN ES, A.J.H. (1972). *Handbruch der Tierernährung, Band II*, pp. 1–54. Hamburg: Verlag Paul Parey

WESTERTERP, K. (1977). *Phys. Zoo.*, **50**, 331–362

ENERGY AND PROTEIN UTILIZATION BY PIGS OF DIFFERENT SEX AND GENOTYPE

M.F. FULLER, J.G. GORDON and R. AITKEN
The Rowett Research Institute, Bucksburn, Aberdeen AB2 9SB, Scotland

Summary

Pigs of genetically improved (I) and control (C) stock, including boars, gilts and castrates, were used in a total of 85 energy- and nitrogen-balance experiments. Preliminary results suggest that boars and gilts had similar rates of nitrogen retention and of heat production, but they produced more heat and retained more nitrogen than castrates. There were no significant differences in these measurements between pigs of the two genotypes.

Introduction

Current estimates of the nutrient requirements of growing pigs (Agricultural Research Council, 1967; National Research Council, 1979) do not distinguish animals of different sex or genotype and assume that an animal's energy metabolism is a function of its body weight, its food and its environment. Yet differences between the sexes in growth, food utilization and body composition are well documented (Blair and English, 1965; Jung and Piatkowski, 1967; Texier, Desmoulin and Dumont, 1970; Fuller and Livingstone, 1978). Similarly, there is now abundant evidence from national pig testing schemes that when compared in identical circumstances pigs from different genetic sources may differ to an important degree in their rates of lean tissue growth and fat deposition (Meat and Livestock Commission, 1975, 1976, 1977, 1978). All these comparisons imply underlying differences between sexes and between genotypes in energy and protein metabolism.

Few measurements have been made to compare the energy metabolism of pigs of different sexes. Ritzman and Colovos (1941) reported some measurements showing a higher rate of heat production by boars than by castrates. When comparing genotypes, Sharma, Young and Smith (1971) estimated, from a comparative-slaughter experiment, that the energy requirements for maintenance of Yorkshire and Lacombe pigs of about 20 kg were 570 and 455 $kJ/kg^{0.75}$ per d respectively: a breed difference of about 25%. In a further experiment Sharma *et al.* (1972) reported that in the Yorkshire breed males had a higher maintenance requirement than females, but that in Hampshire X Yorkshire pigs the reverse was true. Lister, Wood and Perry (1974) compared Pietrain and Large White pigs in a comparative-slaughter trial, the results of which suggested that at the same rate of food intake the heat production of the Pietrains was higher than that of the Large Whites. Subsequently, Fuller *et al.* (1976) were unable to detect any differences in either the rate of heat production or the rate of nitrogen retention between Pietrain and Large White × Landrace pigs when compared at the same rate of food intake.

In the present experiment a comparison has been made of two pig genotypes that might be taken to represent the top and bottom 20 percentiles of the British pig herd. The experiment included boars, gilts and castrates. Values from only 24 animals are so far available and the results are, therefore, preliminary.

Materials and methods

The pigs were of two genotypes:

1. Improved (I). These were Large White × Landrace hybrids from a British pig breeding company, and this stock has consistently ranked highly in the Meat and Livestock Commission's Commercial Product Evaluations (Meat and Livestock Commission, 1975, 1976, 1977, 1978).
2. Control (C). These are hybrids of the Meat and Livestock Commission's Large White and Landrace control herds which have been maintained for 13 years with no genetic selection.

Each group of pigs consisted of a litter-mate trio of a boar, a gilt and a castrate: eight such trios have so far completed the experiment. The diet consisted of (g/kg) barley 744, wheat offal 100, soya bean meal 83, white fish meal 65 and mineral and vitamin supplements 8. It supplied 14.0 MJ metabolizable energy (ME) and 2.8 g nitrogen per kg dry matter (DM). The average daily feed intake supplied 1.22 MJ ME and 2.46 g nitrogen/kg$^{0.75}$ per day.

In each balance period (normally 3–4 for each pig) faeces and urine were collected continuously for 7 d, and bladder catheters were used in the females. Heat loss was measured continuously for 48 h during that period in a gradient-layer calorimeter (Pullar, 1969).

Results

GROWTH RATE

Growth rates are shown in *Table 34.1.* Although the animals were few and the errors consequently high, there were significant differences between the genotypes and between the sexes. The improved pigs grew on average 70 g/d faster

Table 34.1 RATES OF GROWTH (kg/d; 30–90 kg)

Type	*Boars*	*Gilts*	*Castrates*	*Mean*	*Approx. SED*
C	0.72	0.64	0.64	0.67	0.024
I	0.80	0.73	0.70	0.74	
Mean	0.76	0.69	0.67	0.71	
Approx. SED		0.030			

than the control stock. Boars grew 12% faster than gilts and castrates, between which there was little difference.

HEAT PRODUCTION

Rates of heat production were analysed by calculating, for each animal, the regression of log heat production on log body weight. The pooled regression gave the general relationship:

$$\text{Heat production} = a\ W^{0.55}$$

The exponent 0.55 had a standard error of 0.03. Its value is close to that of 0.56 found by Breirem (1939) and that of 0.57 found by Fuller and Boyne (1972) from calorimetric measurements with growing pigs.

Analysis of the coefficients 'a' showed that the difference between genotypes was not significant. The heat production of castrates, however, was significantly ($P<0.05$) lower than that of boars and gilts, between which the difference was not significant. The results, adjusted to a body weight of 60 kg, are given in *Table 34.2*

Table 34.2 RATES OF HEAT PRODUCTION (MJ/d AT 60 kg)

Type	*Boars*	*Gilts*	*Castrates*	*Mean*	*Approx. SED*
C	14.5	14.8	14.0	14.4	0.45
I	15.2	15.6	14.0	15.0	
Mean	14.8	15.2	14.0	14.7	
Approx. SED		0.55			

NITROGEN RETENTION

Rates of nitrogen retention were also expressed per $kg^{0.55}$. Analysis of these values showed that there was no difference between the genotypes, but that the boars and gilts retained significantly more nitrogen than the castrates ($P<0.01$). These results, also adjusted to a body weight of 60 kg, are shown in *Table 34.3.*

Table 34.3 RATES OF NITROGEN RETENTION (g/d AT 60 kg)

Type	*Boars*	*Gilts*	*Castrates*	*Mean*	*Approx. SED*
C	24.6	21.6	18.5	21.6	0.9
I	23.0	23.6	17.9	21.5	
Mean	23.8	22.6	18.2	21.5	
Approx. SED		1.1			

Discussion

These preliminary results are in agreement with much other evidence that when the pigs are given the same amount of food, boars and gilts grow faster and deposit protein at a higher rate than castrates. There is, however, some discrepancy in the comparison of the genotypes: the improved animals grew on average 70 g/d faster than the control stock but retained nitrogen at the same rate. Because urine collections from gilts were made with catheters, it is likely that their nitrogen balances are the most precise and that the difference in nitrogen retention between the genotypes is most accurately estimated by the results for the gilts, which accord well with the growth rates.

The early work of Ritzman and Colovos (1941) and later indirect evidence from measurements of growth and body composition suggested that boars and gilts produce more heat and deposit less fat than castrates; the present results support these suggestions also. These differences in heat production may be attributed, in the current conventions of energy metabolism, to differences in maintenance requirements, implying that the same differences would be observed at any rate of food intake; or to differences in the energy costs of growth, implying that boars and gilts use more ME for each unit of protein or fat deposited than castrates. The present observations do not exclude either interpretation, and it is possible to calculate the consequences of each assumption. These are given in *Table 34.4.*

Table 34.4 ENERGY AND PROTEIN METABOLISM IN BOARS, GILTS AND CASTRATES

1. If the energy expenditures per unit of protein and fat accretion are taken to be 44.7 and 53.4 kJ respectively, the energy required for the maintenance of a 60 kg animal may be calculated:

	Boars	*Gilts*	*Castrates*
Energy retention (MJ/d)	11.5	11.1	12.3
ME associated with			
protein accretion (MJ/d)	6.7	6.3	5.1
fat accretion (MJ/d)	10.8	10.4	12.9
maintenance (MJ/d)	8.9	9.5	8.3
$kJ/kg^{0.75}$ per d	0.41	0.44	0.38
k_f	0.66	0.66	0.68

2. If, however, boars, gilts and castrates are assumed to require for maintenance the same amount of ME (9.5 MJ/d), then the efficiencies with which they use energy for growth diverge rather more:

	Boars	*Gilts*	*Castrates*
ME for maintenance (MJ/d)	9.5	9.5	9.5
k_f	0.68	0.66	0.73

If it is assumed that the ME required for 1 g of protein accretion is 44.7 kJ and for 1 g of fat deposition it is 53.4 kJ (Fowler *et al.*, 1980), and that the energy costs are the same in each sex, then when the heat production, associated with fat and protein accretion, is subtracted from the total ME it gives an estimate of the residual ME which is the maintenance requirement. These calculations lead to the conclusion that boars and gilts have maintenance requirements substantially higher than castrates. The value of k_f, the efficiency of utilization of ME, is calculated to be 0.66 in both boars and gilts and 0.68 in castrates, and this difference is a consequence of the higher ratio of fat to protein gain in the castrates.

The alternative assumptions, that the energy requirement for maintenance is constant and that differences in heat production between the sexes stem from differences in the ME associated with each unit of protein or fat deposited, lead to the conclusion that in boars and gilts the k_f is similar to that calculated with the assumption of constant energy costs of tissue accretion: in castrates the k_f is then 0.73, a value similar to that found in a number of previous studies with castrates in a thermoneutral environment (Lund, 1938; Nehring *et al.*, 1960; Fuller and Boyne, 1972; Holmes, 1974; Close, 1978).

Whichever assumptions are made the results suggest that food utilization by boars and gilts is associated with both higher rates of nitrogen retention and with lower rates of fat deposition than in castrates, and that these differences are accompanied by higher rates of heat production. These cannot be explained without supposing a higher maintenance requirement or higher energy costs of growth. These should be taken into account when considering the nutrient requirements of pigs of different sex.

References

AGRICULTURAL RESEARCH COUNCIL (1967). *The Nutrient Requirements of Farm Livestock, No. 3: Pigs.* London: Agricultural Research Council

BLAIR, R. and ENGLISH, P.R. (1965). *J. agric. Sci., Camb.,* **64**, 169

BREIREM, K. (1939). *Tierernährung,* **11**, 487

CLOSE, W.H. (1978). *Br. J. Nutr.,* **40**, 433

FOWLER, V.R., CLOSE, W.H., FULLER, M.F. and WHITTEMORE, C.T. (1980). In *Energy Metabolism,* pp. 151–156. Ed. L.E. Mount. EAAP Publication No. 26. London: Butterworths

FULLER, M.F. and BOYNE, A.W. (1972). *Br. J. Nutr.,* **28**, 373

FULLER, M.F. and LIVINGSTONE, R.M. (1978). *J. agric. Sci., Camb.,* **91**, 337

FULLER, M.F., WEBSTER, A.J.F., MacPHERSON, R.M. and SMITH, J.S. (1976). In *Energy Metabolism,* p. 177. Ed. M. Vermorel. EAAP Publication No. 19. Clermont-Ferrand: G. de Bussac

HOLMES, C.W. (1974). *Anim. Prod.,* **19**, 211

JUNG, H. and PIATKOWSKI, B. (1967). *Arch. Tierzucht.,* **9**, 399

LISTER, D., WOOD, J.D. and PERRY, B.N. (1974). In *Proceedings of 25th meeting of the European Association for Animal Production, Pigs Commission,* Copenhagen

LUND, A. (1938). *Beretn. Forsøgslab.* No. 180

MEAT AND LIVESTOCK COMMISSION (1975). *Commercial Product Evaluation Report.* London

MEAT AND LIVESTOCK COMMISSION (1976). *Commercial Product Evaluation Report.* London

MEAT AND LIVESTOCK COMMISSION (1977). *Commercial Product Evaluation Report.* London

MEAT AND LIVESTOCK COMMISSION (1978). *Commercial Product Evaluation Report.* London

NATIONAL RESEARCH COUNCIL (1979). *Nutrient Requirements of Domestic Animals,* No. 2, *Nutrient Requirements of Swine.* Washington, D.C.: National Academy of Sciences

NEHRING, K., SCHIEMANN, R., HOFFMANN, L. and KLIPPEL, W. (1960). *Arch. Tierernähr.,* **10**, 275

PULLAR, J.D. (1969). In *International Encyclopaedia of Food and Nutrition, Vol. 17: Nutrition of Animals of Agricultural Importance,* Part 1, p. 471. Ed. D.P. Cuthbertson. Oxford: Pergamon Press

RITZMAN, E.G. and COLOVOS, N.F. (1941). *Tech. Bull. New Hamps. agric. Exp. Stn,* 75

SHARMA, V.D., YOUNG, L.G. and SMITH, G.C. (1971). *Can. J. Anim. Sci.,* **51**, 761

SHARMA, V.D., YOUNG, L.G., SMITH, G.C. and SAISON, R. (1972). *Can. J. Anim. Sci.,* **52**, 751

TEXIER, C., DESMOULIN, B. and DUMONT, B.L. (1970). *Journées Rech. Porcine en France,* pp. 209–216. Paris: INRA-ITP

35

A SYSTEM FOR PREDICTION OF RETENTION OF PROTEIN AND FAT IN GROWING ANIMALS

K.H. MENKE
Institute of Animal Nutrition, University of Hohenheim, 7000 Stuttgart 70, Federal Republic of Germany

Summary

A curvilinear function is proposed for the description of the relationship between metabolizable energy (ME) intake and retention of energy as protein, which permits the calculation of the maximum potential energy retention in protein (A) and the efficiency of ME-utilization for protein deposition (k_w) at different feeding levels. The effect of feed composition and animal characteristics on A and k_w is described in separate functions with A and k_w as dependent variables and body weight, protein:energy ratio and amino-acid pattern as independent variables.

Introduction

Factorial calculations on energy requirements presuppose data on the composition of body-weight gain, because of the differences in efficiency of utilization of ME for deposition of fat (k_f) and protein (k_p). Moreover, it is worthwhile to know the composition of gain for practical and economic reasons. Protein synthesis must be regarded as the primary factor affecting body composition. The amount of fat retained can be regarded as the result of ME-intake minus maintenance requirement and energy consumption for protein retention.

A function for prediction of protein retention must include the effects of diet composition (protein:energy ratio and amino-acid pattern) and feed intake as well as the effects of the endogenous factors of age, sex and breed of the animal. The interactions between these exogenous and endogenous factors can best be studied if feed intake is regarded as the independent variable in curvilinear functions for all sets of experiments with different compositions of rations and different animal characteristics. Thus, the difficulties of conducting and interpreting experiments with isocaloric or isonitrogenous diets can be avoided. The effects of the protein:energy ratio and the amino-acid composition of the feed as well as of age, sex and breed of the animals on the regression coefficients of the function between feed intake and protein retention can be described by separate regression functions.

Definitions

The increase in retention of energy in protein (RPE^x; MJ/kg $W^{0.75}$) with increasing feed intake, expressed as ME-intake of a ration of constant composition (IME^x), is given by a square function:

$$RPE^x = b_0 + b_1\, IME^x + b_2\,(IME^x)^2 \qquad (1)$$

The intercept b_0 is related to the basal value of protein turnover (B_N; g N/W^x) and is expressed in energetical units (B; MJ/W^x):

$$b_0 = -B = -0.148\, B_N \qquad (2)$$

A rough estimate of B_N can be derived from the sum of endogenous losses of nitrogen (Menke, 1977, equation 42).

The regression coefficient b_1 represents the efficiency of utilization of ME for protein retention, which is reached when feed intake tends towards zero ($k_{w.o}$). It is linked to the actual efficiency of utilization measured at a certain proportion (AG) of the maximum potential energy retention of protein (A) by the function:

$$b_1 = k_{w.o} = \frac{2\,k_w\,(1 - \sqrt{1 - AG})}{AG} \qquad (3)$$

$$\text{where } k_w = \frac{RPE^x + B}{IME^x} \qquad (4)$$

$$\text{and } AG = \frac{RPE^x + B}{0.001\,A} \qquad (5)$$

The protein retention capacity (A; kJ/W^x) is defined as the maximum of the feed intake dependent square function, and is calculated from the regression coefficients b_1 and b_2 by the equation:

$$A = -\frac{250\,b_1^2}{b_2}\ ;\ (kJ/W^x) \qquad (6)$$

In most cases the maximum energy retention in protein (A) will not be reached in feeding experiments because of the limitations in feed intake of the animal. Estimates based on extrapolation of measurements made at different levels of feeding therefore presuppose the adoption of a curvilinear function for the description of the changes in protein retention with increasing feeding level. The square function has been chosen not only because of the high correlation coefficients found in experiments of this type but also because of the fact that the constants of the square function can be linked to physiological interpretations. This is difficult or impossible with other types of function (Menke, 1980).

Experimental determination of $k_{w.o}$, A and B

At least three levels of feeding are necessary for the determination of the three constants of interest. It is recommended to choose energy intakes of 0.6 and 0.9 MJ ME^x and one *ad lib.* level (1.2–1.6). The determination of the basal value B can be improved by inclusion of a fourth level at about 0.3 ME^x (below maintenance). If the assumption holds true that the basal value B is independent of feed composition and animal characteristics of a certain species, then the number of levels can be reduced to two, one of which should be in the range of 0.8 ME^x and one *ad lib.* From the existing literature on basal nitrogen metabolism the value of B for ruminants can be taken as 0.045 MJ/W^x (Menke, 1977) and for pigs as 0.034 MJ/W^x (Berschauer, 1977).

An alternative method for determination of the constants $k_{w.o}$ and A is based on the relationship between composition of gain which is expressed as 'relative energy retention':

$$RE_{rel} = \frac{RE^x}{RPE^x + B} \qquad (7)$$

and the degree of exhaustion of protein retention capacity (AG, see equations 5 and 1):

$$AG = \frac{b_1\, IME^x + b_2\, (IME^x)^2}{-\, b_1^2 / 4\, b_2} \qquad (8)$$

This relationship between RE_{rel} and AG gives a sigmoid function. Evaluation of experiments of Burlacu *et al.* (1973) and Burlacu, Iliescu and Stavri (1978), Verstegen *et al.* (1973), Holmes (1974) and Gaus (1980) by Menke (1980) led to the equation:

$$AG = 0.13\, (\sin 12.8\, RE_{rel} + 1)^{2.8} + 0.1 \qquad (9)$$

With the aid of equations 7, 9 and 5 the maximum potential energy retention in protein (A) is calculated from RE^x and RPE^x measured in respiration experiments, using a constant value for B. The coefficient $k_{w.o}$ is derived from the measured values of RPE^x and IME^x with the aid of equations 3, 4 and 9.

Preliminary results

Regression analysis of the data from Walker and Norton (1971) with milk-fed lambs resulted in A-values of 230, 490 and 490 with increasing protein content in the ration (10, 28 and 45% protein energy). The coefficient $k_{w.o}$ increased from 0.20 and 0.30 to 0.38 respectively. With growing pigs Gaus (1979) found A-values of about 450, 250 and 160 at body weights of 45, 70 and 100 kg. With baby pigs of 15–20 kg W Gruber (1979) found nearly linear relationships from which A-values higher than 1000 kJ/W^x could be derived. Calculation of A and $k_{w.o}$ from data of Thorbek (1979) using equation 9 led to the following relationship:

$$A = 1090 - 22.26\,W + 0.1355\,W^2 \qquad r = 0.914;\ n = 381;\ P<0.001 \qquad (10)$$

The coefficient $k_{w.o}$ was also related to body weight:

$$k_{w.o} = 0.232 - 0.00101\,W \qquad r = 0.783;\ n = 381;\ P<0.001 \qquad (11)$$

The amount of protein energy retained (RPE^x) is calculated from equations 1, 10 and 11 with $b_0 = -0.034$, $b_1 = k_{w.o}$ and $b_2 = -250 b_1^2/A$. The amount of fat energy retained (RFE^x) can be calculated from:

$$RFE^x = (IME^x - ME^x_m - \frac{RPE^x}{k_p})\,k_f \qquad (12)$$

using average figures for ME^x_m (470), k_p (0.50) and k_f (0.80), or from $RFE^x = RE^x - RPE^x$ with $RE^x = 0.736\,IME^x + 0.00119\,W - 0.43$ (Berschauer, Gaus and Menke, *see* pages 101–105).

Acknowledgement

This chapter was supported by Deutsche Forschungsgemeinshaft, SFB 142–B.3.

References

BERSCHAUER, F. (1977). *Hohenheimer Arbeiten Nr. 91, Verl. Ulmer. Stgt.*

BURLACU, G., BAIA, G., IONILA, D., MOISA, D., TASCENCO, V., VISAN, I. and STOICA, I. (1973). *J. agric. Sci., Camb.*, **81**, 195–202

BURLACU, G., ILIESCU, M. and STAVRI, J. (1978). *J. agric. Sci., Camb.*, **90**, 165–172

GAUS, G. (1980). *Dissertation Hohenheim.* In preparation

GRUBER, F. (1979). *Diplomarbeit Hohenheim*

HOLMES, C.W. (1974). *Anim. Prod.*, **19**, 211–220

MENKE, K.H. (1977). *Übers. Tierern.*, **5**, 1–46

MENKE, K.H. (1980). *Z. Tierphys. Tierernahr. Futtermittelk.* In press

THORBEK, G. (1975). *424. Beretn. f. Statens Husdyrbrugs forsøg.* 1958 Copenhaven V: Rolighedsvej 25

VERSTEGEN, M.W.A., CLOSE, W.H., START, I.B. and MOUNT, L.E. (1973). *Br. J. Nutr.*, **30**, 21–35

WALKER, D.M. and NORTON, B.W. (1971). *Br. J. Nutr.*, **26**, 15–29

EFFICIENCY OF UTILIZATION OF FOOD ENERGY BY GOSLINGS

D. GROSSU, G. BURLACU and J. STAVRI
Institute of Animal Nutrition, Baloteşti, Ilfov, Romania

Summary

The efficiency of energy utilization of diets based on maize by goslings (0–60 days) was measured. The metabolizable energy (ME), heat production and retained energy were 76.6 ± 1.8%, 42.94 ± 2.73% and 33.66 ± 3.21% respectively, of the gross energy. The ME was utilized with a net efficiency of 82.1 ± 3.76% for maintenance and of 73.7 ± 1.93% for production. The maintenance requirement calculated by regression is 618 kJ/kg$^{0.75}$ per d and 456 kJ NE/kg$^{0.75}$ per d, and when ascertained by starvation experiments after 24 hours it was 539 kJ NE/kg$^{0.75}$ per d and 481 kJ NE/kg$^{0.75}$ per d after 48 hours.

Introduction

Even though the efficiency of energy utilization of food by poultry has been somewhat more extensively studied than by other species, there are still rather few experiments on geese. Some results concerning energy metabolism during starvation and specific dynamic action in goslings were established by Poczopko (1967) and Burlacu and Matei–Vlădescu (1962). There is thus a need to study the efficiency of utilization of food energy by goslings and to establish the energy requirement for maintenance and for production.

Materials and methods

The experiments were made on 64 goslings over 1–60 d. The goslings were fed in the first period (0–3 weeks) with a food consisting of 59.8% maize, 21% soybean meal, 7% sunflower meal, 1.4% meat bone meal, 1.6% fish meal and 2.2% premix, and with a gross energy (GE) of 18 000 kJ/kg dry matter (DM) and 23% crude protein (CP). In the second period (4–8 weeks) the goslings received a food consisting of 64.8% maize, 18% soybean meal, 7% sunflower meal, 1.5% meat bone meal, 1.3% fish meal and 2.2% premix with a GE of 10.4 MJ/kg DM and 20.6% CP. The experiments were made in April–June 1978. Energy- and nitrogen-balance trials were made using an experimental technique described in another paper (Burlacu, Faget and Gubcenco, 1971). The respiratory exchanges were measured over a temperature range of 21–27°C depending on the age of the goslings (Poczopko, 1967).

Results and discussion

In the experiments with goslings (80–4200 g) that were fed *ad lib.*, the energy intake varied from 460 to 4400 kJ GE/d. The ratio of ME:GE in the investigations was 76.6 ± 1.8% and was in accordance with Burlacu, Baltac and Paraschiv (1970), Wenk and van Es (1976) and Chwalibog, Henckel and Thorbek (1978), who found that growing chickens converted GE to ME with an efficiency of approximately 72–76% and 76.9% respectively. The heat production and retained energy were 42.94 ± 2.73% and 33.66 ± 3.21% of the gross energy respectively. The protein and fat energy were 47.6 ± 7.1% and 52.4 ± 15.2% respectively of the energy balance. The digestibility of nutrients (y) varied linearly with the level of intake (x = GE intake; kJ/kg $W^{0.75}$ per d) according to the equations:

DCP%	$y = 91.37 + 0.000079\ x$	sb = 0.0004, n = 68
DEE%	$y = 65.59 + 0.0034\ x$	sb = 0.0021, n = 68
DCF%	$y = 61.35 - 0.0105\ x$	sb = 0.0012, n = 68
DNFE%	$y = 90.70 - 0.0031\ x$	sb = 0.0026, n = 68

(DCP = digestible crude protein; DEE = digestible ether extract; DCF = digestible crude fibre; DNFE = digestible nitrogen-free extract).

The efficiency of ME utilization as net energy and the maintenance requirement calculated with different methods are demonstrated in *Table 36.1.*

Table 36.1

Methods	*Level of feeding*	*Nr. det.*	*Maintenance requirement* (kJ/kg $W^{0.75}$)		*Efficiency* (%)
			ME	NE	
Regression	*Ad lib.* + near maintenance level	66	618	456	73.7
Regression	Near maintenance + starvation 24 h	71	585	535	91.2
Regression	Near maintenance + starvation 48 h	67	585	480	82.1
Starvation	Starvation 24 h after *ad lib.*	21	–	539	–
Starvation	Starvation 24 h after maintenance	22	–	518	–
Starvation	Starvation 48 h after *ad lib.*	20	–	472	–
Starvation	Starvation 48 h after maintenance	19	–	464	–

NE = net energy

There are large differences between values determined by different methods. These results are between the values of 502 and 605 obtained for broilers. The former was determined by Wenk and van Es (1976) and the latter by Johnson and Crownov (1976). In an earlier trial by Poczopko (1967) with goslings of different ages, he found that he obtained values between 481 kJ/kg $W^{0.75}$ per d and 367 kJ/kg $W^{0.75}$ per d after 12 hours of starvation. In the starvation experiments there was an influence on heat production in relation to feeding level before starvation. However, these results differ from the results obtained in growing chickens by Thorbek and Henckel (1976) where there was a difference at 4–5 d of starvation. When the food consumed (y) by the goslings fed *ad lib.* was related to their body weight the following equations were calculated where x is body weight:

y_1 (DM intake g/d) = $-0.014 + 0.214x - 0.048x^2$ n = 39; RSD = ± 4.2
y_2 (GE intake kJ/d) = $0.380 + 2.33x - 0.434x^2$ n = 39; RSD = ± 3.1
y_3 (ME intake kJ/d) = $0.096 + 2.29x - 0.493x^2$ n = 39; RSD = ± 5.1

On the basis of starvation experiments, the total heat production (y; kJ/d) of goslings in starvation of 24 or 48 hours after both *ad lib.* and maintenance levels of feeding was related to their body weight (W) by the following equations:

24 h starvation, after ad lib.
$y = 0.0029\ W^{0.70639 \pm 0.0135}$

24 h starvation, after maintenance level
$y = 5.77\ W^{0.64972 \pm 0.149}$

24 h starvation, after ad lib. and maintenance level
$y = 0.326\ W^{0.69042 \pm 0.0102}$

48 h starvation, after ad lib.
$y = 0.32\ W^{0.67074 \pm 0.0107}$

48 h starvation, after maintenance level
$y = 0.348\ W^{0.65923 \pm 0.0107}$

48 h starvation, after ad lib. and maintenance level
$y = 0.322\ W^{0.6668 \pm 0.0075}$

The b values established in various conditions of starvation in goslings differ from the average value (0.75) established by Kleiber (1961) in adult animals. Poczopko and Kowalszyk (1965) found, as in these results, a value of 0.69 in growing chickens.

References

BURLACU, G., FAGET, G. and GUBCENCO, V. (1971). *Brevet RSR*, 25–978

BURLACU, G. and MATEI–VLĂDESCU, C. (1962). *Probl zoot. veter.*, **XII**, 28–41

BURLACU, G., BALTAC, M. and PARASCHIV, M. (1970). In *Energy Metabolism in Farm Animals*, pp. 209–212. Eds. A. Schürch and C. Wenk. EAAP Publication No. 13. Zurich: Juris Druck and Verlag

CHWALIBOG, A., HENCKEL, S. and THORBEK, G. (1978). *Z. Tierphysiol. Tierernähr. Futtermittelk.*, **41**, (2), 87–99

JOHNSON, D.E. and CROWNOV, J.C. (1976). In *Energy Metabolism in Farm Animals*, pp. 121–124. Ed. M. Vermorel. EAAP Publication No. 19. Clermont-Ferrand: G. de Bussac

KLEIBER, M. (1961). *The Fire of Life.* New York: John Wiley

POCZOPKO, P. (1967). In *Energy Metabolism in Farm Animals*, pp. 361–367. Eds. K.L. Blaxter, J. Kielanowski and G. Thorbek. EAAP Publication No. 12. Newcastle upon Tyne: Oriel Press

THORBEK, G. and HENCKEL, S. (1976). In *Energy Metabolism in Farm Animals*, pp. 117–120. Ed. M. Vermorel. EAAP Publication No. 19. Clermont-Ferrand: G. de Bussac

WENK, C. and VAN ES, A.J.H. (1976). In *Energy Metabolism in Farm Animals*, pp. 189–192. Ed. M. Vermorel. EAAP Publication No. 19. Clermont-Ferrand: G. de Bussac

37

DISCUSSION PAPER: USE OF ENERGY FOR MAINTENANCE AND GROWTH

SIR KENNETH BLAXTER, FRS
The Rowett Research Institute, Bucksburn, Aberdeen AB2 9SB, Scotland

In opening this discussion on the utilization of energy for maintenance, it should be stressed that there are two aspects:

1. The descriptive analysis of energy metabolism – often called the classic approach.
2. The physiological and biochemical analysis of metabolism.

Holmes (Holmes *et al., see* page 97) in his chapter was somewhat apologetic about using the first approach and so too was Farrell (Farrell and Siregar, *see* page 119). In fact these approaches are not separate, and eventually it should be possible to provide an intellectually satisfying model of metabolism that gives sufficient explanation of what we describe and measure in calorimetric work.

At present we have a reasonable framework of ideas about how to describe changes in energy retention with changes in the amount and quality of feed given, and this framework is now well supported by quantitative estimates of its parameters and provides good practical estimates of outcome. These aspects will be illustrated shortly, for they follow on the discussion by van Es on feed evaluation (*see* page 85).

This purely descriptive work generates a need to explain the causal mechanisms. Indeed many of the problems which relate to possible biochemical or biophysical explanations of biological events in growth have arisen from descriptive work, and some of these problems were not self-evident or could not be anticipated from studies on isolated systems.

In addition to dealing with these two aspects we should consider whether modern biochemical and physiological knowledge enables us to manipulate growth or the composition of the body, or the use of nutrients to promote growth.

I wish, in reviewing the 18 preceding chapters and relating them to other work, to begin with the descriptive model and show that it is very reasonable in that it permits the prediction of performance of animals. I apologize for commencing at this elementary level.

The descriptive model is simply the relationship between feed-energy intake and energy retention in animals which have been given amounts of feed over a sufficient time so that an equilibrium can be assumed. Equilibrium may be difficult to attain in some instances as evidenced by Gray and McCracken's demonstration (*see* page 163). Stability of metabolism in young animals is not easy to achieve as we noted long ago, and Webster, Brockway and Smith (1974) avoided it by using their extrapolation procedure. Equally, the demonstration by Fowler *et al.* (*see* page 151) showed the problems associated with defining maintenance, and Walker and Jagusch (1967) showed with lambs that weight equilibrium, fat equilibrium, protein equilibrium and energy equilibrium are different things.

The relationship is a curvilinear one between time rates of retention and of intake, and the model simply expresses it in terms of an intercept on either the retention axis (fasting metabolism) or the feed axis (maintenance) and a slope. The slope is not constant, varies continuously and is greater below maintenance than above.

The slope, wherever it is measured, is affected by the nature of the diet, the nature of the animal, the species and its physiological state. Equally, while the intercept of fasting may be taken to reflect animal factors only, the intercept at maintenance varies both with animal attributes and feed attributes. Van der Merwe and van Rooyen (*see* page 135) demonstrated breed differences in maintenance in cattle, and Fuller, Gordon and Aitken (*see* page 169) also showed it for breeds of pigs, while sex differences were demonstrated indirectly by Ludvigsen (*see* page 115) and by Walach-Janiak, Kotarbińska and Kielanowski (*see* page 157). The model is sufficient to include such variation even though it does not explain why it occurs.

Energy retention reflects deposition of organic material in the body, which is conventionally regarded as consisting solely of lipid and protein. Garrett (*see* Part 1, page 3) describes the relationships between age, weight, breed and sex and the energy content of gain, and so too do Byers and Rompala (*see* page 141). The studies by the Agricultural Research Council Working Party on body composition in ruminants have included data from throughout the world, and these show clear effects of rate of growth, breed, sex and mature size. Such values are of great importance in relating the basic model to practice.

The slope, which varies with the amount and quality of feed, has been comprehensively discussed in the preceding chapters. It is affected by the composition of gain as shown by Berschauer, Gaus and Menke (*see* page 101), and it varies with the type of feed, with metabolizability (as shown by Vermorel, Bouvier and Geay; *see* Part 1, page 49), with protein (as shown by Holmes *et al.*; *see* page 97) and with fat (as indicated by Boyd and McCracken; *see* page 111).

The nature of these complex relationships led me, some 20 years ago, to abandon the idea that there could ever be a concordance between an energy measure of feed and energy retention and to propose, in the metabolizable energy system, a three-component system of meeting animals' energy needs. This separates all those factors that are involved in the slope of the relationship from those that are associated with the animal under consideration.

Knowing the way in which metabolizability changes with intake and the way in which the slope at different levels of intake changes with the nature of the diet, one can compute gain of energy and therefore gain in weight.

The model system I am now using implies continuing curvature rather than constancy of k_f over the whole range, and on numerical integration the model predicts body weight over a very considerable range.

Indeed, I will state with some assurance that the empirical and descriptive approach has been highly justified in terms of its primary objective, namely to predict animal performance under a wide variety of circumstances. For practical purposes I do not think the framework of idea is wrong or in error or useless. This does not mean that it gives us a deep understanding of what is taking place. The descriptive approach simply poses the problems to be solved.

There are many difficulties and I wish to select two that arise from the preceding chapters. The first relates to the origin of the heat increment associated with feeding and particularly with the effect of the nature of the diet on the heat increment, and the second concerns the relationship between protein synthesis rates and energy metabolism.

The first problem is a very old one discovered by Zuntz and Hagemann (1898) in experiments with horses. A given amount of metabolizable energy (ME) from roughages produces more heat than does one from concentrated feed. After almost 100 years we still do not have an unequivocal answer to explain this problem. Ørskov and McDonald's results are very pertinent here (*see* page 147). Armsby (1903) conjectured that the origin of the difference could be due either to work done in prehending and masticating food or to heat arising from the fermentation of the food or to heat arising in intermediary metabolism, where he followed the ideas of Rubner (1902). In this book we have seen the first two jettisoned. Graham (1965) showed that the energy cost of eating could not account for the differences in heat production, while stoichiometric calculation showed that the heat of fermentation was too small a quantity. This work was elegantly confirmed by direct measurement by Webster, Smith and Mollison (1976). This left the heat arising from intermediary metabolism.

Armstrong and I (Armstrong and Blaxter, 1957) showed that in fasted animals the heat increments were proportional to the free energy of ATP released on complete dissimilation of the substrates absorbed. A substitution took place of body lipid by the absorbed metabolite. Acetic and butyric acids were used very efficiently provided C_3 fragments were supplied in the molar ratio 15 C_2 : 1 C_3; if C_3 fragments were not available, then massive metabolic disturbance with ketosis and enhanced glucogenesis occurred and large increases in heat production took place.

Above maintenance we found that increments of heat were very considerable for acetic acid and much smaller for propionate. We could not provide a biochemical explanation for these findings. However, they did accord with the correlation between ruminal concentrations of acetate and heat increment, for roughages are associated with acetate fermentations. Much evidence, however, indicated that this was not so. Slaughter experiments by Ørskov and calorimetric work by Bull and others gave much smaller increments for acetic acid than those we found.

The conclusion of Hovell following his work at the Rowett Research Institute (Hovell and Greenhalgh, 1978) was that above maintenance the heat increment was variable and depended on the ratio of glucose precursor to acetate and butyrate. Ørskov and McDonald (*see* page 147) show that over the effective range in which the acetic acid proportion lies there is relatively little effect of the acetic acid proportion on efficiency or its inverse the heat increment. Yet Moe and Tyrrell (*see* Part 1, page 59) and the workers at Beltsville have amply confirmed our findings that the heat increment of acetic acid is high when added to a roughage diet. They equally confirmed those of Hovell and Greenhalgh (1978) and Bull, Reid and Johnson (1970) in showing that the heat increment of acetic acid was low when added to a concentrate diet.

This problem is not yet solved after a century. I would certainly like to hear a discussion between Dr Ørskov, Dr Moe, Dr Bull and others on this matter and would hope that the biochemical model builders will take part.

The second problem of growth relates to protein deposition and lipid deposition, the discrepancy between biochemical explanation and calorimetric observation and the nature of animal differences in metabolic rate. Following the demonstration of this by Kielanowski (1976) there have been many estimates made of the cost of protein deposition, and additional ones have been presented here. Webster *et al.* (*see* page 125) give efficiencies of 44 and 74% for protein and fat, Gray and McCracken (*see* page 163) 44 and 89%, Holmes *et al.* (*see* page 97) 53 and 92%. There are, however, some curious findings. Thus, Garrett (*see* Part 1, page 3) stated that he had found no difference in efficiency with stilboestrol, which augments protein deposition, nor had van Es (*see* Part 1, page 85), while Vermorel, Bouvier and Geay (*see* Part 1, page 49) found no differences between adult sheep and young calves in the efficiency with which they used the same feed.

Stoichiometric calculation suggests the efficiency with which glucose can be used to synthesize fat is about 74%, a figure that agrees with experimental findings. For protein, assuming four ATP equivalents for the formation of a peptide bond, this does not agree. If the messenger RNA has to be synthesized for each peptide bond, then the ATP requirement might be 10 ATP and maximally it might be 16 ATP if the DNA turns over at such a rate, which is highly unlikely: the experiments indicate about 30 ATP. The answer most people propose is that this reflects the turnover of protein. However, Edmunds, Buttery and Fisher (*see* page 129) conclude that this cannot be the sole reason. Here we have to be careful, because in equating nitrogen retention with protein retention we are in slight error for growth consists of replication of a DNA unit, that is protein plus nucleic acid. Certainly amino-acid movement and the maintenance of electrochemical gradients must be considered as these authors imply.

Webster *et al.* (*see* page 125) have taken the argument further to adduce that the metabolic rate of an animal is directly proportional to protein turnover. If we think about the reasons why the metabolic rates of animals differ and hence their maintenance requirements, then we have to postulate differences in work done in the purely thermodynamic sense of that term. It seems highly improbable that the efficiency of production of ATP or reduced coenzymes varies between species. Work done in a fasting animal consists of pressure/volume work, the work of polymerization of molecules (protein synthesis, lipid synthesis and glycosidic bond synthesis – all of which turn over) and the maintenance of ion gradients. All these are energy-demanding processes and differences between animals in fasting metabolism must reflect these functions. Webster's approach (Webster *et al., see* page 125) identifies the turnover component of protein as being a virtually constant proportion of metabolic rate which implies proportionality of the other terms. However, questions arise about deviations of metabolic rate from such a general concordance. Does this reflect species differences in determinants, leaky membranes, futile cycles or curious species-specific demands for synthesis of particular molecules or their rate of turnover?

Finally (and related), Christensen, Henckel and Thorbek (*see* page 107) commenced with the idea that oxidative phosphorylation was disrupted by deficiency of the essential fatty acids (EFAs) and that membrane structures were generally disturbed. They also expected, together with Grete Thorbek, changes in oxygen consumption when there was a deficiency of EFAs, that is an increase in oxidation without ATP production: it did not happen. Nevertheless, this sort of approach – the linking of biochemical studies on isolated systems to calorimetry – should surely be uppermost in our minds.

References

ARMSBY, H.P. (1903). *The Principles of Animal Nutrition.* New York: John Wiley and Sons

ARMSTRONG, D.G. and BLAXTER, K.L. (1957). *Br. J. Nutr.,* **11**, 247–272

BULL, L.S., REID, J.T. and JOHNSON, D.E. (1970). *J. Nutr.,* **100**, 262–276

GRAHAM, N. McC. (1965). In *Energy Metabolism of Farm Animals*, p. 231. Ed. K.L. Blaxter. EAAP Publication No. 11. New York and London: Academic Press

HOVELL, F.D.DeB. and GREENHALGH, J.F.D. (1978). *Br. J. Nutr.,* **40**, 171–183

KIELANOWSKI, J. (1976). In *Protein Metabolism and Nutrition*, p. 207. Eds. D.J.A. Cole, K.N. Boorman, P.J. Buttery, D. Lewis, R.J. Neale and H. Swan. EAAP Publication No. 16. London: Butterworths

RUBNER, M. (1902). *Die Gesetze des Energieverbrauchs bei der Ernährung.* Leipzig: Deuticke

WALKER, D.M. and JAGUSCH, K.T. (1967). In *Energy Metabolism of Farm Animals*, p. 187. Eds. K.L. Blaxter, J. Kielanowski and G. Thorbek. Newcastle upon Tyne: Oriel Press

WEBSTER, A.J.F., BROCKWAY, J.M. and SMITH, J.S. (1974). *Anim. Prod.,* **19**, 127–139

WEBSTER, A.J.F., SMITH, J.S. and MOLLISON, G.S. (1976). In *Energy Metabolism of Farm Animals*, p. 221. Ed. M. Vermorel. EAAP Publication No. 19. Clermont-Ferrand: G. de Bussac

ZUNTZ, W. and HAGEMANN, O. (1898). 'Untersuchungen über die Stoffwechsel des Pferdes bei Ruhe und Arbeit' *Landw. Jbr.,* 27, Ergänzungsband III, p. 1

38

GENERAL DISCUSSION

The discussion was concerned with three questions:

1. Is the relation between metabolizable energy intake (MEI) and energy retention (ER) linear or curvilinear?
2. Why is protein deposition such an inefficient process in a growing animal?
3. In the ruminant animal why is the heat increment associated with a roughage diet higher than that with a diet providing the same metabolizable energy (ME) intake from concentrates?

First, it was stated that in energy-balance studies with animals a knowledge of total ER alone is insufficient, and that the proportions of energy retained as fat and protein must also be known. It was suggested that in mature animals and in animals given a high MEI, where fat makes up a high proportion of ER, one might expect an upward turn in the line relating ER to MEI due to the high efficiency of fat deposition relative to that of protein (assuming a constant maintenance need). Conversely, where protein deposition exceeds fat deposition a downward turn might be expected. Many workers consider that the main curvilinearity that is to be expected in the relation between MEI and ER is the result of differences between k_m and k_{pf}. For example, the relative changes in body protein and fat that are associated with a change in MEI are thought to be of considerable importance in determining the slope of the line around the point of energy equilibrium (that is k_m). There would seem to be insufficient evidence to conclude that ruminants differ from non-ruminants with respect to this relation between MEI and ER. However, a number of factors need to be considered, including the metabolizability of the diet, the level of MEI, the age of the animal, the environmental conditions and the value of the exponent used for calculating metabolic body weight.

Secondly, calculations based on the energy cost of peptide-bond formation indicate that this factor alone cannot explain the inefficiency with which protein is deposited during growth, and may, in fact, only account for about one-tenth of the total energy loss. On present evidence the energy losses associated with the sodium flux, with the synthesis of DNA and RNA, with the absorption of amino acids, with the replacement of cells in the intestinal mucosa and with the turnover of protein are insufficient to explain the inefficiency of protein deposition. Furthermore, it is to be expected that k_p will vary according to the composition of the diet and of the tissues deposited, to the proportion of dietary protein digested in the hind gut and to the amount of protein formed as wool or lost in milk.

Thirdly, the present consensus of opinion is that there is no single factor responsible for the high heat increment associated with the feeding of roughage when compared with concentrates. It is thought that heat increment is the sum of a number of inefficiencies – among which may be listed the energy associated with the processes of eating and digestion, the costs of fermentation, heat production within the gut and the proportions of volatile fatty acids in the rumen – and that the relative contribution of each of these factors to the total heat increment will be affected by the composition of the diet.

PART 3

SIMULATION MODELS FOR ENERGY EXCHANGE AND GROWTH

39

MODELS OF TISSUE AND ANIMAL METABOLISM

N.E. SMITH, R.L. BALDWIN and W.M. SHARP
Department of Animal Science, University of California, Davis, California 95616, USA

Summary

Several stages in the development of a model of metabolism in the intact ruminant are discussed to illustrate the interplay between modelling and experimental research. The types of data used in such models and the advantages of and the continuing nature of modelling analyses when these are incorporated into a research programme are also considered here. Examples drawn from models and example results from several stages of modelling analyses of energy metabolism are presented to illustrate the technique.

Introduction

Consideration of the complex of metabolic, nutritional and physiological data that must be accommodated in analyses of animal regulatory functions indicates a need for utilization of dynamic mathematical modelling techniques as a research tool. Objectives in the use of dynamic modelling techniques in support of studies of energy metabolism include evaluations of available concepts and data for adequacy as explanations of the behaviour of metabolic systems, evaluation of alternative hypotheses for probable adequacy and identification of critical experiments. Within the context of these objectives, one can visualize the necessity for close interactions between modelling and experimental analyses. In developing this discussion, emphasis will be placed upon illustration of the interplay between research modelling and experimentation.

The first stage of model development which requires collection, summarization and evaluation of a wide range of energetic, physiological and metabolic data will be considered only briefly because this has been reported elsewhere (Smith, 1971; Baldwin and Smith, 1971a,b; Smith *et al.*, 1975; Baldwin *et al.*, 1976).

The basic characteristics of a dynamic model of whole animal metabolism will be discussed briefly with emphasis on a generalized mathematical representation of the metabolism of a tissue to illustrate the fundamental characteristics of the model. Solutions of models of ruminant metabolism led to identification of specific research problems that required both experimental work and the construction and evaluation of detailed models of tissue metabolism. One of the reasons for construction of detailed models was that they were required for interpretation of radioisotope data. The latter stages of the research programme will be discussed in relation to current status and future emphasis.

Whole animal models

The first stage of model development requires definition of the basic characteristics of the animal in steady state to be modelled. In the case of our model, a lactating cow, her weight, production, diet and energy status were set at 550 kg, 30 kg, 60% concentrate and 40% alfalfa and energy balance. Then, organ weights and energy expenditures, extracellular and intracellular pool volumes and metabolite pool sizes, nutrient absorption rates, metabolite transport rates to tissues, turnover and oxidation rates of major metabolites and metabolite utilization rates were established, and they were based on an extensive literature search and data summarization and integration (Smith, 1971; Smith *et al.*, 1975). The results of this search and integration are depicted diagrammatically in *Figure 39.1.* In the diagram major state variables (for example blood triacylglycerol (TG) and pool sizes, both of which must be defined) and net fluxes in moles per day are identified. This diagram serves, together with specific concepts of metabolic regulation, as the basis for model construction. A mathematical representation of tissue metabolism requires consideration of energy requirements for basal metabolism and for variable biosynthetic and work functions, affinity of the tissue for extracellular metabolites, biosynthetic and oxidative capacities and specific (hormonal) regulatory processes. A simplified set of equations for a typical tissue in modified FORTRAN format would be:

$$\mathrm{DADP} = \mathrm{ATPUSE} - \sum_{i=1}^{n} \mathrm{METOX}_i{*}\ \mathrm{SATP}_i \tag{1}$$

$$\mathrm{ATPUSE} = \mathrm{BASATP} + \mathrm{VARATP} \tag{2}$$

$$\mathrm{BASATP} = \text{a constant or a function of organ weight} \tag{3}$$

$$\mathrm{VARATP} = \sum_{i=1}^{n} \mathrm{VARFUN}_i{*}\ \mathrm{SVAR}_i \tag{4}$$

$$\mathrm{METOX}_i = \mathrm{APPCAP}_i * \mathrm{METC}_i \,/\, (\mathrm{METC}_i + \mathrm{APPAFF}_i) \tag{5}$$

$$\mathrm{APPCAP} = \mathrm{f}(\mathrm{MAXCAP}_i, \mathrm{ADPC}, \ldots.) \tag{6}$$

$$\mathrm{APPAFF}_i = \text{a constant} \tag{7}$$

where DADP is the differential equation for ADP; METOX_i is moles of metabolite 'i' oxidized; SATP_i is moles ATP formed per mole METOX_i; BASATP is the basal rate of ATP use; VARATP is ATP used in variable functions (VARFUN_i * SVAR_i; for example, SVAR_i * VARFUN_i = 2 * moles lactose synthesized); APPCAP_i is the apparent capacity to oxidize MET_i; METC_i is the concentration of MET_i; APPAFF is the apparent affinity for MET_i; MAXCAP_i is the maximum capacity to oxidize MET_i; and ADPC is ADP concentration. Bases for the definition and the computation of ATP use in terms of basal and variable functions and the use of Michaelis–Menten kinetics to compute metabolite oxidation were documented by Baldwin and Smith (1971a,b; 1979).

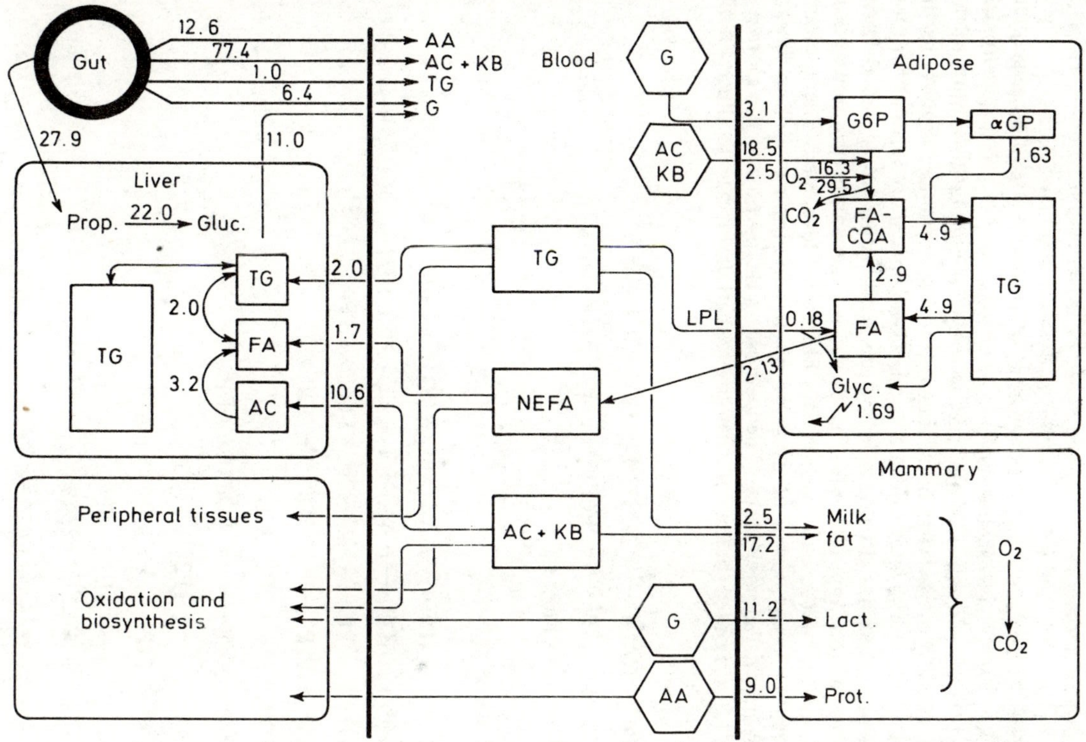

Figure 39.1 Simplified block diagram of energy metabolism in a lactating dairy cow

Adipose metabolism

Evaluation of the behaviour of the first version of the lactating cow model indicated that the adipose element was inadequate, that is the concepts and data available and used in model construction were not sufficient to explain reality (Smith, 1971). This observation led to the performance of a series of studies of ruminant adipose metabolism (Baldwin *et al.*, 1973; Yang and Baldwin, 1973a,b; Grichting, Smith and Baldwin, 1977; Yang, Baldwin and Garrett, 1978). A portion of these studies involved measurements of relative rates of glucose-u-^{14}C, -1-^{14}C, -2-^{14}C, -6-^{14}C and acetate-1-^{14}C conversions to CO_2, fatty acids and glyceride glycerol. A number of approaches to the interpretation of data of this type have been proposed for non-ruminant tissues. Several assumptions are essential to these methods of interpretation. These appear to be valid for most non-ruminant tissues but do not hold for ruminant tissues (Yang and Baldwin, 1973b). Thus a new approach which does not require simplifying assumptions for interpretation of the radioisotope data had to be developed. The approach adopted was the construction of dynamic models for the metabolism of each tracer in ruminant adipose tissue. The pentose cycle portion of the glucose-2-^{14}C model is presented below in chemical equation format to illustrate the approach:

$$\text{G6P2* + NADP} \xrightarrow{K_1} \text{NADPH}_2 + \text{CO}_2 + \text{RU5P1*}$$
$$\text{3RU5P1* + TP} \xrightarrow{K_2} \text{2F6P1* + F6P3* + TP}$$
$$\text{G6P1*} \underset{K_4}{\overset{K_3}{\rightleftharpoons}} \text{F6P1*}$$
$$\text{G6P3*} \underset{K_4}{\overset{K_3}{\rightleftharpoons}} \text{F6P3*}$$
$$\text{G6P1* + NADP} \xrightarrow{K_1} \text{NADPH}_2 + \text{CO}_2\text{*} + \text{RU5P}$$
$$\text{G6P3* + NADP} \xrightarrow{K_1} \text{NADPH}_2 + \text{CO}_2 + \text{RU5P2*}$$
$$\text{3RU5P2* + TP} \xrightarrow{K_2} \text{2F6P3* + F6P2* + TP}$$
$$\text{F6P1* + ATP} \longrightarrow \text{F16DP1* + ADP}$$
$$\text{F16DP1*} \longrightarrow \text{TP1* + TP}$$
$$\text{F6P2* + ATP} \longrightarrow \text{F16DP2* + ADP}$$
$$\text{F16DP2*} \longrightarrow \text{TP2* + TP}$$
$$\text{F6P3* + ATP} \longrightarrow \text{F16DP3* + ADP}$$
$$\text{F16DP3*} \longrightarrow \text{TP3* + TP}$$
$$\text{RU5P + TP1*} \longrightarrow \text{F6P4* + TP}$$
$$\text{RU5P + TP2*} \longrightarrow \text{F6P5* + TP}$$
$$\text{RU5P + TP3*} \longrightarrow \text{F6P6* + TP}$$
$$\text{G6P4*} \rightleftharpoons \text{F6P4*}$$
$$\text{G6P5*} \rightleftharpoons \text{F6P5*}$$
$$\text{G6P6*} \rightleftharpoons \text{F6P6*}$$
$$\text{G6P4* + NADP} \longrightarrow \text{NADPH}_2 + \text{FCO}_2 + \text{RU5P3*}$$

and so on.

Where G6P2* is glucose-6-P labelled (*) in position 2; RU5P(–*) is ribulose-5-phosphate labelled in various (–) positions; F6P is fructose-6-phosphate; TP is triose-phosphate; F16DP is fructose-1,6-diphosphate; K_1–K_n are the rate constants. Note that stoichiometries are only indicated for labelled species and rate constants were only indicated to the extent required to illustrate the approach.

When chemical equations are input to KINSYM (Baldwin and Smith, 1971a) in a form similar to that above, then the chemical equations are converted to differential equations of the general form:

$$DG6P1^* = + K_2(RU5P1^*)(TP)(2) - K_3(G6P1^*) + K_4(F6P1^*) \ldots$$

If methods of computing fluxes (for example $+ K_3(G6P1^*)$) other than the mass action forms indicated are desired, these can be inserted. The differential equations can then be solved to obtain estimates of steady state rates of labelled CO_2, fatty acid and glyceride-glycerol production. The initial set of rate constants and pool sizes used were derived from a 'cold' model of ruminant adipose metabolism described previously (Baldwin *et al.*, 1976). Outputs from the model can be compared with experimental data and errors of prediction calculated. Rate constants can then be manually adjusted in all models (glucose-1-^{14}C, -2-^{14}C, -6-^{14}C, -u-^{14}C and acetate-1-^{14}C) or they can be corrected using a computer program based on error minimization techniques (Koong *et al.*, 1975) to obtain an optimum fit to the data. This process is prohibitively expensive because five dynamic models must be solved to steady state for each iterative step. An alternative approach which achieves the same objective at much lower costs may be realized.

References

BALDWIN, R.L. and SMITH, N.E. (1971a). *Fed. Proc.*, **30**, 1459–1465

BALDWIN, R.L. and SMITH, N.E. (1971b). *J. Dairy Sci.*, **54**, 583–595

BALDWIN, R.L., REICHL, J.R., LOUIS, S., SMITH, N.E., YANG, Y.T. and OSBORNE, E. (1973). *J. Dairy Sci.*, **56**, 340–349

BALDWIN, R.L., YANG, Y.T., CRIST, K. and GRICHTING, G. (1976). *Fed. Proc.*, **35**, 2314–2318

BALDWIN, R.L. and SMITH, N.E. (1979). In *Advances in Nutritional Research*, Vol. 1, pp. 1–24. Ed. H.H. Draper. New York: Plenum Publishing Corporation

GRICHTING, G., SMITH, N.E. and BALDWIN, R.L. (1977). *Fed. Proc.*, **36**, 1141 (abstr.)

KOONG, L.J., BALDWIN, R.L., ULYATT, M.J. and CHARLESWORTH, T.J. (1975). *Comput. Progr. Biomed.*, **4**, 209–213

SMITH, N.E. (1971). *Quantitative Simulation Analyses of Ruminant Metabolic Functions: Basal Lactation and Milk Fat Depression.* PhD Thesis, University of California, Davis

SMITH, N.E., ELY, L., BALDWIN, R.L., DOLKAS, C., KOONG, L.J., MOBERG, G. and PLUCINSKI, T. (1975). *Comput. Biol. Med.*, **5**, 315–323

YANG, Y.T. and BALDWIN, R.L. (1973a). *J. Dairy Sci.*, **56**, 350–365

YANG, Y.T. and BALDWIN, R.L. (1973b). *J. Dairy Sci.*, **56**, 366–374

YANG, Y.T., BALDWIN, R.L. and GARRETT, W.N. (1978). *J. Anim. Sci.*, **47**, 686–690

40

DEVELOPMENT OF A COMPUTER PROGRAM FOR SIMULATION OF ENERGY METABOLISM IN THE WHOLE ANIMAL

A.R. SCHULZ
Department of Biochemistry, School of Medicine, Indiana University, Indianapolis, Indiana 46202, USA

Summary

A computer program is described that represents the animal as a pseudo-steady-state system in which the metabolism can be expressed as a series of stoichiometric expressions. Input for the program includes feed intake, respiratory exchange and urinary urea excretion.

Introduction

The genesis of this model for simulation of energy metabolism lies in a conclusion expressed by Rubner (1902) on the basis of his incisive investigations of energy metabolism during the latter part of the nineteenth century. Rubner stated, 'The energetic concept of the process of nutrition is not based upon the fact that someone started to express nutritional materials in calories, but upon the knowledge and proof that the animal organism depends upon the energy value of the nutritional materials and body substance. In spite of the varying chemistry of catabolism which apparently occurs during changes in the form of nourishment, the energy metabolism is the determining factor and focal point around which everything else revolves.' The wisdom of Rubner's conclusion was confirmed by the subsequent investigations of many scientists, but notably those of Lohmann, Meyerhof, Warburg and Lipmann (Florkin and Stotz, 1975). This concept culminated in studies that have established that the obligatory generation of 'energy-rich' phosphate bonds during metabolism and respiration and the utilization of this 'phosphate-bond energy' during endergonic physiological functions (for example nerve transmission, muscle contraction and biosynthesis) determine the amount of respiration. Thus, the ability of the living organism to couple the generation of ATP to metabolism and respiration defines the flameless combustion, of which Lavoisier spoke, and constitutes the mechanistic basis for Pflüger's premise that the rate of oxygen consumption was regulated within the living cell (Florkin and Stotz, 1972).

An algorithm has been developed for simulation of energy metabolism in the whole animal which is independent of both kinetic considerations and estimates obtained by regression analysis. This model is based on the assumption that the animal can be represented as a pseudo-steady-state system, that is the intermediary metabolites remain constant but the body substance and/or animal products may accumulate. The model is further based on the representation of the metabolic transformations that make significant contributions to energy metabolism as series of stoichiometric expressions which summarize the major metabolic pathways. This algorithm was developed originally for the mongastric animal (Schulz, 1978), and it has been extended to the ruminant, to energy balance in the mammalian animal during lactation and to the avian. The algorithm has also been extended to provide for energy balance on the basis of feed consumption, respiratory exchange and urinary-urea excretion. The integration of the latter algorithm into the program for simulation of energy metabolism will be the subject of this chapter and only the application to monogastric animals will be discussed.

Program description

The computer program for this simulation algorithm was written in BASIC for use with a microcomputer. The dietary energy is calculated as the total moles of ATP which would be generated if the digestible nutrients were oxidized completely to CO_2, H_2O and urea. The moles ATP can be converted to their equivalent in kilojoules by multiplying moles ATP by 77.8. Estimates of the available energy obtained in this manner are in close agreement with estimates of metabolizable energy. The stoichiometry for glucose metabolism is given by the following expression:

$$\text{glucose} + 6\ O_2 \longrightarrow 6\ CO_2 + 36\ \text{ATP}$$

The yield in moles ATP from fats is calculated on the assumption that dietary fats can be represented as triglycerides, and it can be calculated from the following expression:

$$\text{moles ATP} = 25.5*N - 1 - 6*U,$$

where 'N' is the number of carbon atoms and 'U' is the number of double bonds per mole of fatty acid. The stoichiometries for the complete oxidation of 21 amino acids by mammals have been presented elsewhere (Schulz, 1978).

Since the algorithm provides for the calculation of oxygen consumption and carbon dioxide production during simulated energy metabolism, it is possible to simulate energy metabolism on the basis of feed consumption, respiratory exchange and urinary-urea excretion. When energy metabolism is simulated by this portion of the algorithm, protein metabolism is evaluated by comparing the grams of dietary nitrogen with the grams of urinary-urea nitrogen. If these quantities are equal net protein is not synthesized and all of the dietary amino acids are metabolized. If the dietary nitrogen is less than the urinary-urea nitrogen, a net degradation of body must be simulated and the moles of each amino acid metabolized are given in the following expression:

$$\text{moles amino acid}_i = \text{moles dietary amino acid}_i + \text{body protein degraded (g)} \times \frac{\text{moles amino acid}_i}{\text{body protein (g)}}$$

Where the dietary nitrogen exceeds the urea nitrogen excreted, the moles of each amino acid required for protein synthesis are given by the following expression:

$$\text{moles amino acid}_i = \text{body protein synthesized (g)} \times \frac{\text{moles amino acid}_i}{\text{body protein (g)}}$$

The stoichiometry for incorporation of each amino acid into protein is then:

$$\text{amino acid} + \text{peptide}_{n-1} + (4 + Z)\ \text{ATP} \longrightarrow \text{peptide}_n$$

where Z is the transport factor for amino acids.

The amount of fat and carbohydrate synthesis or degradation is determined by the carbon balance. The moles of carbon available for incorporation into additional body substance are given below:

moles C to be incorporated = moles C in acetyl-CoA + moles C in glucose + moles CO_2 produced to this point in metabolism – total moles CO_2 to be produced

If the carbon available is less than the CO_2 to be produced, then the acetyl-CoA and glucose are oxidized by simulation and additional body substance must be oxidized. The respiratory exchange for oxidation of body substance is estimated as follows:

moles CO_2 to be produced = total moles CO_2 to be produced – moles CO_2 produced to this point
moles O_2 to be consumed = total moles O_2 to be consumed – moles O_2 consumed to this point

The moles of body fat (assumed to be triolein) and/or moles of glucosyl units of glycogen oxidized to provide the necessary respiratory exchange are calculated by solution of the following equations:

moles CO_2 to be produced = 57 (moles triolein oxidized) + 6 (moles glucosyl units oxidized)
moles O_2 to be consumed = 80 (moles triolein oxidized) + 6 (moles glucosyl units oxidized)

The oxidation of each mole of triolein results in the production of 452 moles of ATP, and the oxidation of each glucosyl unit of glycogen is associated with the formation of 37 moles of ATP.

If the carbon available as acetyl-CoA and glucose exceeds that required for carbon dioxide production, the excess is incorporated into body substance. Triolein is synthesized from acetyl-CoA and minimal glucose according to stoichiometry:

$$44.33\ \text{acetyl-CoA}_{\text{mito}} + 0.5\ \text{glucose} + 11.67\ O_2 + (51 + Z)\ \text{ATP} \longrightarrow 34.67\ CO_2 + \text{triolein}$$

where Z is 43.33 times the transport factor for a two-carbon unit. Any acetyl-CoA not needed for fat synthesis is oxidized:

$$\text{acetyl-CoA} + 2\,O_2 \longrightarrow 2\,CO_2 + 12\,\text{ATP}$$

At this point in the simulated metabolism, glucose is the only remaining source of carbon. A test is conducted, once again, to determine the moles of CO_2 remaining to be produced. The remaining respiratory exchange results from either fat synthesis or glucose oxidation. The amount of glucose utilized in each of these metabolic processes is determined by solution of these equations:

moles CO_2 to be produced = 27 (moles glucose converted to triolein) + 6 (moles glucose oxidized)
moles O_2 to be consumed = 4 (moles glucose converted to triolein) + 6 (moles glucose oxidized)

Glucose is incorporated into body fat according to the stoichiometry:

$$14\,\text{glucose} + 4\,O_2 + (70 + ZZ)\,\text{ATP} \longrightarrow 27\,CO_2 + \text{triolein}$$

where ZZ is 14 times the transport factor for glucose.

The respiratory exchange and urea excretion have been accounted for at this point in the simulated metabolism. There may be dietary glucose remaining or there may be a deficit of glucose. If a surplus of glucose remains, it is converted into glycogen:

$$\text{glucose} + (2 + ZZ)\,\text{ATP} + \text{glucosyl}_{n-1} \longrightarrow \text{glycosyl}_n$$

If there is a deficit of glucose, the deficit is removed by degradation of glycogen to provide the necessary glucose.

The computer program described above appears to simulate energy metabolism accurately over a wide range of feed intakes, and its value lies in the interpretation of respiratory exchange data which represent anabolic as well as catabolic reactions.

Acknowledgement

This work was supported in part by the Grace M. Showalter Residuary Trust.

References

FLORKIN, M. and STOTZ, E.H. (1972). In *Comprehensive Biochemistry*, Vol. 30, pp. 215–249. Amsterdam: Elsevier
FLORKIN, M. and STOTZ, E.H. (1975). In *Comprehensive Biochemistry*, Vol. 31, pp. 159–184. Amsterdam: Elsevier
RUBNER, M. (1902). In *The Laws of Energy Consumption in Nutrition*. Translated by R.J.T. Joy (1968). Washington, DC: US Army Research Institute of Environmental Medicine, Natick, Mass.
SCHULZ, A.R. (1978). *Br. J. Nutr.*, 39, 235

41

SIMULATION MODELS FOR ENERGY METABOLISM IN MONOGASTRIC AND RUMINATING ANIMALS

J.R. REICHL
Institute of Animal Nutrition, University of Hohenheim, 7000 Stuttgart 70, Federal Republic of Germany

Summary

A set of six models may be used for a stepwise system analysis of energy metabolism. These comprise the following. Whole animal: (1) *general outputs* for general relationships between caloric nutrients and the heat production; (2) *Metabolic pathways* for calculation of stoichiometries for metabolism of single components, for example amino acids. Individual tissues: (3) *flux rates* for calculation of fluxes of concerned metabolic pathways; (4) *pool sizes* for a steady-state adjustment of metabolite pools. Subcellular fractions: (5) *redox states* for calculation of redox potentials for metabolite pools; and (6) *membranes* concerning energy transformations within the membrane.

Introduction

Every one of these models contains over 100 variables and approximately the same number of equations, and each of the different levels of model yields final results about energy metabolism which are comparable with specific experimental measurements. The continuation of an execution of models on the next level is optional. It is also possible to carry out the whole analysis on a 16 K desk computer, because of division of these models into smaller subprograms.

The only information necessary for the execution of the whole set of models (which should be submitted to the computer before an execution of the first model) are the intakes of potentially digestible protein, digestible fat and digestible carbohydrates (which could be precalculated from the crude analysis of feed rations) and the quantity of retained protein and fat (submitted either as an experimental value or as desired goals). Protein retention, as well as the intake of digestible nutrients, is a fixed variable during an execution of the whole set of models; the fat retention is a goal which should be reached by repeating runs of simulation. If specific analyses of amino acids, fatty acids and various carbohydrates (cellulose, starch, pentoses and hexoses) are available, these could also be submitted to the computer before execution of the first model.

The simulations on the second and third level are dynamic; nevertheless, it is not necessary to submit information to the computer about the experimental flux rates and metabolite pools. All of these parameters or variables will be calculated during a simulation simply from the intake of nutrients and from the adjusted proportionality coefficients, which in this chapter are called stoichiometries and distribution coefficients. A comparison of metabolite pools, fluxes and redox potentials measured in experiments with results of simulation is useful for checking the necessity of repeating the simulation. For this purpose it is not necessary to have complete analyses of all metabolites and fluxes considered in simulation. This characteristic of the models that are presented here allows absolute distinction of simulation results from experimental results.

Models on the level of the whole animal

MODEL 1 – 'GENERAL OUTPUTS'

In contrast to a multiple regression analysis, used for determination of regression coefficients from a system of measured inputs and outputs in a system of relations between feed and animal, simulation analysis with this model provides a system of measurable general outputs by combining a set of theoretically calculated coefficients. In both cases the proper coefficients must not be affected by associated effects. (For example, a coefficient for transformation of glucose to fat must not include an excess of moles of ATP which are not used for biosynthesis of fat.) When this condition is fulfilled, the coefficients can be arranged in a linear matrix form in a fashion similar to that found in preliminary work (Reichl, 1976). A set of equations resembling the equations previously published (Reichl and Baldwin, 1975) for microbial digestive functions as well as equations for the metabolism of volatile fatty acids in ruminants are also included in this model.

Model inputs: Measured intakes of digestible protein, digestible fat and digestible carbohydrates; measured or desired retentions of protein and fat.

Model outputs: Heat production, ATP balance, O_2 consumption, CO_2, CH_4 and urea production, yield of microbes of the digestive tract and series of efficiencies.

Computing algorithm: Stepwise solution of a system of algebraical equations. Calculation of byproducts by setting limits for turnover of particular equations. Forward to model 2, then optional return to this model; and repeated execution by changing the limits.

MODEL 2 – 'METABOLIC PATHWAYS'

The more general results of the first model could be improved by using this model, which has a possibility for achievement of a balance of single amino acids and fatty acids between feed and animal body. This balance shows how many units of each component can be decomposed or have to be synthesized. If analytical values of these components are not available then the standard compositions of body protein and fat will automatically be used, but for the feed components a use of tabulated values for single feedstuffs is recommended. This model works in a similar fashion to the model of Schulz (1978) with some differences, for example, not only the ATP balance but also the heat balance are simultaneously calculated, and not only the final but also the intermediate products, for example, pyruvate, acetyl-CoA, acetacetate, α-ketoglutarate, NADH and NH_3 are calculated and used later as stoichiometries for protein and fat decomposition or for biosynthesis in the model 3. Equations for the calculation of composition of microbes in a manner used previously by Reichl and Baldwin (1976) are also included in this model.

Model inputs: In addition to general inputs used by model 1, analysis of single components of protein and fat in the feed and the animal body are also evaluated in this model.

Model outputs: In addition to final products as in model 1, a balance of intermediary products is also stored and printed by the computer.

Computing algorithm: Solution of a system of algebraical equations. Calculation of intermediary products by setting the balance of intakes and depositions to zero. Optional return to model 1, stop, or forward to model 3, where calculated stoichiometries will be used.

Models on the level of individual tissues

MODEL 3 – 'FLUX RATES'

An execution of this model uses the same inputs of digestible nutrients as in the first two models, but in amounts corresponding to one meal. In model 2 calculated stoichiometries for protein and fat decomposition and synthesis are necessary prerequisities of this model. The goal of this model is a calculation of single fluxes for individual reactions including decomposition of feed compounds, growth of rumen and gut microbes and transfer of nutrients to the blood, their distribution to the individual tissues (liver, muscle, adipose tissue and other organs) and finally their metabolism in the tissues. A recycling of some compounds is also considered in this model, but only net forward fluxes by reversible reactions are calculated. The main property of this model is control of the distribution of metabolites to the individual tissues and to the individual metabolic pathways. First execution of this model usually uses the same distribution coefficients of metabolites to all tissues or pathways. These coefficients will be changed manually by repeated executions of this model in seeking a particular goal, which is usually a requested deposition of fat. Information about actual fluxes determined in experiment is not needed, but could be considered by choosing new distribution coefficients. The main criterion for an appropriate result of this model is an output of the next model. Formally this model is essentially an extended version of a model published previously (Reichl, Karras and Jansen, 1976).

Model inputs: In addition to general inputs used by model 1 and the stoichiometries calculated in model 2, a set of changeable distribution coefficients.

Model outputs: Final products of energy metabolism as in model 1 but reflecting metabolism of individual tissues. Calculated individual fluxes are also stored and printed by the computer.

Computing algorithm: Stepwise solution of a system of algebraical equations. Calculation of individual fluxes by setting intermediary products stepwise to zero. Forward to model 4 and then optional return to this model, and repeated execution by changing the distributions.

MODEL 4 – 'POOL SIZES'

A set of equations used in model 3 is employed in this model. However, their execution is now in the form of first order differential equations. This model is fully dynamic. As in model 3, this model uses the same inputs of digestible nutrients as model 1, but in amounts that correspond to one meal. A simulation time corresponds to a time between two feedings. Simulation of this model is carried out with uniform rate constants but adjusted to the time of simulation. Execution of this model starts with uniform pools of all metabolites, but these will vary during an execution of the model until in repeated executions they reach a steady state. Information about the actual experimental pools is not needed as an input to the model, but it is necessary for comparison with the adjusted steady-state pools. If a series of similar problems has to be solved with this model, the pools adjusted for a preceding problem could be used as starting pools in the first execution of the next problem. The only tools able to change the pools are the distribution coefficients which are adjustable in model 3. To improve the result of model 4, it is necessary to repeat simulation with model 3. Also, in model 4 only forward net fluxes and pools are used without distinction of mitochondria and cytosol. This model is also an essentially extended version of a model published previously (Reichl, Karras and Jansen, 1976).

Model inputs: In addition to inputs as for model 3, adjusted rate constants.

Model outputs: A time table of individual pools of metabolites.

Computing algorithm: Integration of a system of differential equations. Repeated executions until steady state of pools is reached. Optional return to model 3, stop, or forward to model 5, where calculated pools will be used.

Models on the level of subcellular fractions

MODEL 5 – 'REDOX STATES'

The same inputs of digestible nutrients as in model 4 are used in this model. This model is also fully dynamic and employs a system of first order differential equations. Flux rates and pool sizes calculated in the last two models are used in this model, but, with the exception of one tissue which is to be examined in detail, the rest of the tissues are represented in this model with only minimum equations. Completely reversible reactions and differentiation to cytosol and mitochondria are used in the elected tissue (it could be chosen from any of the three tissues mentioned before). Using the concept of Krebs and Veech (1969), from ratios between some adjusted steady pools and with a corresponding equilibrium constant, redox states of the nicotinamide nucleotides are calculated. The next step in the execution of this model is the calculation of redox potentials and a summarizing of free energy. Simulated redox potentials could then be compared with those experimentally measured in isolated mitochondria, but as well as by simulation on the preceding level, experimental data are also not used as inputs in this model.

Model inputs: Fluxes calculated in model 3, pools calculated in model 4 and a set of equilibrium constants.

Model outputs: Time table of individual pools and potentials.

Computing algorithm: Integration of a system of differential equations. Repeated executions until steady state of pools is reached. Forward to model 6 and then optional return to this model, and repeated execution by changing some fluxes.

MODEL 6 – 'MEMBRANES

A coupling of redox systems to the synthesis of ATP will be studied in this model as well as partitioning of an excess of ATP in other than biosynthetic processes. In this model the concept of Wilson, Erecinska and Dutton (1974) is used. This model is at present in the course of development.

Under consideration are two other models, which are based on the level of environmental conditions, that is heat loss and adaption.

Acknowledgement

This paper was supported by Deutsche Forschungsgmeinschaft, SFB 142–B.4.

References

KREBS, H.A. and VEECH, R.L. (1969). *Adv. Enzyme Regul.,* **7**, 397–413

REICHL, J.R. and BALDWIN, R.L. (1975). *J. Dairy Sci.,* **58**, 879–890

REICHL, J.R. and BALDWIN, R.L. (1976). *J. Dairy Sci.,* **59**, 439–454

REICHL, J.R. (1976). In *Energy Metabolism of Farm Animals*, pp. 161–164. Ed. M. Vermorel. EAAP Publication No. 19. Clermont-Ferrand: G. de Bussac

REICHL, J.R., KARRAS, K. and JANSEN, R. (1976). In *Energy Metabolism of Farm Animals*, pp 153–156. Ed. M. Vermorel. EAAP Publication No. 19. Clermont-Ferrand: G. de Bussac

SCHULZ, A.R. (1978). *Br. J. Nutr.*, **39**, 235–254

WILSON, D.F., ERECINSKA, M. and DUTTON, P.L. (1974). *Ann. Rev. Biophys. Bioeng.*, **3**, 203–230

SIMULATION OF NUTRITIONAL–ENVIRONMENTAL INTERACTIONS IN GROWING SWINE*

D.P. STOMBAUGH and A. OKO
The Ohio State University and Ohio Agricultural Research and Development Center, Columbus, Ohio, USA

Summary

A mathematical model was designed to evaluate production alternatives and nutritional-environmental interactions in growing pigs. The model describes feed intake, absorption from the intestine, glucose metabolism and the anabolism and catabolism of fat and protein. The model solves 25 simultaneous differential equations in parallel every 0.125 minutes to provide an interconnected and dynamic description of growth.

Introduction

A dynamic model that can accurately predict carcass quality, rate of gain and feed efficiency in response to dynamic changes in nutritional, environmental and biological inputs would be a useful engineering and economic tool. The objective of this project was to develop a comprehensive and dynamic model that would 'interact' with its environment as a real animal does. It was desired to describe the physiological and metabolic processes and their control as realistically and completely as possible.

The modelling technique converted the differential equations describing each of the controlled physiological and metabolic processes to a dynamic solution which could be integrated and solved in parallel, using an IBM developed language (Continuous System Modelling Program III, CSMP III). The CSMP language permits the digital computer to operate in much the same fashion as an analog computer to solve simultaneous differential equations. The program uses 25 integrators to solve 25 simultaneous differential equations once every 0.125 minutes.

The simulation model

The simulation model contains several major interconnected components which are listed below:

1. Feed formulation and composition.
2. The control of feed intake.

*This paper was approved as Journal Article No. 131–79 by the Ohio Research and Development Center, Wooster.

3. The adsorption of proteins, fats and glucose from the intestine.
4. The basic metabolic pathways for glucose, glycogen, amino acids and triglycerides.
5. The utilization and production of chemical energy (adenosine triphosphate).
6. The thermoregulatory system. The user must indicate the ration, how it is to be fed, the initial pig weight, ambient temperature, wind speed, humidity, the number of days to be simulated and the output variables desired.

FEED INTAKE

The feed intake component of the program controls both the rate and frequency of feed intake. When feed is freely available the frequency of feed intake is regulated by volume of feed in the stomach, the blood glucose level, a combined thermoregulatory signal from the body core and the skin surface and a signal which prevents eating for six hours each night. In addition to these signals, which act almost immediately, a long-term adjustment occurs based on the levels of adipose and protein tissue in the body.

ABSORPTION

Foodstuff flows from the stomach to the small intestine at a rate proportional to the square root of the volume of mixture in the stomach. Glucose, fat and protein are removed from the intestine at rates proportional to their concentrations in the intestine. The percentage of individual components entering the bloodstream is based on the digestibilities of each component.

GLUCOSE METABOLISM

The blood glucose level is determined by continuously integrating the flow rates of glucose both into and from the blood. The flow of glucose entering the blood comes from the following:

1. Adsorption from the intestine.
2. Formation of glucose from glycogen through glycogenolysis.
3. Formation of glucose from glycerol and lactate through gluconeogenesis.
4. The breakdown of certain groups of 'surplus' amino acids through gluconeogenesis.
5. When the glucose level is very low (for example, during starvation) the breakdown of protein tissues supplies amino acids for glucose synthesis.

The flow rates of glucose from the blood include:

1. Glucose consumed by body cells as fuel when the glucose level is high.
2. Glucose consumption by organs such as the brain which can use only glucose.
3. The use of glucose to form glycogen whenever the glycogen level falls below its setpoint.
4. The formation of triglycerides.
5. Glucose utilization for any required synthesis of non-essential amino acids.

In this model hormone levels in the blood are not included and consequently, functions based on the blood glucose level are used as regulators for many metabolic reactions.

FAT SYNTHESIS AND BREAKDOWN

Fat deposition is determined by continuously integrating flow rates of fat both to and from fat deposits. The rate of catabolism is controlled by the blood glucose level, the ATP level and animal weight. The flow rate of fat to storage comes from the intestine and by synthesis from glucose and other substances.

PROTEIN SYNTHESIS AND BREAKDOWN

The rate of protein synthesis is modelled as the sum of two components: a slowly changing component and a second more dynamic component. The slowly changing component is influenced by the actual protein mass in the body and the protein mass at maturity (Whittemore and Fawcett, 1976). The more dynamic synthesis rate is influenced by the insulin level (represented by a function of the blood glucose level). This rate is limited by the availability of the individual amino acids needed to form protein molecules. The model provides quantitative dynamic information for each amino acid and optimizes the utilization of the limiting amino acids.

Protein catabolism is a function of the actual protein tissue mass and is influenced by the blood glucose level.

ADENOSINE TRIPHOSPHATE AND HEAT PRODUCTION

Within each metabolic pathway the rates of production and utilization of chemical energy were calculated in terms of adenosine triphosphate (ATP) and acetyl-CoA. The need for ATP controls the breakdown of acetyl-CoA and fat catabolism. Thermal energy production in each process is continuously monitored and used in the thermoregulatory model.

TEMPERATURE REGULATION

In the thermoregulatory portion of the model, the animal's body was modelled as two cylinders (one represented the trunk and one represented the extremities). Each cylinder contained three layers (core, fat and skin). Heat losses from the animal's surface and from the respiratory tract were determined by complex heat-transfer equations. Thermoregulatory responses were initiated and controlled by error signals generated from the difference in the actual temperature of the trunk core and the skin surface from their set points.

Simulation results and discussion

Typical results from simulation runs are shown in *Figures 42.1* and *42.2* and in *Table 42.1.* For all simulation results presented, the conductive heat transfer section of the thermoregulatory model was not used, wind speed was maintained constant at 1.0 m/s and relative humidity was fixed at 55%.

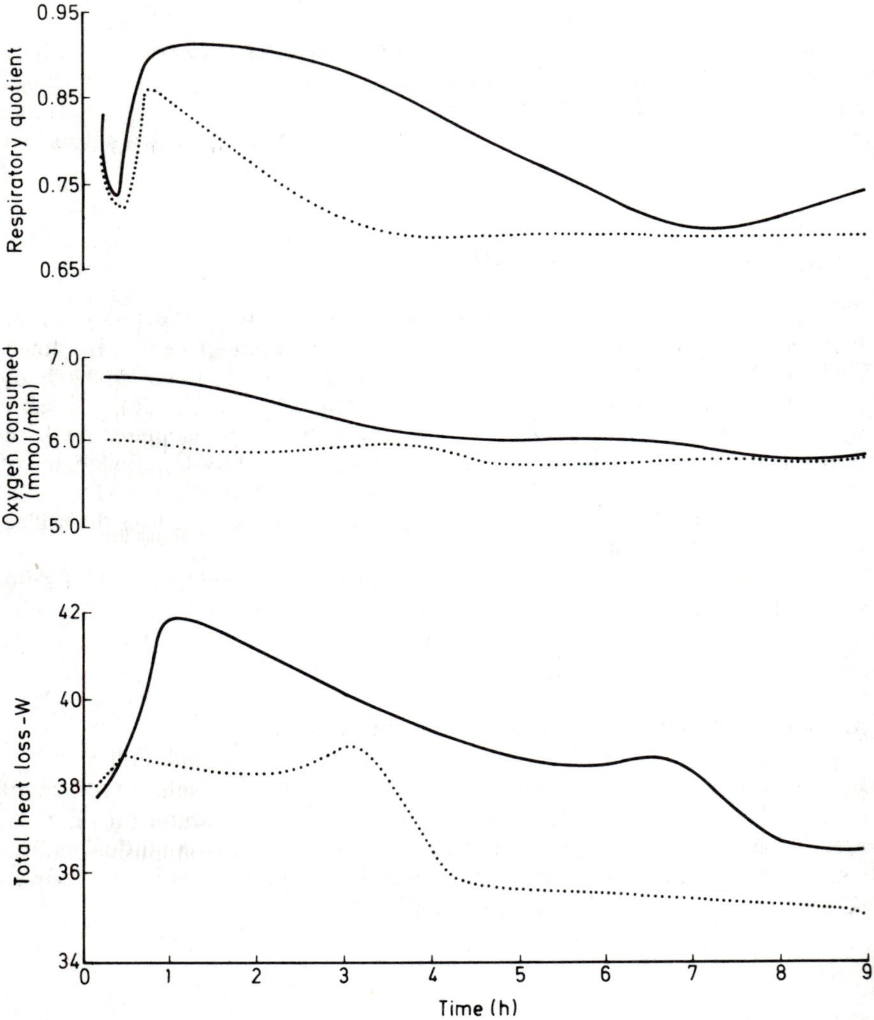

Figure 42.1 Simulation results for a 20 kg pig showing respiratory quotients, rates of oxygen consumption (oxygen consumed; mmol/min) and total heat loss rates (total heat loss; W) as functions of time (h). Results show the effects of ingesting either one 50 g (dotted lines) or one 200 g (solid lines) meal between the 20 and 25 minute readings on the time axis.

To illustrate the dynamic nature of the model, *Figures 42.1* and *42.2* show variations in selected nutritional parameters following a brief period of feed intake. Changes in the respiratory quotient (RQ), oxygen consumption and heat loss, similar to those measured in previous calorimeter trials (Stombaugh and

Grifo, 1977), are shown in *Figure 42.1.* These dynamic changes resulted from complex adjustments in the synthesis and breakdown rates of chemical substances within the body. A few of these adjustments are illustrated in *Figure 42.2.*

Data presented in *Table 42.1* illustrate the use of the model to evaluate feeding and management alternatives. Computer costs required to operate this model

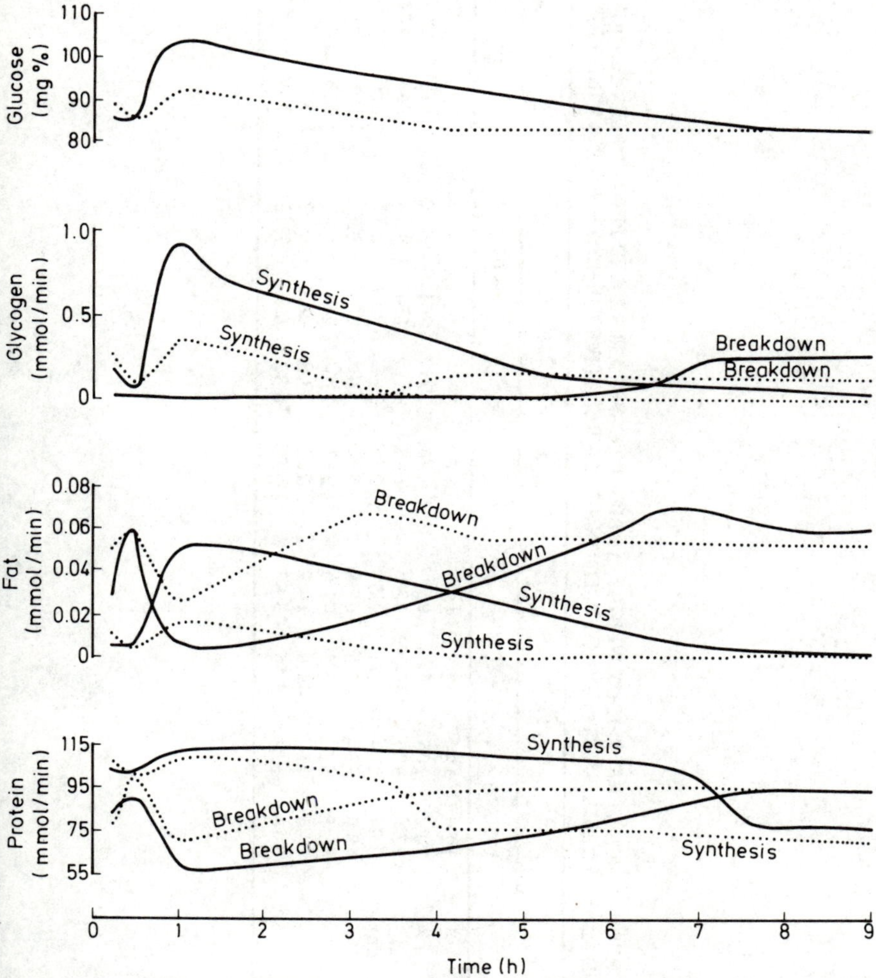

Figure 42.2 Simulation for a 20 kg pig showing blood glucose levels (glucose; mg %), synthesis and breakdown rates of glycogen (glycogen; mmol/min), fat (fat; mmol/min) and protein (protein; mmol/min) as functions of time (h). Results show the effects of ingesting either one 50 g (dotted lines) or one 200 g (solid lines) meal between the 20 and 25 minute readings on the time axis.

are quite large; consequently, each of these 12 simulations covered only a one-day time period. The model simulated three body weights (20, 50 and 80 kg) at two ambient temperatures (15 and 30°C) and with rations containing 15.9 MJ/kg of gross feed energy. These corn and soybean-meal rations were formulated to obtain either 12% or 18% crude protein.

Table 42.1 PREDICTED PIG PERFORMANCE VERSUS BODY WEIGHT, AMBIENT TEMPERATURE AND PROTEIN LEVEL. PIGS WERE FED A 15.9 MJ/kg RATION *AD LIB.* AND MAINTAINED IN A STANDING POSITION AT 55% RELATIVE HUMIDITY WITH A WIND VELOCITY OF 1 m/s

Pig weight (kg)	*20*				*50*				*80*			
Ambient temperature (°C)	*15*		*30*		*15*		*30*		*15*		*30*	
Protein (%)	12	18	12	18	12	18	12	18	12	18	12	18
Feed intake (g/d)	1069	1063	957	959	2041	2057	2011	2012	3178	2916	1895	1895
Rate of gain (g/d)	377	520	354	477	685	864	691	850	895	824	350	322
Feed conversion ratio (kg feed/kg gain)	2.84	2.04	2.71	2.01	2.98	2.38	2.91	2.37	3.55	3.54	5.42	5.88
Net protein deposition (g/d)	48.0	78.4	41.8	68.8	89.0	130	89.0	126	126	130	64.0	65.0
Net fat deposition (g/d)	86.8	62.2	95.3	71.0	157	116	165	121	168	95.8	−5.3	−33.0
Nitrogen excreted (g/d)	13.5	15.0	12.9	14.7	26.5	29.9	26.4	30.0	39.4	52.1	41.7	52.9
Total heat loss (MJ/d)	6.16	6.19	3.55	3.64	8.28	8.33	6.94	7.10	10.6	10.8	9.63	9.77

The major discrepancy between these simulation results and those in the literature resulted from the failure of the low protein ration to reduce feed consumption (Sharda, Mahan and Wilson, 1976). If each simulation run had extended over a longer time period, this reduction would have occurred through the actions of long-term regulators contained in this model.

Effects of a 30°C ambient temperature on feed intake and feed efficiency were most pronounced on the 80 kg pig (heat stress). The 15°C temperature (slight cold stress) increased the feed consumption slightly (through changes in the metabolic processing of the nutrients) for the 20 kg pig.

References

SHARDA, D.P., MAHAN, D.C. and WILSON, R.F. (1976). *J. Anim. Sci.*, **42**, 1175–1181

STOMBAUGH, D.P. and GRIFO, Jr., A.P. (1977). *Trans. Am. Soc. agric. Engrs*, **20**, 954–960

WHITTEMORE, C.T. and FAWCETT, R.H. (1976). *Anim. Prod.*, **22**, 87–96

43

SUFFICIENT STATISTICS FOR FEED EVALUATION IN ADDITION TO THE DESCRIPTION AND CONTROL OF GROWTH

C.Z. ROUX and H.H. MEISSNER
Animal and Dairy Science Research Institute, Irene, Republic of South Africa

Summary

Repeated measurements on the logarithms of cumulative feed intake, cumulative heat production, body protein and fat of a growing animal on *ad lib.* feed intake result in a covariance matrix which can be approximated by a matrix of rank one. The resulting theory in terms of sufficient statistics is described with reference to miscellaneous supporting observations. Such observations include the tight fit to data of a log-linear description and the predictable modification of the growth pattern and body composition by restricted feed intake. The relationship of the derived statistics to energy concentration is explicable in terms of control of feed intake by energy demand or physical capacity; therefore, it is concluded that they can be important in feed evaluation.

Introduction

The need for the analysis of multiple successive measurements on the same individual originates from the operation of the control mechanisms of growth and feed intake on an individual level. The established body of statistical knowledge for analysing successive measurements is commonly called time series analysis. To fix ideas in a growth and nutrition context, let:

$$\begin{aligned} x_1(t) &= \ln \text{(cumulative feed intake, say in MJ metabolizable energy; ME)} \\ x_2(t) &= \ln \text{(cumulative heat production in MJ)} \\ x_3(t) &= \ln \text{(amount of body protein in MJ)} \\ x_4(t) &= \ln \text{(amount of body fat in MJ)}, \end{aligned} \qquad (1)$$

all measured at time t. It is assumed that some estimate of pre-experimental feed intake dating back to conception is included in $x_1(t)$. (Often this need not be very accurate since the rapid increase in cumulative feed intake soon causes the relative error to be small.)

Ignore, for the moment, the possible effect of measurement error and assume that the covariance matrix with elements,

$$\Sigma_t (x_i(t) - \overline{x}_i)(x_j(t) - \overline{x}_j), \text{ for } i, j = 1,2,3,4,$$

is singular with rank one. It then follows from Roux (1979) that the change of any $x_i(t)$, over time, is described by the following equation:

$$x_i(t) - \alpha_i = \rho(x_i(t-1) - \alpha_i) + \epsilon_i(t) \tag{2}$$

where ρ and α_i are constants and $\epsilon_i(t)$ is a random disturbance term. It also follows from the unity rank of the covariance matrix that:

$$x_i(t) = a_i + b_i\, x_1\,(t) \tag{3}$$

for i = 2,3,4, where the designation of the variables follows from equation (1).

The solution of equation (2) is given by:

$$x_i(t) = \alpha_i - (\alpha_i - x_i\,(0)\,\rho^t + \sum_{j=0}^{t-1} \rho^j\, \epsilon_i(t-j) \tag{4}$$

It follows from Roux (1976, 1979) that:

$$\gamma = -\ln \rho = c(1 + b_2^2 + b_3^2 + b_4^2) \tag{5}$$

where the b_i are defined in equation (3).

A consequence of equations (4) and (5) is that growth rate in protein and fat, and by implication body mass, is determined by the way feed energy is partitioned into the synthesis of protein and fat and into heat production. Write $y_i(t)$ = exp $x_i(t)$, with the subscripts retaining the meaning given to them in equation (1). The rank one assumption on the covariance matrix then also gives:

$$dy_1 = y_1 \sum_{i=2}^{4} b_i\, dy_i/y_i \sum_{i=2}^{4} b_i^2 \tag{6}$$

where the b_i are defined in equation (3). Equation (6) is fundamental to the understanding of the control mechanisms associated with equations (2), (3) and (4), since it portrays the way in which an animal obeys the law of conservation of energy. This portrayal is substantiated from equation (3) because when dy_i/y_i = $b_i\, dy_i/y_i$ is substituted an identity results in equation (6).

The operation of the control mechanisms

Assume now that after some perturbation, perhaps due to a change in diet, an animal's growth again settles down to a regular pattern. The perturbation may be due to, say, different energy concentrations in the diets, and therefore there may, initially, be a lack of energy for protein or fat synthesis. The return to a regular pattern may be due to a change in the relative values of the y_i (i = 1,2,3,4) in equation (6) in such a way that movement along the lines of equation (3) can be resumed without a change in the b_i. The change in the relative values of the y_i will then be reflected in changes in the intercepts a_i, as can be described in terms of initial values ($x_i(0)$) in the solution of equation (4) to the difference equations, where t = 0 is, say, the starting point of the regular pattern. A change in intercept was obtained, for example, for the log-linear relationship between body protein and cumulative ME, by the inclusion of increased amounts of sawdust in pig diets (*see* Roux and Kemm, 1979).

In some situations the new diet might enforce a change in the b_i before equation (6) can be obeyed. In equation (5) this will result in a change from γ to $\gamma' = c'(1 + (b_2')^2 + (b_3')^2 + (b_4')^2)$. An example is again from the experiment reported on by Roux and Kemm (1979) where some increases in percentages of sawdust resulted in changes from ρ to ρ', which are due to changes in c and b_4, the slope for ln (body fat).

There is also a control mechanism associated with each $x_i(t)$. This is evident from the term involving $\epsilon_i(t)$ in equation (3), from which it is clear that the effect of a perturbation at t = 0 will be eliminated for large enough t if $|\rho| < 1$. From theory (*see* Roux, 1976, 1979) it is clear, however, that this mechanism only involves movement along the lines of equation (3). Movement away from this equation is controlled by changes in the intercepts, if the growth structure defined by the slopes remains the same, as follows from the discussion on equation (6).

Experimental evidence

THE RANK OF THE COVARIANCE MATRIX

Almost all essential properties of the present theory follow from the assumption of a covariance matrix with rank one. Meissner (1977) reports an experiment with three lambs permanently kept in confinement from near birth, with measurement for present purposes from 25 kg (13 weeks) to nearly 50 kg (30 weeks), with their heat production measured in an open system calorimetric unit and their body composition obtained by the tritium dilution technique. For each lamb the dominating root in a principal component analysis had a value of over 99%. In a slaughter experiment involving 63 pigs on six treatments (*see* Roux and Kemm, 1979) with data from 11 weeks (20–27 kg) to 31 weeks (83–121 kg), and with both body protein and fat levels obtained by chemical analysis and heat production by subtraction, values of the same magnitude were obtained for the dominating root for each treatment, despite the fact that here individual variation is added to measurement error. In the light of accepted knowledge about the roots of matrices with small perturbations added to them, it seems fair to conclude that the 1% deviation from a single non-zero root is due to measurement error or individual variation, as the case may be, instead of being indicative of a lack of fit.

The adequacy of equation (2) can also be tested by testing h_2 for significance in the model:

$$x_1(t) = g + h_1 x_1(t-1) + h_2 x_1(t-2) + \epsilon(t)$$

where $x_1(t)$ is ln (cumulative feed intake) as ascertained at time t.

It is general experience with many sets of data that h_2 is seldom significant (*see* Roux, 1979). Hence the conclusion from this line of evidence is also that the model equation (2) fits the data well.

RESTRICTED FEEDING

In an experiment on the feeding of baconers Siebrits (1979) found that pigs fed with a $\rho = 0.94$ grew in body mass with a $\rho = 0.94$ on a low protein (14%) diet and a $\rho = 0.92$ on a high protein (16%) diet, in both cases with a standard error of 0.003. The deviant value of ρ was accompanied by slight curvilinearity in the system, as one might expect. The value of 0.92 for ρ is in the vicinity of the value of 0.91 found by Roux and Kemm (1979) for a diet with 19% protein. The ρ used for feed intake was, actually, more extreme than a $\rho = 0.93$ obtained for pigs on *ad lib.* feed intake with their feed energy concentration diluted by 32% sawdust. Hence the high protein diet allowed some reversion in the direction of faster normal growth at lighter body masses.

There seems to be general agreement that body composition can only easily be modified through the fat component. Hence it may be possible to choose a desired b_4 for fat deposition in equation (5) and to feed animals accordingly to a desired fat percentage at a given body mass. For values of γ near to *ad lib.* feed intake, this may be done on the assumption of the constancy of c, b_2 and b_3, as is shown by an experiment reported on by Meissner (1977). Here $\gamma'/\gamma = 0.88$ induced a ratio $b_4'/b_4 = 0.86$, as the slopes b_2 and b_3 for heat production and protein did not differ significantly and $c = c'$, approximately. Meissner (1977) also calculated feed intake and rates of gain in body mass for different (assumed) fat slopes and obtained values which are, for some comparable situations, in good agreement with tables of the Agricultural Research Council (1965).

Feed evaluation

An advantage in experimental design and analysis is that the a_i, b_i (i = 2,3,4 and 5, say, for body mass), $x_1(0)$, α_1 and ρ for feed intake are sufficient statistics for normal distributions under the usual regularity assumptions. Working with sufficient statistics allows analysis of estimates of parameters for individual animals instead of the actual observations. Hence rations can be quantified in relation to *ad lib.* feed intake in terms of percentages of $\gamma = -\ln\rho$, and the behaviour of the slopes and intercepts can be described with reference to this percentage. This approach is currently under investigation at the Animal and Dairy Science Research Institute and may improve on the procedure of assuming approximate constancy of some of the slopes, intercepts, $x_1(0)$ and α_1.

Evidence is available from Kemm's experiment on pigs (*see* Roux and Kemm, 1979) on the behaviour of $[a_i, b_i; x_1(0), \alpha_1, \rho]$ on diets with different energy concentrations, which are explicable in terms of concepts of control of feed intake by level of energy demand or by physical capacity. For instance, ρ remained constant from 0 to 8% sawdust ($\rho = 0.91$), it increased linearly between 8 and 24% and remained constant again from 24 to 32% ($\rho = 0.93$). This behaviour of ρ is consistent with estimates of Meissner (1977) on ruminants on concentrates and roughages. In Kemm's experiment all $[a_i, b_i]$ in equation (3), in addition to body mass, remained approximately constant for both 0 and 8% sawdust which seems to indicate this as the domain of control by energy demand. The delineation of the domain of control by physical capacity in terms of ρ awaits further experimentation, since it is unclear if in the sawdust situation it should include the region from 8 to 24%.

References

AGRICULTURAL RESEARCH COUNCIL (1965). *The Nutrient Requirements of Farm Livestock, No. 2: Ruminants.* London: ARC

MEISSNER, H.H. (1977). *An Evaluation of the Roux Mathematical Model for the Functional Description of Growth.* PhD Thesis, University of Port Elizabeth

ROUX, C.Z. (1976). *Agroanimalia,* **8**, 83–94

ROUX, C.Z. (1979). *Animal Growth as a Single Dimensional First Order Autoregressive Time Series.* Unpublished

ROUX, C.Z. and KEMM, E.H. (1979). *The Influence of Dietary Energy Concentration on the Growth Pattern, Body Composition and Feed Utilization of the Growing Pig, 1: A Suitable Mathematical Model and Data Analysis.* Submitted to *Animal Production*

SIEBRITS, F.K. (1979). *Die kwantifisering van die effekte van nat voeding op die spekvark.* MSc thesis submitted to the University of Pretoria

44

QUANTIFICATION OF ANIMAL AND DIETARY DIFFERENCES WITH THE AID OF THE ROUX MATHEMATICAL MODEL

H.H MEISSNER, C.Z. ROUX and H.S. HOFMEYR
Animal and Dairy Science Research Institute, Irene, Republic of South Africa

Summary

Quantification is mainly achieved by repetitive measurement on the same animal and accurate description of the data by the proposed model whereby measurement error becomes negligible. The growth theory involved predicts growth performance to be a function of the parameters in the growth equation which would change in a systematic fashion depending on level of energy intake. The model has been successfully applied to studies of biological variability in animals, to effects of different feeding levels and compensatory growth and to estimates of voluntary intake.

Introduction

In evaluation programmes of the productive merit of farm animals, a mathematical model which reliably approximates growth curves could be a valuable asset. The same applies to a model which could effectively differentiate between the intakes of diets with small differences in quality.

The essence of the Roux mathematical model (Roux, 1976; Roux and Meissner, *see* pages 217–221) is based on repetitive measurements on the same animal, and combines the log form of the Gompertz equation ($x_i(t) = \alpha_i - \beta_i \exp(-\gamma t)$) to approximate growth with time and the log form of the allometric equation ($y_i = a_i + b_i x_i$) to approximate growth of one component relative to another, for description of the interrelationships of various growth variables. The merit of the model is discussed using growth data of several experiments.

Accuracy of fit

Growth experiments were conducted in which voluntary feed intake, mass changes and composition with the aid of the tritium-dilution technique were continuously measured from about birth to maturity in every single animal. Two breeds of lambs and three breeds of bull calves, which had access to either a concentrate or a roughage diet representing extremes in energy concentration, were used. Some of the results following weaning covering a mass range of 15–60 kg for lambs and 110–600 kg for calves are presented.

The Roux model was fitted to sets of growth data for every lamb or calf and the data substituted for the calculated parameters of the growth equations. It is a valid procedure to describe the growth of a particular animal by its statistical parameters (for example slope and intercept) because under regularity assumptions all information from the data is contained in them (Roux, personal communication).

As shown in *Table 44.1* the Roux model approximates most sets of growth data satisfactorily, the exception being fat on the roughage diet where due to inadequate energy supply continual positive growth in this component was not always possible.

Table 44.1 ACCURACY OF FIT OF THE ROUX MODEL TO SETS OF GROWTH DATA

Growth of one component relative to another in terms of r^2 (mean ± s.d.)

Cumulative feed intake	*Mass*	*Protein*	*Fat*
Concentrate diet: lambs (n = 51)	0.991 ±0.007	0.964 ±0.036	0.971 ±0.025
Roughage diet: lambs (n = 26)	0.992 ±0.006	0.965 ±0.025	0.894 ±0.055

Growth with time in terms of mean error of estimate of predicted vs observed growth (%; mean ± s.d.)

	Cumulative feed intake	*Mass*	*Protein*	*Fat*
Concentrate diet: lambs (n = 51)	1.42 ±0.68	1.61 ±0.92	3.01 ±1.21	4.53 ±1.83
Roughage diet: lambs (n = 26)	1.35 ±0.55	1.65 ±1.12	2.89 ±1.30	18.9 ±6.54

Furthermore, 99% confidence limits suggest very little deviation in conceptual replications of the experiment on the same animals. It could be deduced that extrapolation, within the obvious limits imposed by the model, could be done with confidence since the limit values of the confidence interval calculated for mean values were only marginally less than the limit values calculated for values at the extremes.

Quantification of small differences

INDIVIDUAL DIFFERENCES

Intrabreed differences in factors relevant to growth and efficiency are often exceptionally small, for example, the coefficient of variation in cumulative feed intake and the amount of protein at 40 kg mass in the above-mentioned lamb experiment were only 7.9% and 2.7% respectively. Nevertheless, by means of covariance analyses on the slopes and intercepts, where every individual represents a cell, these differences could be quantified as shown in *Table 44.2.*

Table 44.2 F-VALUES FOR DIFFERENCES IN SLOPES AND INTERCEPTS BETWEEN INDIVIDUALS, BREEDS AND FEEDING LEVELS

Ln (cumulative feed intake) = X

Component Y	*Differences between*	*Slope*	*Intercept*
	Individuals		
ln (mass)	Mutton Merino males (n = 11)	4.54**	14.23**
	Karakul males (n = 15)	4.28**	1.76
ln (protein)	Mutton Merino males	1.58	2.11*
	Karakul males	1.38	6.02**
	Breeds (Afrikaner, Hereford and Simmental)	– §	5.76**
ln (mass)	Concentrate vs roughage diet	–	29.6***
	Breed – diet interaction	–	0.14
	Feeding levels (1–7)		
ln (mass)	Before 30 kg mass (n = 28)	1.50	14.04***
	After 30 kg mass (n = 14)	7.98**	8.39**

* $P < 0.05$; ** $P < 0.01$; *** $P < 0.001$
§ See text for explanation (*see* Breed differences)

BREED DIFFERENCES

Differences between the Afrikaner, Hereford and Simmental bull calves on both the concentrate and roughage diets for example, were effectively demonstrated by a one-way analysis of variance procedure on all individual equations as shown in *Table 44.2.* An interesting feature in this experiment was the uniformity of slopes within diet regardless of breed. Therefore, a common slope per diet could be calculated and the intercepts adjusted which implies that the differences between breeds were approximately proportional throughout the growth interval studied. Also, there was no indication of breed X diet interaction which means that the relative differences in efficiency were the same on both diets.

FEEDING LEVEL DIFFERENCES

In another experiment, groups of six lambs were fed one of seven feeding levels within the mass interval of 22–30 kg. The feeding levels varied from *ad lib.* for level 1 to about 50% *ad lib.* for level 7. Four lambs per feeding level were slaughtered on reaching a weight of 30 kg, while the remaining were put on level 1 for the interval 32–45 kg and then slaughtered at the weight of 45 kg. Despite the very limited numbers, differences between feeding levels could be effectively established as shown in *Table 44.2.* Prior to changeover the slopes did not differ significantly, so a common slope was calculated and the intercepts adjusted accordingly. The exponents of the adjusted intercepts are given in *Table 44.3* as proportions of one another, with feeding level 1 taken as 100.

The slopes did differ significantly following the change to level 1 (*Table 44.2*), and in fact they showed a highly significant positive correlation with feeding levels 5–7, with levels 1–4 being essentially the same. The same

Table 44.3 PROPORTIONAL PRODUCTION OF MASS PER UNIT CUMULATIVE FEED INTAKE

Feeding level	*1*	*2*	*3*	*4*	*5*	*6*	*7*
Before 30 kg mass	100	99.8	96.9	96.7	96.4	86.9	81.5
After 30 kg mass	100	94.5	101	102	108	123	117
Total	100	96.1	99.8	100	104	108	103

determination was done with levels 1–4 prior to the changeover, while the progressive change in slope from level 5 to level 7 was recognized in the calculation of the proportional production of mass per unit cumulative feed intake as shown in *Table 44.3.*

Apart from the interesting aspect of very similar total efficiency, the main object of *Table 44.3* is to show that the model enables systematic categorization of the effects of different feeding levels and possibly different dietary energy concentrations (preliminary results), which would presumably allow reliable extrapolation beyond the limits of experimentation.

Prediction according to growth theory

The parameter γ in the log form of the Gompertz equation plays a central role in the growth theory (Roux, 1976). It can be partitioned into components which represent the proportional allocation of consumed energy to heat production (2), protein (3) and fat (4) according to:

$$\gamma = c\,(1 + b_2^2 + b_3^2 + b_4^2)$$

where b_i is the proportional allocation to 2, 3 and 4 and c is constant.

Additional postulates are that $b_2' \cong b_2$, $b_3' \cong b_3$, but $b_4' < b_4$ for any given feeding level relative to *ad lib.* (the b_i' notation represents the conceptual feeding level). Furthermore, the reduction in b_4 with decreasing feeding level would be systematic up to a point where the system would presumably break.

In a growth experiment with lambs where (2), (3) and (4) were continually measured in addition to mass and feed intake, in which one group received feed *ad lib.* and the other 80% *ad lib.*, the above hypothesis was substantiated. Thus, the Roux model would probably be effective in predicting animal performance from the basis of *ad lib.* intake instead of maintenance, which is the conventional point of departure but it is at the other end of the scale of interest and therefore liable to ineffective prediction of levels of feeding near *ad lib.*

References

ROUX, C.Z. (1976). *Agroanimalia,* **8**, 83–94.

45

ENERGETICS OF GROWTH IN CATTLE

H.F. TYRRELL and P.W. MOE

Ruminant Nutrition Laboratory, Animal Science Institute, Agricultural Research Science and Education Administration, US Department of Agriculture, Beltsville, Maryland 20705, USA

Summary

Fifty-nine energy-balance measurements were made on eight Hereford heifers during the fattening period in order to estimate the maintenance requirement at body weights between 220 and 430 kg. Analysis of covariance indicated that partial efficiency of metabolizable energy (ME) use for tissue deposition was not affected by diet, but maintenance requirement on the same diet decreased by 54–75 kJ/$W^{0.75}$ for each 100 kg increase in body weight.

Introduction

The results of a study comparing the energetics of growth and lactation were reported by Tyrrell, Moe and Oltjen (1974) and included 59 energy-balance measurements on eight Hereford heifers over a 55-week interval of time. Webster, Brockway and Smith (1974) reported that while the fasting metabolic rate of cattle appeared to decline as the cattle grew, the basal metabolism, estimated from metabolic measurements on fed animals, appeared to remain constant as the cattle grew. Webster, Smith and Mollison (1976) presented additional data that indicated variation in the estimated maintenance requirement was associated not only with age or body size of growing cattle but also with the physiological state of the animal. They referred to this as an individual's 'impetus to grow'. Our data (Tyrrell, Moe and Oltjen, 1974) appeared suitable for further analysis using a similar approach to that suggested by Webster, Brockway and Smith (1974).

Materials and methods

Two corn silage based rations (70% vs 40% of ration dry matter; DM) were fed to Hereford heifers (average initial weight 200 kg) continuously throughout the 55-week study. Complete energy-balance measurements were made using open-circuit respiration calorimeters at approximately six-week intervals. Feed intake was restricted to 120% of maintenance during the first 12 weeks (trials 1 and 2) of the study followed by 24 weeks of *ad lib.* intake (trials 3, 4, 5 and 6). Intake was then restricted to calculated maintenance (trials 7 and 8) for 12 weeks followed by measurement of fasting metabolic rate (trial 9). The experiment was originally designed to end at this point, but it became evident that heifers were in significant positive tissue energy balance during trials 7 and 8. Therefore, heifers were realimented for five weeks following fasting and trial 10 was conducted with the heifers much closer to zero tissue energy balance than had been achieved previously.

Results

Animal performance and other details have been reported previously (Tyrrell, Moe and Oltjen, 1974) and will not be repeated here. The relationship between total energy retention and distribution of energy between fat and protein was not affected by diet, animal or time (body weight) of measurement as the intake level changed, which indicated that partial efficiency of ME use for tissue synthesis was constant in this experiment. This hypothesis was tested by subjecting the 59 energy-balance and heat production observations to analysis of covariance on ME intake with diet, trial, diet × trial interaction and animal on the same diet analysed as fixed effects within the least squares model. The unbiased estimates of regression coefficients within diet were not different ($P>0.05$), which substantiated the use of the common regression coefficient (constant k_f) to estimate differences among fixed effects adjusted to equal ME intake. Trial (time or body weight) was found to be a significant ($P<0.01$) source of variation in heat production ($kJ/W^{0.75}$) adjusted to equal ME intake ($kJ/W^{0.75}$). This indicates that the intercept of the regression of heat on ME intake, i.e. maintenance requirement, is different among trials. Orthogonal single degree of freedom comparisons were used to test hypotheses about trial differences.

Orthogonal comparisons among trials indicated that heat production between trials 1 and 2 was not different. Therefore we can conclude that adaptation to the chamber environment was not a significant factor affecting maintenance requirement. Heat production between trials 7 and 8 was not different which indicated that the metabolic rate during trial 7 was not affected by the previous high plane of nutrition. Heat production during trial 10 was greater ($P<0.05$) than during trials 7 and 8, and indicated that subjecting the cattle to fast and realimentation changed the metabolic rate. Heat production (a) was lower during trials 5, 6, 7 and 8 than during trials 1, 2, 3 and 4; (b) was lower during trials 7 and 8 than during trials 1 and 2; and (c) was lower during trials 5 and 6 than during trials 3 and 4. These three comparisons indicate a significant ($P<0.01$) decline in heat production as Hereford heifers grow from approximately 200 kg to 400 kg in live weight.

The estimate of the common regression of retained energy on ME, together with least square means within diet and trial, were used to calculate the ME required for zero retained energy (*Table 45.1*). These calculated values lead to two conclusions. First, the ME required for maintenance by animals consuming the 70% corn silage diet is greater than for the animals consuming the 40% corn silage diet. Secondly, within both diets the ME required for maintenance declines by 54–75 $kJ/W^{0.75}$ for each increase in 100 kg body weight. This indicates that incorporation of a decreasing metabolic rate ($kJ/W^{0.75}$) in association with

Table 45.1 APPARENT MAINTENANCE REQUIREMENT OF HEREFORD HEIFERS

Trial number[a]	*70% Silage diet*		*40% Silage diet*	
	Body weight (kg)	*Maintenance* (kJ/$W^{0.75}$)	*Body weight* (kg)	*Maintenance* (kJ/$W^{0.75}$)
1[b]	220	552	222	502
2[b]	231	565	246	485
3[c]	284	565	311	536
4[c]	328	577	354	494
5[c]	357	531	387	473
6[c]	385	477	418	460
7[d]	389	435	410	397
8[d]	398	431	431	393
10[e]	390	494	420	356
Average	331	515	355	456

[a] Trials carried out in chronological sequence at approximately six week intervals
[b] Feed restricted to 120% maintenance
[c] Fed *ad lib.*
[d] Feed restricted to 100% maintenance
[e] Feed restricted to 100% recalculated maintenance

increasing body weight of growing cattle in the computation of ME for maintenance should improve the accuracy of predicting the energy requirements and growth rates of cattle.

References

TYRRELL, H.F., MOE, P.W. and OLTJEN, R.R. (1974). In *Energy Metabolism of Farm Animals*, pp. 205–208. Eds. K.H. Menke, H.J. Lantsch and J.R. Reichl. EAAP Publication No. 14. Stuttgart: Universität Hohenheim Dokumentationsstelle

WEBSTER, A.J.F., BROCKWAY, J.M. and SMITH, J.S. (1974). In *Energy Metabolism of Farm Animals*, pp. 135–138. Eds. K.H. Menke, H.J. Lantsch and J.R. Reichl. EAAP Publication No. 14. Stuttgart: Universität Hohenheim Dokumentationsstelle

WEBSTER, A.J.F., SMITH, J.S. and MOLLISON, G. (1976). In *Energy Metabolism of Farm Animals*, pp. 221–224. Ed. M. Vermorel. EAAP Publication No. 19. Clermont-Ferrand: G. de Bussac

46

EFFECTS OF BODY WEIGHT ON PARTITIONING OF AVAILABLE ENERGY TO FAT AND LEAN GAINS AND ON PARTIAL EFFICIENCIES

L.J. KOONG
Department of Animal Science, University of California, Davis, California 95616, USA

Summary

A method to study the effects of body weight on partitioning of available energy for growth (MEA) to fat and lean-tissue gains and on partial energetic efficiencies is proposed. A system of two simultaneous non-linear equations was used to calculate fat gain (FG) and lean gain (LG) as a function of MEA and body weight. The partial efficiencies for FG and LG were estimated from data obtained from energy-balance trials on pigs with body weights ranging from 20 to 80 kg. The results showed that the efficiency for FG is constant at 72% for all pigs; however, the efficiency for lean-tissue gain decreases from 55.5% for pigs at 20 kg to 49.9% for pigs at 80 kg. The results also showed that at a given level of energy intake, the older animal utilizes more of its available energy for fat synthesis.

Introduction

A new method for estimating energetic efficiencies of utilization of metabolizable energy (ME) for FG and LG was first introduced by Koong, Garrett and Canolty (1976). The method partitioned MEA into two portions: one used for fat synthesis and the other for lean-tissue synthesis. The basic relationship was expressed quantitatively in the following set of equations:

$$FG = P \cdot E_f \cdot MEA \quad (1)$$
$$LE = (1-P) \cdot E_\ell \cdot MEA \quad (2)$$

The parameter P represents the fraction of MEA which is used for fat synthesis and (1–P) is the fraction of MEA used for lean-tissue synthesis. These fractions of MEA are then converted to FG and LG with efficiencies E_f and E_ℓ, respectively. In the previous work, only the effect of MEA on the partition coefficient P was examined.

The purpose of this study was to extend the method to incorporate the possible effects of body weight (LW) on P as well as on the partial efficiencies for FG and LG.

Materials and methods

The trials using barley and skim milk as the major energy and protein sources from studies on energy metabolism in growing pigs reported by Thorbek (1975) were used for this study. The data set contains 94 animals with ages ranging from 80 to 190 d and LW ranging from 20 to 80 kg. Metabolizable energy intake (MEI) varied from 10.0 to 37.7 MJ/d.

Equations (1) and (2) were used as the basic model for estimating the partial efficiencies. MEA was calculated in this study by subtracting maintenance requirement (MR) from MEI where MR was estimated by using the equation reported by Thorbek (1975):

$$\mathrm{MR\ (kJ)} = 7029 + 33.5\ \mathrm{LW\ (kg)}$$

As in the previous study, the Michaelis–Menten equation was assumed to express the effect of MEA on the partition coefficient:

$$P = \frac{P_{max} \cdot \mathrm{MEA}}{K + \mathrm{MEA}} \tag{3}$$

where P_{max} is the maximum fraction of MEA that can be used for fat synthesis and K represents the level of MEA at which 50% of maximum fraction of MEA is used for fat synthesis.

It is well known that the young growing animal will use more of its available energy for LG whereas the more matured animal will direct most of its available energy for fat deposition. This effect of maturity on partition coefficient was introduced into equation (3) by letting

$$K = \frac{A}{\mathrm{LW}} \tag{4}$$

where A is a constant to be estimated.

The possible effects of LW on partial efficiencies for FG and LG were investigated by assuming that E_f and E_l are linear functions of LW:

$$E_f = B + C \cdot \mathrm{LW} \tag{5}$$
$$E_l = D + E \cdot \mathrm{LW} \tag{6}$$

Substituting equations (3), (4), (5) and (6) into (1) and (2) results in:

$$\mathrm{FG} = \left(\frac{P_{max} \cdot \mathrm{MEA}}{\frac{A}{\mathrm{LW}} + \mathrm{MEA}}\right)(B + C \cdot \mathrm{LW})\,\mathrm{MEA} \tag{7}$$

$$\mathrm{LG} = \left(1 - \frac{P_{max} \cdot \mathrm{MEA}}{\frac{A}{\mathrm{LW}} + \mathrm{MEA}}\right)(D + E \cdot \mathrm{LW})\,\mathrm{MEA} \tag{8}$$

These equation sets then became the basic model to be used for estimating partial efficiencies with experimental data.

In equations (7) and (8), MEA, LW, FG and LG are experimentally measured variables for each animal and P_{max}, A, B, C, D and E are unknown parameters. Since the magnitudes and variabilities of FG and LG range several-fold and are strongly correlated, a weighted non-linear least-squares method must be used to estimate the values of these parameters. The iterative simplex algorithm for function minimization developed by Nelder and Mead (1965) was used to minimize the weighted sum of squares of residuals.

Results and discussion

The parameter values estimated by fitting equations (7) and (8) to experimental data from growing pigs fed barley and skim milk diet are:

P_{max} = 0.846
A = 13210.3
B = 0.715
C = 0.00014
D = 0.573
E = –0.00092

The value of P_{max} indicates that the maximum fraction of MEA that could be used for fat synthesis is 85%. Since the other five parameters are all parts of the effects of LW on partition coefficient and partial efficiencies in the form of K, E_f, and E_l, they are summarized at different LWs (*Table 46.1*) to give a more meaningful interpretation.

The values for K decrease as the animal becomes older. The lower value for K indicates that at a given level of MEA older animals utilize more MEA for fat synthesis than younger animals. The effect of MEA and LW on the partitioning of available energy for fat synthesis is illustrated in *Figure 46.1*.

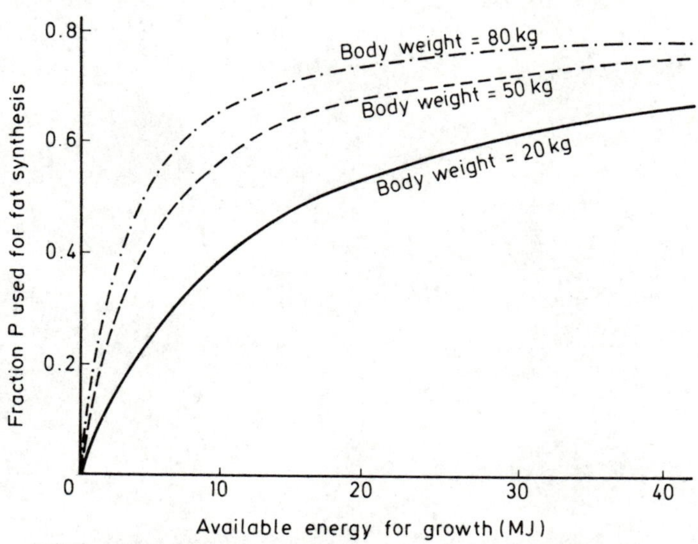

Figure 46.1 The effects of MEA and LW on the partition coefficient, P

Table 46.1 ESTIMATED VALUES FOR PARAMETERS K, E_f, AND E_ℓ FOR GROWING PIGS AT DIFFERENT LIVE WEIGHTS (LW).

LW (kg)	E_f (%)	E_ℓ(%)	*K* (MJ)
20	71.8	55.5	11.5
50	72.2	52.7	4.60
80	72.6	49.9	2.89

The results of this study also show that energetic efficiency for FG is relatively constant at 72% for growing pigs from 20 to 80 kg of LW. This is very similar to the value of 75% reported by Thorbek (1975) for the same data set. The energetic efficiency for protein deposition decreases from 55.5% for pigs at 20 kg to 49.9% for pigs at 80 kg. The question of whether energy cost for protein deposition increases has been raised by Kotarbinska and Kielanowski (1969). The present findings tend to support their belief that the costs of protein deposition are higher in older animals.

References

KOONG, L.J., GARRETT, W.N. and CANOLTY, N.L. (1976). In *Energy Metabolism of Farm Animals*, pp. 149–152. Ed. M. Vermorel. EAAP Publication No. 19. Clermont-Ferrand: G. de Bussac

KOTARBINSKA, M. and KIELANOWSKI, J. (1969). In *Energy Metabolism of Farm Animals*, pp. 299–310. Eds. K.L. Blaxter, J. Kielanowski and G. Thorbek, EAAP Publication No. 12. Newcastle upon Tyne: Oriel Press

NELDER, J.A. and MEAD, R. (1965). *Comput. J.*, 7, 308–313

THORBEK, G. (1975). *Beretn. Forsøgslab., København*, **424**, 193

47

DISCUSSION PAPER: PRINCIPLES UNDERLYING MODELLING ANALYSES IN ANIMAL ENERGETICS

R.L. BALDWIN
Department of Animal Science, University of California, Davis, California 95616, USA

Introduction

The application of mathematical modelling techniques in analyses of energy transactions in animals is not in its infancy. On the other hand, it has not yet been developed to the point where a number of unique and definitive contributions arising from the approach can be cited. Many believe that the modelling approach has great promise as an analytical tool and have even gone so far as to state that this may be the only approach that will enable rigorous quantitative evaluation of our understanding of the complex systems we call animals *and* that will allow effective or maximul use of our knowledge in the solution of animal and food production problems. Contributions to date have been formalization of current knowledge, evaluations of current knowledge and hypotheses for probable adequacy as explanations of the quantitative and dynamic behaviour of animal systems, and identification of critical areas of experimentation. Also, a rational array of philosophies, approaches and techniques has begun to emerge. These advances are well illustrated by the chapters in this part of the book. In this discussion paper, an attempt to identify several aspects of philosophy, approach and technique will be made in order to illustrate how these relate to these chapters. For clarity, chapters will not necessarily be referred to in the order in which they appear; all named references are to the chapters presented in Part 3 of this book.

Philosophy, approach and technique

Modelling studies are highly analogous to experimental studies in that the most important step is the setting of an objective. The setting of a specific objective for a modelling exercise is essential for the selection of elements of the system and interactions among these which must be included, and they comprise construction of a block diagram, selection of the mathematics and data used in the model, and, most importantly, evaluation and validation of model adequacy. The latter is achieved when the objective is attained. Models or modelling papers should be evaluated according to several criteria: was the objective set rational and appropriate; are the mathematics and data incorporated rigorous and appropriate; and was the objective attained? Sometimes, a mass of experimental data collected without a specific objective in mind can be analysed, after the fact, and yield new insight. This does not happen in modelling; the modelling objective is paramount.

Two types of equation are commonly used in modelling: empirical equations which are devised using statistical techniques to describe a given data set, and theoretical equations which represent a theory or concept about the fundamental nature of the system. Almost all models have both empirical and theoretical aspects but vary greatly in the degree of each. Models consisting of empirical equations can only be applied with confidence within the range limited by the data to which the empirical equations were fitted. Models based upon sound theoretical equations should provide better projections over a broader range, but are more difficult to devise and validate.

A common goal in research modelling is to drive toward causal models based on theoretical equations because these reflect true understanding of causal relationships which are determinants of system behaviour. Two approaches to this goal are evident among the chapters presented in this section. The first approach is to start with basically empirical equations and attempt assignment of theoretical or causal meaning to terms in these equations. The analyses of Koong (*see* page 231). Roux and Meissner (*see* page 217), Meissner, Roux and Hofmeyr (*see* page 223) and Tyrrell and Moe (*see* page 227) are examples of this approach. Koong (*see* page 231) explored the properties of the Michaelis–Menten type equation as a possible improved, as compared to multiple regression, method of deducing partial efficiencies and determinants of lean and fat gain. In this case, he explored effects of live weight upon estimated parameters. The goal is reduction of unexplained error through additions of defined measurable elements to the equation. Similarly, improvements in the applicabilities of the Gompertz and allometric equations for growth to varying nutritional circumstances are presented by Roux and Meissner (*see* page 217). Tyrrell and Moe (*see* page 227) sought improvements in estimates of animal maintenance requirements. These approaches might be generally characterized as proceeding from the intact animal inward and as emphasizing deductive estimation of parameters not directly measurable as separate entities. One can argue differently, but I believe that this first approach yields the greatest immediate practical benefit. Over the longer term the second approach will yield greater benefit to research and possibly greater general practical benefit. Both are appropriate.

The contributions of Smith, Baldwin and Sharp (*see* page 193), Schulz (*see* page 199), Reichl (*see* page 203) and Stombaugh and Oko (*see* page 209) represent a second approach which might be characterized as proceeding from knowledge of parts to descriptions of the whole animal. Objectives vary among these chapters but they are all concerned with evaluation of our knowledge of the parts for adequacy as explanations of the whole. A second goal is identification of theoretical relationships which should be added to empirical models used to evaluate and optimize animal performance and management systems.

Important concepts, particularly important in this latter type of modelling of energy metabolism, are that validated stoichiometric relationships be developed and that model complexity be held to the minimum essential for achievement of the modelling objective. (I might note, as an aside that it took some time for me to recognize that incorporation of elements into models simply because they were interesting and satisfying is inappropriate.) Let failure of a simple model to simulate reality force the addition of elements, not academic attractiveness. The work of Schulz (*see* page 199) emphasized evaluation of stoichiometric relationships based on knowledge of biochemical pathways for adequacy. The simultaneous algebraic equations used can be and have been solved by hand, but computers greatly increase the speed and accuracy of solution. This is essential to the validation analyses presented which look very promising. The work of Smith, Baldwin and Sharp (*see* page 193) represents the next stage of metabolic analysis: the dynamic stage. Identification of the major determinants of patterns of nutrient use among tissues is sought along with the development of deductive models for the interpretation of tissue metabolic data. The studies of Smith, Baldwin and Sharp (*see* page 193) and Stombaugh and Oko (*see* page 209) appear to emphasize patterns of nutrient use and are based on stoichiometric relationships. However, they differ in focus. Smith, Baldwin and Sharp (*see* page 193) emphasize properties and interactions among tissues, whereas Stombaugh and Oko (*see* page 209) emphasize central regulatory elements and interactions with the environment.

Reichl (*see* page 203) emphasizes the concept of modelling at several levels of organization and the concept of selecting only elements appropriate to scale. He presented a hierarchy of compatible models ranging from consideration of general, input/output relationships to evaluations of intracellular redox states and membrane functions. At each level appropriate mathematics and elements explicitly represent changes, for example, details regarding membrane function are inappropriate (out of scale) within input/output models. Over the long term, one can visualize the use of hierarchical sets of models in the evaluation and optimization of animal production systems. Single models incorporating a large proportion of theoretical elements would probably either be too complex (costly) to optimize or too simple to be applied generally.

Conclusion

The development of efficacious, rigorous, validated, generally applicable models is a time-consuming process. Models evolve slowly and often require the conduct of experiments for collection of essential data and for testing of alternative hypotheses: a strength of the technique. Validation is a difficult process, and publications providing full documentation of models are generally very long. These properties of modelling research compromise communication. Short overview papers and papers focusing on specific, often theoretical, aspects of a model are feasible as illustrated by these chapters. However, they are not fully satisfying. Complete models with full documentation and validation results are required for satisfaction. Patience.

48

GENERAL DISCUSSION

The chief subjects discussed in the 10 preceding chapters include the nature of the relation between energy intake and energy utilization, the question of maintenance requirement and the problems of validation of metabolic models.

The relations between energy intake and utilization may be apparently curvilinear, but it has been argued that the relation could be resolved into a set of linear functions. Curvilinearity could be related to changes in diet and physiological state, and to shift in maintenance requirements. The average body maintenance requirement could in turn be dependent on variations between animals in respect of high metabolic density organs such as the liver, which in 40 kg sheep varied from 400 to 1600 g in weight.

In connection with model systems, the possible effects of variations in rates of transfer of metabolites were considered in conditions that were not steady-state, and the use of enzyme kinetics to validate equations was also discussed. The animal could be modelled at different levels, such as tissue, organ and whole animal, with variables introduced at the minimum required. A good test of a model was whether it could simulate transition states such as changes in blood metabolite patterns in response to eating.

PART 4

COMPARATIVE ASPECTS OF ENERGY METABOLISM

49

THERMOGENESIS IN PIGS

N.J. ROTHWELL* and M.J. STOCK*
Department of Physiology, Queen Elizabeth College, Campden Hill Road, London W8 7AH, England
M.I. GURR**, R. MAWSON and C. FISHER†
Nutrition and Agriculture Division, Unilever Research Laboratory, Colworth House, Sharnbrook, Bedford MK44 1LQ, England

Summary

Groups of four animals in each experiment received a 25% protein diet restricted to maintain body weight at 6 kg or 20 kg and were compared with similar groups receiving a 1–2% protein diet fed *ad lib.* A three-fold difference in metabolizable energy (ME) intake was maintained. In 20 kg animals most of the 'excess energy' was used for carcass fat deposition, but in 6 kg pigs 80% of the 'excess energy' was accounted for by increased heat production as measured by indirect calorimetry. Differences in blood thyroid hormone levels, adipose tissue lipolysis and liver enzyme activity were noted.

Introduction

Debate continues about the existence of diet-induced thermogenesis in man or in farm animals. Miller and Payne (1962) described two pigs that maintained similar weights for long periods yet whose energy intakes differed by four- or five-fold. They ascribed the 'energy gap' to thermogenesis, yet did not measure heat production or carcass composition. The intention here was to repeat Miller and Payne's experiment, comparing animals of different ages, comparing energy expenditure as estimated from metabolic-balance data with indirect calorimetry and making careful measurements of carcass composition at the beginning and end of the experiments.

Materials and methods

All pigs were Large White (LW)-Landrace (LR) crossbreeds as follows:

LW♂ × (LW × LR) ♀

Each experiment involved pairs of pigs, one of which was fed a low protein (LP) diet (protein 2.1–2.4%; ME 20 MJ/kg DM) *ad lib.* while the other received

* Present address: Physiology Department, St. George's Hospital Medical School, Tooting, London SW17 0RE
** Present address: Nutrition Department, National Institute for Research in Dairying, University of Reading, Shinfield, Reading RG2 9AT
† Present address: Nutrition Department, Poultry Research Centre, King's Buildings, West Mains Road, Edinburgh EH9 3JS

a high protein (HP) diet (protein 21–27%; ME 15 MJ/kg DM) in restricted amounts that were sufficient to maintain weight.

The nitrogen and energy contents of the diets were determined by Kjeldahl analysis and ballistic bomb calorimetry respectively. Carcasses were analysed for water by drying and for fat content by acid hydrolysis. Initial body energy was estimated from analyses of separate groups of baseline animals.

ENERGY EXPENDITURE

This was calculated from the ME intake (E_{in}) and the change in body energy content (ΔBE):

$$E_{out} = E_{in} - \Delta BE$$

HEAT PRODUCTION

The production of heat was estimated by indirect calorimetry. Oxygen was analysed by paramagnetism and carbon dioxide by infrared, using the equation:

$$\text{Heat production (kJ)} = 16.18\ O_2 + 5.02\ CO_2 + 2.17\ CH_4 - 5.97\ \text{(urinary nitrogen)}$$

EXPERIMENT 1

Eight gilts (20 kg) were divided into two groups of four (HP/LP). Measurements included five five-day-balance periods in metabolic cages with one day in the calorimeter at the start of each balance period. The experiment lasted for 42 days.

EXPERIMENT 2

Six castrate males (6 kg) were studied in three pairs (HP/LP) for which the protocols differed slightly. Measurements included one or two four-day balance periods and calorimetry for two or four days.

EXPERIMENT 3

Eight castrate males (6 kg) were divided into three groups. A group of four (two HP/LP pairs) was put on metabolic balance for 21 days. Balance and heat production measurements were carried out continuously on the other two pairs in the calorimeters, one for a five-day period and the other for an 11-day period.

Table 49.1 ENERGY BALANCE: COMPARISON OF 20 kg AND 6 kg PIGS

		Final body weight (kg)	*Metabolizable energy intake* (MJ)	*Change in body energy* (MJ)	*Calculated energy expenditure* (MJ)	*Calculated energy for fat deposition* (MJ)
Experiment 1	HP (n=4)	24.0 ± 0.7	182.20 ± 23.89	+3.19	179.01 ± 19.23	1
(42 days)	LP (n=4)	26.7 ± 13.0	585.77 ± 72.10**	+285.76	300.01 ± 19.86**	101
Experiment 2	HP (n=3)	7.05 ± 0.22	51.24 ± 3.00	+6.04	45.20 ± 2.02	2.1
(23–29 days)	LP (n=3)	6.63 ± 0.16	143.08 ± 11.1	+43.31	99.77 ± 4.23	15.2
Experiment 3	HP (n=2)	6.1	34.14	−9.09	42.96	0
(21 days)	LP (n=2)	5.7	89.12	+11.45	87.67	4.0

Values are means ± s.e.m. ** $P<0.01$
Starting weights: Experiment 1:20 kg
Experiment 2 & 3:6 kg

Table 49.2 ENERGY EXPENDITURE: COMPARISON OF BALANCE METHOD AND INDIRECT CALORIMETRY[1]

		Days on balance	*Calculated energy expenditure (A)*	*Days in calorimeter*	*Heat production* (*B*; MJ/d)	$\frac{B}{A} \times 100$
Experiment 1	HP (n=4)	5 × 5 d	4.26 ± 0.46	5 × 1 d	5.64 ± 0.07	132
	LP (n=4)	5 × 5 d	7.14 ± 0.47	5 × 1 d	7.45 ± 0.19	104
Experiment 2	HP (n=3)	1 × 4 d or 2 × 4 d	1.76 ± 0.18	4 × 1 d or 1 × 4 d	1.70 ± 0.05	97
	LP (n=3)	1 × 4 d or 2 × 4 d	3.71 ± 0.06	4 × 1 d or 1 × 4 d	2.21 ± 0.01	60
Experiment 3	HP (n=1)	1 × 5 d	1.87	1 × 5 d	1.73	93
	LP (n=1)	1 × 5 d	2.59	1 × 5 d	2.58	100
	HP (n=1)	1 × 11 d	1.94	1 × 11 d	1.59	82
	LP (n=1)	1 × 11 d	2.67	1 × 11 d	2.15	81

[1] Mean ± s.e.m.

Results

COMPARISON OF CALCULATED ENERGY BALANCE

This is shown in *Table 49.1* for 6 kg and 20 kg pigs. A much larger proportion of the extra energy expenditure of the LP pigs was attributable to fat deposition in the 20 kg pigs. A three-fold difference in energy intake was achieved in both groups.

COMPARISONS OF CALCULATED ENERGY EXPENDITURE AND HEAT PRODUCTION (*Table 49.2*)

There was better correlation between the two methods when the animals were acclimatized to the conditions in the calorimeter for extended periods.

NITROGEN BALANCE AND BODY COMPOSITION (*Table 49.3*)

Fat deposition was significantly increased with LP diets when compared with pigs fed HP diets.

Table 49.3 NITROGEN BALANCE AND BODY COMPOSITION[1]

		Nitrogen gained (g/d)	*Water (%)*	*Fat (%)*	*Protein (%)*
Experiment 1	HP (n=4)	+8.76	70.2 ± 0.5	6.3 ± 0.8	19.9 ± 0.3
	LP (n=4)	–0.04	48.6 ± 0.8	35.6 ± 0.5	13.3 ± 0.4
Experiment 2	HP (n=3)	+1.84 ± 0.25	70.2 ± 1.2	11.0 ± 0.2	16.3 ± 0.1
	LP (n=3)	+0.24 ± 0.01	53.3 ± 3.6	30.7 ± 3.5	11.9 ± 0.9
Experiment 3	HP (n=2)	+1.21	75.0	4.4	17.2
	LP (n=2)	–0.22	64.5	13.8	16.1

[1] Mean ± s.e.m.

CHARACTERISTICS OF ADIPOSE TISSUE (*Table 49.4*)

Some characteristics of subcutaneous adipose tissue of 6 kg animals are shown in *Table 49.4*. In pigs on LP diets the fat content of adipose tissue was increased by enlargement of fat cells but without change in their number. The tissue of these pigs was more sensitive to the lipolytic effect of noradrenaline.

Table 49.4 CHARACTERISTICS OF THE ADIPOSE TISSUE[1]

	Adipocyte mass (μg/cell)	*Number of adipocytes per pig* (× 10^8)	*Lipid content of adipose tissue* (mg/g)	*Basal lipolysis* (μmol FA/cell per 2 h) (× 10^4)	*Increase in lipolysis in presence of noradrenaline* (%)
HP[2]	0.05 ± 0.01	104 ± 54	415 ± 44	4.93 ± 0.18	32 ± 8
LP[2]	0.11 ± 0.02*	101 ± 69	655 ± 46**	6.00 ± 0.42	100 ± 23*

[1] Subcutaneous adipose tissue (backfat) from pigs in experiment 3
[2] Mean ± s.e.m.; n = 4
* $P<0.05$; ** $P<0.01$ compared with HP

Discussion

In all experiments a roughly three-fold difference in energy intake between LP and HP fed animals was maintained. In the 20 kg animals almost all the excess energy intake of LP fed pigs was directed into deposition of extra carcass fat, but it was difficult to prevent the animals from gaining weight. These pigs gained approximately 7 kg over the period of the experiment and the high energy density of this gain resulted in a large increase in total body energy content. Animals weighing 6 kg maintained weight, yet only a small proportion of the excess energy was directed into fat deposition. There were no significant differences in excreted energy or physical activity. It seems, therefore, that the elevated heat production of LP fed animals was due to a large increase in diet-induced thermogenesis.

Differences in estimation of energy expenditure from balance and calorimetric measurements arose in large part from the fact that calorimetry results were unrepresentative of the metabolic conditions existing over the whole period of the balance trial. Good agreement was obtained when pigs were maintained in the calorimeter for the duration of the experiment.

The most striking finding, the difference between 6 kg and 20 kg pigs in terms of the response to over-feeding, remains to be explained. An understanding of the physiological and biochemical events whereby the young pig is transformed from an essentially lean animal with a high thermogenic capacity to one with a large propensity to fatness would be of scientific as well as commercial value.

Acknowledgements

We are grateful to the SRC and Unilever Ltd for a CASE studentship for N.J.R.

Reference

MILLER, D.S. and PAYNE, P.R. (1962). *J. Nutr.*, 78, 255–262

50

ENERGY UTILIZATION OF A CARBOHYDRATE-FREE PROTEIN-FAT DIET BY SOWS

H.L. MÜLLER and M. KIRCHGEßNER
Institut für Ernährungsphysiologie, Technische Universität München, D-8050 Freising-Weihenstephan, Federal Republic of Germany

Summary

The effect of a carbohydrate-free protein-fat diet (PF diet), when compared with a mixed diet (CPF diet), on energy metabolism was studied in a change-over respiration trial with 10 mature non-pregnant sows. Under the conditions of undernutrition, neutral temperature and isocaloric supply of metabolizable energy (ME), no differences were found in the nitrogen, carbon and energy retention. Also, the mean heat production (HP), the time pattern of HP and the physical activity were the same in both dietary groups. The maintenance requirement for ME at zero retained energy (RE) was estimated at 420 kJ/kg$^{0.75}$ under both feeding conditions. However, the calculated ATP formation was different, yielding a turnover of 304 mol/d for the mixed diet and 275 mol for the PF diet.

Introduction

The utilization of nutrients as energy sources for animals was studied systematically for the first time by Rubner (1902). His work was complemented and refined by other investigations (Blaxter, 1967; Armstrong, 1969) and, moreover, provided with biochemical evidence (Krebs, 1960). This led to the commonly accepted concept of the necessary energy expenditure for ATP synthesis. Observations in human nutrition regarding the weight-reducing effect of low-carbohydrate diets (JAMA, 1973; Kasper, Thiel and Ehl, 1973) and the finding of equally high oxygen consumption after ingestion of protein and carbohydrate (Garrow and Hawes, 1972) can, however, not be brought into harmony with this concept. The present study was conducted to test these findings by exact determination of energy balance. For this purpose the influence of a carbohydrate-free protein-fat diet, when compared with a mixed diet, on the heat production and energy balance was assessed in a model study with sows (Kirchgeßner and Müller, 1980).

Materials and methods

The experiment was conducted with 10 adult non-gravid sows (German Landrace) according to the change-over design AB-BA. Each 10-day-preliminary period was succeeded by an eight-day collection period, of which the animals spent the last four days in the respiration chamber. Thereafter, the change-over to the other diet followed with the same length of periods. The major dietary constituents were 16% cadaver meal, 6% fish meal, 5% casein and 67% starch meal (= CPF diet) in treatment A and 53% cadaver meal, 21% fish meal, 16% casein and 3% mixed fat (= PF diet) in treatment B. Dietary nitrogen contents were 2.9 and 9.5%, and the gross energy (GE) contents were 18 and 20 kJ/g dry matter (DM). These feeds were offered in two meals daily; the amounts fed were rationed so as to result in a slightly negative energy balance and to allow the same ME intake in both treatments. The gaseous exchange of each animal was measured in four successive 24-hour periods. Energy balance was computed by the carbon-nitrogen method. For the presentation of the time course of energy exchange the respiratory quotients (RQ) method was applied. Climatic conditions were 20°C and 60% relative humidity.

Results and discussion

The balance data are presented in *Table 50.1.* The high nitrogen intake with the PF diet is compensated by an increased nitrogen excretion via faeces and urine, so that nitrogen retention is the same and about zero for both dietary groups. Similarly, there is no significant difference in the carbon balance, because the elevated carbon excretion via faeces and urine in response to the PF diet is offset by a 12% lower carbon dioxide production. The RQ, however, is markedly different.

The intermediary energy exchange was not affected by diet. Heat production averages 23.6 MJ/d in both treatment groups, and RE from the PF diet is just 0.8 MJ/d more negative than with the CPF diet. This amount corresponds numerically to the low difference in ME intake. The negative RE is due to the breakdown of body fat in both treatments. According to the comparable body weights, the data expressed on the basis of metabolic body size show the same relationships.

The 24-hour response curves for carbon dioxide production, oxygen consumption and HP, determined by continuous measurement of the gaseous exchange and by application of the RQ method, show a close parallelism between the two diets. Differences in physical activity were not observed. The time spent standing amounted to 205 min/d on the mixed diet and 194 min/d on the PF diet.

The results show that the carbohydrate-free diet did not have a weight-reducing effect in terms of a stimulated degradation of body fat. This agrees with balance studies by Prabucki and Schürch (1977), who found no differences in HP by rats fed isocaloric diets with 10% sugar and varying fat and starch contents.

Regarding the question whether the energy exchange on the basis of the dietary components followed the possible ATP formation according to known biochemical schemes, a calculation of the ATP turnover was performed for both diets (*Table 50.2*). Since the conditions of the experiment were thermally neutral, the total ME and RE were included in this calculation of ATP formation. The energy quantity per mole of ATP formed was taken from Krebs (1960).

The calculation resulted in an ATP turnover of 304 mol/d on the mixed diet and in 275 mol/d on the PF diet, which corresponds to a difference of 10%. This result obviously conflicts with the classic concept of nutrient efficiency in catabolism unless the ATP requirement or the metabolic intensity of the animals were different under the conditions of the two diets. Possibilities for such an

Table 50.1 NITROGEN BALANCE, CARBON BALANCE, GAS EXCHANGE AND ENERGY BALANCE DATA

	CPF diet	*PF diet*	*Difference*	± *s.e. of difference*	*F value* ($F_{0.05}$ = 5.3)
Nitrogen balance (g/d)					
Intake	34.3	113.3	79.0	3.6	488
Faeces	4.4	10.8	6.4	0.4	315
Digestible nitrogen	29.9	102.5	72.6	3.4	462
Urine	28.3	102.7	74.4	7.1	109
Retention	1.6	−0.2	−1.8	5.6	0.1
Carbon balance (g/d)					
Intake	569	570	1	23	0.0
Faeces	103	148	45	10	20.0
Urine	19	56	37	4	79.2
Carbon dioxide	546	482	−64	13	22.6
Retention	−99	−116	−17	19	0.7
Gas exchange (ℓ/d)					
Oxygen	1178	1204	26	33	0.5
Carbon dioxide	1019	899	−120	25	22.6
Respiratory quotient	0.868	0.746	−0.121	0.009	173
Energy balance (MJ/d)					
Intake	23.2	25.9	2.7	1.0	6.8
Faeces	4.1	5.9	1.8	0.4	19.9
Urine	0.7	2.4	1.7	0.2	67.1
ME	18.4	17.6	−0.8	0.8	1.1
Heat production	23.6	23.6	−0.0	0.7	0.0
Retention	−5.2	−6.0	−0.8	1.0	0.7
Protein energy	0.2	−0.0	−0.2	0.8	0.1
Fat energy	−5.4	−6.0	−0.6	1.2	0.2
Energy balance (kJ/d and $kg^{0.75}$)					
Intake	392	434	42	18	5.3
DE	323	334	11	15	0.5
ME	311	295	−16	14	1.4
Heat production	399	395	−4	13	0.1
Retention	−88	−100	−12	16	0.5
Protein energy	4	−0	−4	14	0.1
Fat energy	−92	−100	−8	20	0.1

DE = digestible energy

Table 50.2 YIELD OF ATP CALCULATED FROM THE INTAKE OF THE CPF VERSUS THE PF DIET

Source of energy	*CPF diet*		*PF diet*	
ME (MJ/d)				
Protein	3.93		12.56	
Fat	0.78		5.03	
Carbohydrate	13.74	Σ 18.45	–	Σ 17.59
RE (MJ/d)				
Protein	0.24		−0.03	
Fat	−5.42		−6.00	
Total energy for ATP formation				
Protein	3.69		12.59	
Fat	6.20		11.03	
Carbohydrate	13.74	Σ 23.63	–	Σ 23.62
Yield of ATP (mol/d)	304		275	

adaptation of the metabolic pattern to the nutritional differences could, perhaps, be related to effects on protein turnover.

The results also lend themselves to the deduction of the energy requirement for maintenance when defined as the ME intake at zero-energy retention. The estimate is 420 MJ/kg$^{0.75}$ for either diet. This energy equilibrium is somewhat higher than the maintenance energy requirement estimated for sows fed practical rations (Müller and Kirchgeßner, 1979).

References

ARMSTRONG, D.G. (1969). 'Cell Bioenergetics and Energy Metabolism.' In *Handbuch der Tierernährung*, Vol. I. Ed. W. Lenkeit *et al.*, p. 385. Berlin and Hamburg: Parey

BLAXTER, K.L. (1967). *The Energy Metabolism of Ruminants.* London: Hutchinson

GARROW, J.S. and HAWES, S.F. (1972). *Br. J. Nutr.*, **27**, 211

JAMA (1973). Council on Food and Nutrition, American Medical Association. *J. Am. Med. Ass.*, **224**, 1415

KASPER, H., THIEL, H. and EHL, M. (1973). *Am. J. clin. Nutr.*, **26**, 197

KIRCHGEßNER, M. and MÜLLER, H.L. (1980). *Nutr. Metab.* **24**, 1

KREBS, H. (1960). *Arzneimittel-Forsch.*, **10**, 369

MÜLLER, H.L. and KIRCHGEßNER, M. (1979). *Z. Tierphysiol. Tierernährg. Futtermittelk.* **42**, 271

PRABUCKI, A.L. and SCHÜRCH, A. (1977). *Bibliotheca Nutr. Dieta*, **25**, 78

RUBNER, M. (1902). *Die Gesetze des Energieverbrauchs bei der Ernährung.* Leipzig und Wien: Deuticke

51

HEAT PRODUCTION WHEN SINGLE NUTRIENTS ARE GIVEN TO FASTED COCKERELS

I. TASAKI and M. KUSHIMA
Laboratory of Animal Nutrition, Faculty of Agriculture, Nagoya University, Nagoya 464, Japan

Summary

Fasting heat production of Single Comb White Leghorn cockerels acclimatized to 20 ± 1°C was measured, and further measurements were done after ingestion of single nutrients (470–500 kJ). Heat production increased soon after corn starch ingestion, but this increase ceased within 24 h and 5% of ingested gross energy (GE) was eliminated as heat. Heat increment after casein-gelatin mixture ingestion was 50% of ingested GE during three days, and this increase lasted for more than three days. No heat increment was observed when corn oil was ingested.

Introduction

Since Rubner's work, there have been many investigations concerning specific dynamic action (SDA) of nutrients given to mammals. However, there is little information on SDA in poultry (Tasaki and Sakurai, 1967; Burlacu, 1967). In order to investigate the time course of changes in heat production after a single ingestion of individual nutrients in fasted cockerels, this experiment was carried out.

Materials and methods

Single Comb White Leghorn cockerels of 1–1.5 years and weighing about 2 kg were used in this experiment. The birds were kept in an environmentally controlled room for at least two weeks (Shannon and Brown, 1969) to acclimatize them to an environmental temperature of 20 ± 1°C. To avoid heat production that would result from picking feed and other activities involving physical movements, the birds were restrained in wire cages. Inside the environmentally controlled room the caged birds were fasted for two days, after which they were placed in a respiration chamber kept at 20 ± 1°C. Ventilation was provided at the rate of about 2.5 ℓ/min. The out-going air was introduced into the oxygen and carbon dioxide analysers, and the fasting heat production was calculated using the formula devised by Romijn and Lokhorst (1961) which was based on the oxygen absorbed and the carbon dioxide produced. Immediately after the basal heat production had been measured, each of the nutrients was forced-fed to the experimental birds and the heat production was measured continuously for three successive days.

The experimental diets used here were 30 g of corn starch, 24 g of casein-gelatin mixture (2:1) and 12 g of corn oil. The GE of these diets, as given to the birds, were 470, 500 and 470 kJ respectively. Two methods of forced-feeding were employed. One was the tube-feeding method, which involved the insertion of a silicon tube of 5 mm inner diameter through the mouth of the crop sac. The outer end was attached to a syringe that was outside the respiration chamber. This enabled the nutrients to be given to the birds at any time without the birds having to be removed from the chamber. This method was used for feeding corn oil and corn starch suspended in water. The other method was the dumpling-feeding technique. Here the diets were mixed with warm water and kneaded into several slender dumplings that were then inserted into the crop sac through the oral cavity. This method was rather unsatisfactory because the birds had to be withdrawn from the chamber every time they were fed. Nevertheless, this method was used for feeding the casein-gelatin mixture because it could not be given by using the tube-feeding method.

Results and discussion

Figure 51.1 shows a typical time course of changes in heat production when single nutrients were given to the cockerels. This figure indicates that heat production increased soon after 470 kJ of corn starch was administered (Bird 1), and this increase might be due to the SDA that resulted from the starch ingestion. Heat increment on a daily basis was calculated by subtracting the total heat production on day 1, which represented the basal heat production (Tasaki and Sakurai, 1964), from the total heat production on day 2, day 3 and, in some cases, day 4. The average value for three birds is shown in *Figure 51.2*. This illustrates that the heat increment ceased within 24 hours after the starch ingestion and that the increment was about 5% of the ingested GE. Since the digestibility of corn starch was almost 90% in cockerels (Tasaki and Sakurai, 1967) then approximately 6% of the absorbed energy (DE) was eliminated as heat.

Heat increment on day 3 and day 4 was found to be negative. This was because the basal heat production per bird seemed to decrease gradually day by day (Tasaki and Sakurai, 1964). Assuming that the line passing through day 1, day 3, and day 4 is the level of the basal metabolism (*Figure 51.2*), then the actual heat increment appearing on day 2 may be considered to be more than 6% of the ingested DE the assumed figure for this being about 8%.

When the birds were fasted, the liver glycogen decreased remarkably (Hazelwood and Lorenz, 1959) and the body fat was used for basal heat production (Tasaki and Sakurai, 1964). It is assumed, therefore, that the heat increment that appeared is mainly due to the heat produced in the biosynthesis of liver glycogen from the absorbed glucose.

As was found to be the case in starch ingestion, heat production increased soon after the casein-gelatin mixture was administered (Bird 2), and the heat increment was extremely large on day 2 (*Figure 51.2*). Such an increase in heat production decreased gradually, although it lasted for more than three days. The

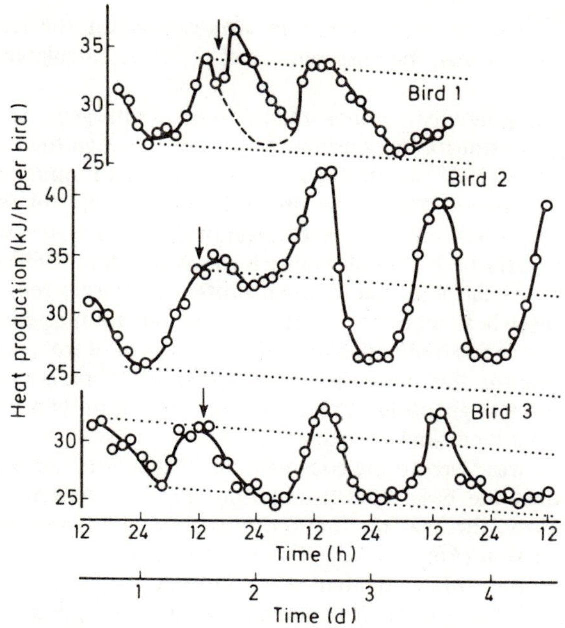

Figure 51.1 Time course of changes in heat production after single ingestion of individual nutrients. Day 1 shows basal heat production and the arrows show the time at which 470 kJ of corn starch (Bird 1), 500 kJ of casein-gelatin mixture (Bird 2) and 470 kJ of corn oil (Bird 3) were singly ingested. The broken lines indicate the assumed basal heat production, and the dotted lines show ranges of assumed diurnal rhythm of basal heat production.

average heat increment for four birds during the three days following protein ingestion was about 50% of the ingested GE. The total heat increment could be calculated through extrapolation to be almost 52%. Since the digestibility of the casein-gelatin mixture was reported to be about 90% in cockerels (Tasaki and Sakurai, 1967), approximately 57% of the supplied DE was eliminated as heat.

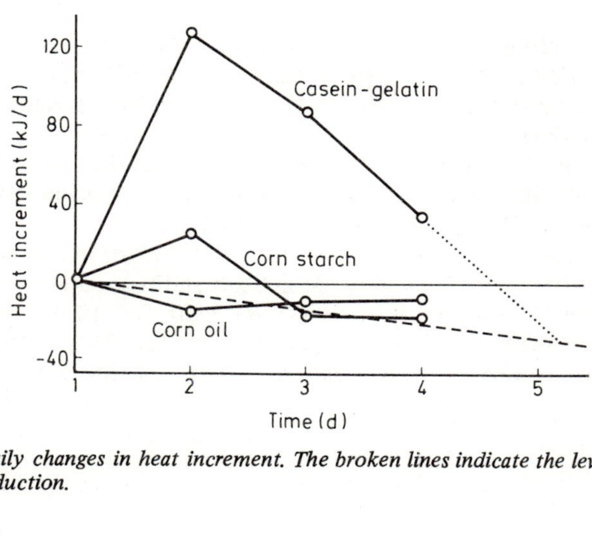

Figure 51.2 Daily changes in heat increment. The broken lines indicate the level of assumed basal heat production.

However, as discussed in connection with starch ingestion, the actual heat increment should be larger than the observed value, i.e. it was calculated as 61% of the ingested DE.

Adult cockerels generally require a small amount of nitrogen, and the quantities ingested in this experiment are considered to be greater than the requirement. As a result, as Burlacu (1967) reported, a certain amount of amino acids remained in the body, and some of these were probably used rapidly for the synthesis of protein that had been catabolized endogenously for basal metabolism. The remainder would still be stored in its free form. Ashworth (1969) suggested that a large amount of heat is produced when protein is synthesized. This could be the rapid and large heat increment that appeared on the first day after protein ingestion. Amino acids remaining in their free form were perhaps oxidized gradually or used for the synthesis of new proteins to replace those that had been catabolized endogenously. This could be one reason why heat increment was observed for a long time.

When corn oil was given to the cockerels the heat production was found to be a little lower than the basal heat production (Bird 3). As has been discussed previously, however, the heat production might not be different from the actual basal heat production (*Figure 51.2*). The fatty acid composition of corn oil was reported to be rather similar to that of chicken fat when compared with other fat sources such as lard (Scott, Nesheim and Young, 1976). When lard was given to the fasted cockerels a heat increment was observed, although it was lower than that observed in corn starch or casein-gelatin mixture feeding (Tasaki and Sakurai, 1967). Approximately 90% of corn oil was absorbed from the chicken intestine (Tasaki and Sakurai, 1967), which resulted in the amount of energy ingested being almost the same as the basal heat production. Therefore, the absorbed corn oil probably compensated for the body fat that was oxidized for basal heat production with a little loss of heat for the synthesis of fatty acids characteristic in chicken fat. This might be a reason why no heat increment was observed when corn oil was ingested.

References

ASHWORTH, A. (1969). *Nature, Lond.*, **233**, 407–409

BURLACU, G. (1967). In *Energy Metabolism*, pp. 149–157. Eds. K.L. Blaxter, J. Kielanowski and G. Thorbek. EAAP Publication No. 12. Newcastle upon Tyne: Oriel Press

HAZELWOOD, R.L. and LORENZ, F.W. (1959). *Am. J. Physiol.*, **197**, 47–51

ROMIJN, C. and LOKHORST, W. (1961). In *Energy Metabolism*, pp. 49–58. EAAP Publication No. 10. The Netherlands: Wageningen

SCOTT, M.L., NESHEIM, M.C. and YOUNG, R.J. (1976). *Nutrition of the Chicken, 2nd edn* p. 25. Ithaca, New York: M.L. Scott and Associates

SHANNON, D.W.F. and BROWN, W.O. (1969). *Br. Poult. Sci.*, **10**, 13–18

TASAKI, I. and SAKURAI, H. (1964). *Jap. J. zootech. Sci.*, **35**, 15–18

TASAKI, I. and SAKURAI, H. (1967). *Jap. J. zootech. Sci.*, **38** 201–211

52

EFFECTS OF AMBIENT TEMPERATURE ON THE HEAT PRODUCTION OF GROWING TURKEYS

M.G. MacLEOD, S.G. TULLETT and T.R. JEWITT
Agricultural Research Council Poultry Research Centre, King's Buildings, West Mains Road, Edinburgh EH9 3JS, Scotland

Summary

Respiration calorimetry was used to study the acute metabolic-rate responses of growing turkeys to temperatures between 20°C and 40°C and the chronic effects of ambient-temperature change on acclimatized turkeys. Acutely-exposed fasting birds showed a zone of minimal metabolism extending from 30°C to 40°C. The variability of data made it difficult to determine whether there was a similar zone in birds fed *ad lib.* As well as giving an acute (<1 d) response to temperature change, the turkeys transferred from a low to a high temperature showed a continuing decrease (>1 d<15 d) in metabolic rate as they became acclimatized.

Introduction

Despite the increasing commercial interest in the turkey, there is a lack of information about the energy metabolism of the species. For example, little is known about its metabolic responses to the ambient temperature, which are important in specifying the thermal environment at which production efficiency is optimal. Both acute and chronic responses are of interest, the latter because they give a guide to the effects of prolonged exposure to a specific temperature and the former because they are important in relation to rapid changes in ambient temperature, including circadian fluctuations. This chapter describes preliminary work on young growing turkeys.

Materials and methods

BIRDS AND INITIAL MANAGEMENT

Broad-breasted White (Ross 'Super Midi') turkeys were reared, from hatching, in a brooder in which they had access to an infra-red radiator. The brooder was kept in a room maintained at about 20°C and lit continuously. Water and a mash feed were supplied *ad lib.*

HEAT-PRODUCTION MEASUREMENTS

Metabolic rate was measured using the calorimetry system and methods of Lundy, MacLeod and Jewitt (1978). Each period of measurement lasted for 22 h, starting at 11.00 h. The lighting regimen was synchronous with that to which the birds were accustomed: continuous for acute experiments and 14 h light:10 h dark for chronic experiments.

ACUTE EFFECTS

Acute responses were measured in 21-day-old turkeys (mean fasted weight = 235 g) fasted for 18 h before fasting heat production measurements and fed *ad lib.* up to and during the measurement of fed heat production. The birds were removed from the brooder and observed in groups of three per calorimeter, at controlled temperatures between 20°C and 40°C and at a water vapour pressure of 2.1 kPa (16 mmHg). Each group was studied at one temperature only. Water was supplied *ad lib.*

CHRONIC EFFECTS

Twenty-four turkeys were removed from the brooder at seven days of age; 12 were transferred to cages in a room held at about 20°C and 12 to similar accommodation at 35°C. Food and water were provided *ad lib.* After acclimatization to these conditions for 28 days, groups of five birds were randomly selected from 20°C (Groups L1 and L2) and 35°C (Groups H1 and H2). The fasting heat production of L1 and H1 was measured at their acclimatization temperatures (20°C and 35°C respectively); groups L2 and H2 were transferred to 35°C and 20°C respectively and their fasting heat production was measured during their first day at these temperatures (*Table 52.1*). The mean fasted weight

Table 52.1 CHRONIC RESPONSE OF HEAT PRODUCTION TO AMBIENT TEMPERATURE (MEAN ± s.e.m.; n = 5).

	Acclimatization temp. (°C)	*Observation (Obs.) temp.*	*Period at temp.* (d)		*Fasting heat production H* (kJ/kg $W^{0.75}$ per d)		*Between-obs. differences in H*
			Obs. 1	*Obs. 2*	*Obs. 1*	*Obs. 2*	
L1 (20°C control)	20	20	30	44	612 ± 23	540 ± 19	−72 ± 12[a]
L2 (Test)	20	35	1	15	540 ± 30	392 ± 9	−148 ± 23[b]
H1 (35°C control)	35	35	34	47	455 ± 12	400 ± 12	−55 ± 14[a]
H2 (Test)	35	20	1	15	603 ± 36	541 ± 17	−62 ± 19[a]

[a b] Valves with different superscripts are significantly different ($P<0.02$)

was 0.540 kg. A further fasting measurement was made after the birds had been held at these temperatures for a further 14 days (mean fasted weight = 1.090 kg).

Results

ACUTE EFFECTS

Fasting 21-day-old turkeys had a zone of minimal metabolic rate extending from about 30°C to about 40°C (*Figure 52.1*), and within this zone the heat production was 490 kJ/kg $W^{0.75}$ per d. Between 20°C and 30°C the fasting heat production, H (kJ/kg $W^{0.75}$ per d), was related to ambient temperature, T_a(°C), by the equation:

$$H = -12.6T_a + 871.$$

Heat production therefore increased by 2.5% of the minimal value for every 1°C decrease in environmental temperature. About 38°C there was some indication of a hyperthermal rise in heat production, but this cannot be verified statistically with the available data.

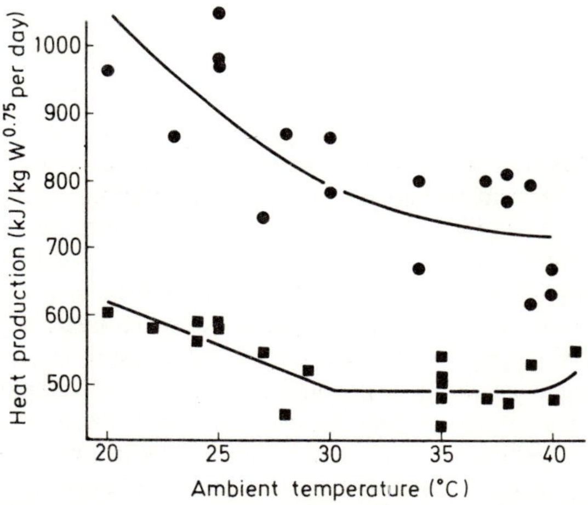

Figure 52.1 Relationship between heat production and ambient temperature in acutely-exposed fasted ▪ and fed • three-week-old turkeys. Each point represents the mean of a group of three birds.

Heat production by the 21-day-old turkeys tended to fall continuously from 20°C to 40°C (*Figure 52.1*). Although the rate of decrease in heat production became less as the temperature increased, the variability in the heat production of fed birds (associated with variation in food intake) was such that it became difficult to determine whether a critical temperature existed within the temperature range examined.

CHRONIC EFFECTS

As would be expected from the acute results, there were marked changes in the heat-production rates of growing turkeys on their first day at temperatures 15°C different from those to which they had been acclimatized for 30–34 days

(*Table 52.1*). Turkeys transferred from 35°C to 20°C had a mean H 32% higher than that of controls kept at 35°C and an identical H to that of birds acclimatized to 20°C, which indicated that their response was complete within one day. Those transferred from 20°C to 35°C, however, showed a decrease of only 9% on the first day and had a metabolic rate that remained significantly higher than that of birds acclimatized to 35°C. After 14 days of acclimatization to 35°C, the birds taken from 20°C had attained a mean H that was the same as that of birds acclimatized to 35°C for 47 days which indicated that adjustment was now complete. The metabolic rates of birds that had been at 20°C for 44 days and 13 days remained similar.

Fasting metabolic rate per unit of metabolic body size of the control groups decreased between the two measurement periods: the statistical test used to determine whether significant adaptation occurred took this into account by comparing the differences between the two measurements made on each bird. The adaptive change that occurred over at most 15 days in birds transferred from 20°C to 35°C was statistically significant (*Table 52.1*): no similar effect occurred in birds transferred from 35°C to 20°C where the full response occurred within the first day.

Discussion

The chronic metabolic-rate response to temperature change was significant only in the case of birds transferred from 20°C to a higher temperature, but this does not preclude the possibility that there may be long-term adaptations (physiological, morphological or behavioural) to temperatures below 20°C.

Fed individuals, unlike the fasted birds, could not be shown to have a zone of minimal metabolism within the temperature range studied; a decline in food intake above 25°C may have allowed a decrease in total heat production additional to that which occurred in fasting heat production. If food intake were to be fixed at a range of levels, one might expect the critical temperature to be lower at a higher feeding level (Close and Mount, 1978), but when food intake is allowed to vary the animal is able to reduce the food intake as the temperature increases, initially because of lower energy requirements and finally to reduce heat load. A comparison of the acute results with the initial results from birds acclimatized to 20°C and 35°C suggests, however, that temperature-acclimatized fasted birds would also show a continuing decrease in H to at least within the temperature range studied.

The critical temperature indicated by the present acute exposure experiment is in agreement with that (30°C) found by Nichelmann *et al.* (1976) in turkey poults of the same age but of a different strain exposed to the experimental temperature for only 2 h; in their case, however, heat production increased just above this temperature to the extent that heat production at 40°C was considerably higher than at 20°C. The shorter period of exposure to changed ambient temperature, as well as between-strain differences, may provide some explanation for the difference between these results.

Adaptation of fasting heat production to high temperature has been shown previously to occur in the domestic fowl (Berman and Snapir, 1965; Shannon and Brown, 1969), and is probably thyroid mediated (Hahn, Ishibashi and

Turner, 1966; Collins and Weiner, 1968). There may be endocrinological implications in the rapid response of 35°C-acclimatized birds to 20°C.

A decrease in fasting-metabolic rate may offer the acclimatized bird a choice of strategems in a hot environment; it can either increase its food intake and hence its rate of production above that of the unacclimatized animal without increasing its total fed heat production beyond the same level (i.e. it has a greater scope for heat increment of feeding), or it can maintain the same food intake as the unacclimatized animal but with a lower total-heat production and hence a reduced risk of heat stress. The former alternative is given support by the data of Davis, Hassan and Sykes (1972) and the latter by Berman and Snapir (1965).

Acknowledgement

We are grateful to Mr A. Raza for technical assistance.

References

BERMAN, A. and SNAPIR, N. (1965). *Br. Poult. Sci.*, **6**, 207–216
CLOSE, W.H. and MOUNT, L.E. (1978). *Br. J. Nutr.*, **40**, 413–421
COLLINS, K.J. and WEINER, J.S. (1968). *Physiol. Rev.*, **48**, 785–839
DAVIS, R.H., HASSAN, O.E.M. and SYKES, A.H. (1972). *J. agric. Sci., Camb.*, **79**, 363–369
HAHN, D.W., ISHIBASHI, T. and TURNER, C.W. (1966). *Poult. Sci.*, **45**, 31–33
LUNDY, H., MacLEOD, M.G. and JEWITT, T.R. (1978). *Br. Poult. Sci.*, **19**, 173–186
NICHELMANN, M., ELLERKAMP, S., HERNTRICH, I. and LYHS, L. (1976). *Mh. Vet. Med.*, **31**, 213–217
SHANNON, D.W.F. and BROWN, W.O. (1969). *Br. Poult. Sci.*, **10**, 13–18

53

OXYGEN CONSUMPTION, CARBON DIOXIDE PRODUCTION AND HEAT PRODUCTION IN MAN

G. THORBEK, S. HENCKEL* and A. NORDÉN†

National Institute of Animal Science, Department of Animal Physiology and Chemistry, 25 Rolighedsvej, DK-1958 Copenhagen V, Denmark

Summary

Oxygen consumption and carbon dioxide production were measured by means of respiration chambers in six males (22–63 years) and six females (21–67 years) and the heat production (HP) based on indirect calorimetry was calculated. In the first experiment (Ia) all subjects were on a 'standard' diet and measurements were performed during 6 h (0900–1500). An identical experiment (Ib) was carried out after a period of 2–3 weeks. This experiment was extended for another hour during which a step test (18 cm × 200 steps) was performed from 1505 h to 1515 h, and the gas exchange was measured from 1500 to 1600 h (II). In a third experiment (III) measurements were undertaken during the last 4 h of a 25-hour fasting period (0900–1300 h).

In experiments with the subjects on a 'standard' diet (Ia,Ib), regressions of gas exchange and HP on metabolic weight were calculated. With no significant differences between the two periods of measurement the data could be pooled as:

$$O_2\ (\ell/h) = 0.865 \times \text{kg}\ W^{0.75}$$
$$CO_2\ (\ell/h) = 0.634 \times \text{kg}\ W^{0.75}$$
$$HP\ (kJ/h) = 17.0 \times \text{kg}\ W^{0.75}$$

After a 10 minute step test (II) the gas exchange increased for the next hour by 39% for oxygen consumption and 47% for carbon dioxide production while the HP increased by 41%, and all the differences were highly significant. With oxygen consumption related to work load the regressions are expressed on body weight as follows:

$$O_2\ (\ell/h) = 0.412 \times \text{kg}\ W$$
$$CO_2\ (\ell/h) = 0.347 \times \text{kg}\ W$$
$$HP\ (kJ/h) = 8.41 \times \text{kg}\ W$$

After 24 h of fasting (III) the gas exchange decreased by 14% for oxygen consumption and 18% for carbon dioxide production while the HP decreased by 15%, all differences being highly significant. The regressions were found as:

$$O_2\ (\ell/h) = 0.741 \times \text{kg}\ W^{0.75}$$
$$CO_2\ (\ell/h) = 0.523 \times \text{kg}\ W^{0.75}$$
$$HP\ (kJ/h) = 14.5 \times \text{kg}\ W^{0.75}$$

* Regnecentralen A/S, Copenhagen

† Current address: Department of Internal Medicine, University of Lund and the Unit for Community Care Sciences, Dalby, Sweden

From the experiment it may be concluded that indirect calorimetry using respiration chambers with open-circuit ventilation can be applied to experiments in man. Further, it was found that metabolic body weight (kg $W^{0.75}$) in this trial could be used to estimate oxygen consumption, carbon dioxide production and HP independent of sex and age.

Introduction

Since the classic experiments of Atwater in man using the respiration calorimeter built by Atwater and Benedict (1905), few experiments based on direct calorimetry have been carried out. Twenty-two years ago Benzinger *et al.* (1958) designed a gradient layer calorimeter for man in which the heat flow across the walls was measured. In a symposium on 'The application of human and animal calorimetry' (University of Aberdeen, 1977) the advantages and disadvantages of direct or indirect calorimetry were discussed and Durnin (1978) indicated that methodological errors by indirect calorimetry have usually been in the range of 10–12%.

Beaton (1975) has pointed out that there are no valid estimates of the adequate intake of protein in relation to energy in human diets. Durnin (1975) has stressed the technical difficulties in estimating energy requirements by the use of Douglas bags, Max Planck respirometers or heat rate recordings.

It is obvious that more experimental work in human subjects concerning their energy metabolism is required. A trial in man using the method of indirect calorimetry and the respiration unit built for cattle in Copenhagen has been carried out in co-operation with the Dalby-Group from Sweden.

Materials and methods

Out of a group of 18 volunteers controlled by a routine clinical and laboratory health screening programme (Bengtsson *et al.*, 1979), 12 healthy subjects were selected (six males and six females) to represent the greatest range in age (21–67 years) and body weight (51–78 kg). The gas exchange in each subject was measured by means of the respiration unit originally built for cattle as described by Thorbek and Neergaard (1970). The respiration chamber (10.3 m^3) was furnished with a chair, a table, a couch, bottles for collection of urine, a washing stand and a telephone. The temperature was kept constant at 22.5°C with a relative humidity of 60%. The units were calibrated with carbon dioxide during the experiments and deviations of ±0.4% and ±1.2% were found for chambers A and B respectively.

In experiment I all subjects were given the same basal diet, the 'standard' diet, for dinner, breakfast and lunch, and it contained 6530 kJ (Gross energy; GE) and 100 g protein (nitrogen (%) × 6.25). In order to eliminate any stress and to obtain the normal degree of satiety for each subject extra sandwiches for lunch and biscuits for tea or coffee were provided. Thereby the intake of energy varied from 6530 to 9960 kJ and from 100 to 119 g protein.

The experiment started at home with dinner at 1800–1900 h. Breakfast was served the following morning at the laboratory at 0800–0830 h, and at 0830 h the subjects entered the respiration chambers, were seated at the desk and occupied themselves with reading, writing, knitting or sewing. At 0900 h measurements of the gas exchange started. At 1030 h tea or coffee was passed in through the closed locker system of the unit, and at 1230 h lunch was passed in to the subject.

From 1230 to 1330 h the subjects were resting or sleeping on the couch and at 1330 h they returned to the desk and tea or coffee was served. At 1500 h measurements of the gas exchange were finished and the subjects left the chambers.

After a period of 2–3 weeks the experiments were repeated (Ib), but this time the subjects remained in the chamber from 1500 to 1600 h in order to perform a step test (II) from 1505 to 1515 h by stepping up and down 18 cm with a speed of 20 steps/min. Thereby a work load from 1836 to 2808 kg, depending on the body weight of the different subjects, was performed. The values of oxygen and carbon dioxide concentration in the outgoing air from 1400 to 1500 h (resting) and from 1500 to 1600 h (including the step test), together with correction figures from changes in the air composition in the chamber, were used for calculating the gas exchange and the HP.

After the second measurements (Ib), including the lunch meal at 1200–1230 h, no food was served for the following 24 h (III) but free access to water was provided. Next day the subjects entered the chambers, as in previous experiments, and the gas exchange was measured during continued fasting from 0900 to 1300 h.

Results

'STANDARD' DIET (EXPERIMENTS Ia AND Ib)

The results obtained by measuring the oxygen consumption and the carbon dioxide production twice over a period of six hours in 12 adults on a 'standard' diet were used to calculate the HP. The data were statistically analysed in relation to body weight or metabolic weight (kg $W^{0.75}$), and it was found that the latter gave the best statistical fit. The statistical analyses showed that the intercepts

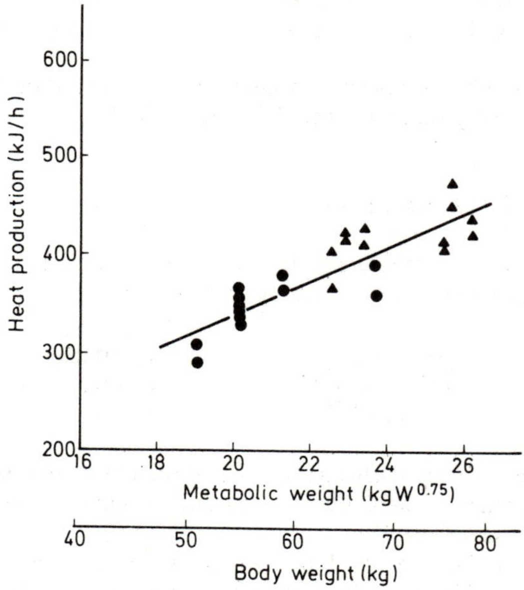

Figure 53.1 Heat production in man on a 'standard diet' (▲ = male; ● = female)

were not significant and the regressions are, therefore, calculated without intercept through the origin. The results concerning HP in relation to metabolic body weight are demonstrated in *Figure 53.1.*

No significant differences between the two periods of measurements were found and the data could be pooled in the following regressions:

O_2 (ℓ/h)	$= 0.865 \times$ kg $W^{0.75}$	$s_b = 0.011$	$r = 0.88$	$n = 24$
CO_2 (ℓ/h)	$= 0.634 \times$ kg $W^{0.75}$	$s_b = 0.008$	$r = 0.87$	$n = 24$
HP (kJ/h)	$= 17.0 \times$ kg $W^{0.75}$	$s_b = 0.21$	$r = 0.88$	$n = 24$

Using the equation for adults an HP of about 343 kJ/h could be expected under the given conditions for a person of 55 kg body weight increasing to 410 kJ/h at 70 kg body weight.

STOP TEST (EXPERIMENT II)

The values for gas exchange and HP from 1500 to 1600 h, which included the step test from 1505 h to 1515 h were compared with the values from 1400 h to 1500 h where the subjects were seated at the desk. As the work load is represented by the body weight times the number of steps and the height, regressions on body weight would be preferred. For comparison all regressions were based on body weight as follows:

O_2, 14–15 (ℓ/h)	$= 0.296 \times$ kg W	$s_b = 0.010$	$r = 0.71$	$n = 12$
O_2, 15–16 ℓ/h)	$= 0.412 \times$ kg W	$s_b = 0.008$	$r = 0.88$	$n = 12$
CO_2, 14–15 ℓ/h)	$= 0.236 \times$ kg W	$s_b = 0.006$	$r = 0.83$	$n = 12$
CO_2, 15–16 ℓ/h)	$= 0.347 \times$ kg W	$s_b = 0.009$	$r = 0.79$	$n = 12$
HP, 14–15 (kJ/h)	$= 5.97 \times$ kg W	$s_b = 0.17$	$r = 0.76$	$n = 12$
HP, 15–16 (kJ/h)	$= 8.41 \times$ kg W	$s_b = 0.16$	$r = 0.89$	$n = 12$

The increments of 39% for oxygen consumption, 47% for carbon dioxide production and 41% for heat production were highly significant.

24-HOUR FASTING (EXPERIMENT III)

The results obtained by measuring the gas exchange over the period of 20–24 hours of fasting gave the following regression equations:

O_2 (ℓ/h)	$= 0.741 \times$ kg $W^{0.75}$	$s_b = 0.017$	$r = 0.86$	$n = 12$
CO_2 (ℓ/h)	$= 0.523 \times$ kg $W^{0.75}$	$s_b = 0.011$	$r = 0.88$	$n = 12$
HP (kJ/h)	$= 14.5 \times$ kg $W^{0.75}$	$s_b = 0.32$	$r = 0.87$	$n = 12$

The fasting values for gas exchange and HE were about 15% below the values obtained in experiment I with the subjects on a 'standard' diet and the differences were highly significant.

The basal metabolic rate of adults lying on a couch is given by Durnin and Passmore (1967) as 230 and 306 kJ/h for females (55 kg) and males (70 kg), respectively, which is 21% and 13% below the values obtained in the present experiment on fasting with the subjects seated at a desk occupied by reading, writing or handiwork.

Conclusions

The investigation has shown that indirect calorimetry using respiration chambers with open-circuit ventilation can be applied to experiments in man for measurement of gas exchange and heat production under different conditions. A statistical analysis of the data showed that oxygen consumption, carbon dioxide production and HP could be calculated under the present conditions with an acceptable accuracy by using metabolic body weight (kg $W^{0.75}$) independently of sex and age.

References

ATWATER, W.O. and BENEDICT, F.G. (1905). *A Respiration Calorimeter with Appliances for the Direct Determination of Oxygen.* Carnegie Institution of Washington, Publication No. 42

BEATON, G.H. (1975). *Limitation of Present Knowledge about Protein and Energy Needs.* Proceedings of the Ninth International Congress on Nutrition, Mexico, 1972, Vol. 1, pp. 31–34. Basel: Karger

BENGTSSON, M.-A., EJLERTSSON, G., HOGLUND, G., KAIJ, K., MELANDER, A., NORDÉN, Å., SAMUELSSON, G. and SCHERSTÉN, B. (1979). *Scand J. Gastroenterol.,* **52**, Suppl. 14, p. 58

BENZINGER, T.H., HUEBSCHER, R.G., MINARD, D. and KITZINGER, C. (1958). *J. Appl. Physiol.,* **12**, Suppl. 1, pp. 1–28

DURNIN, J.V.G.A. (1975). *Unresolved Problems in Establishing International Standards for Energy Requirements.* Proceedings of the Ninth International Congress on Nutrition, Mexico, 1972. Vol. 1, 35–42. Basel: Karger

DURNIN, J.V.G.A. (1978). *Proc. Nutr. Soc.,* **37**, 5–12

DURNIN, J.V.G.A. and PASSMORE, R. (1967). Energy, Work and Leisure. London: Heinemann Educational Books

THORBEK, G. and NEERGAARD, L. (1970). In *Energy Metabolism*, pp. 243–246. Eds. A. Schürch and C. Wenk. EAAP Publication No. 13. Zurich: Juris Druck

54

DIET-INDUCED THERMOGENESIS DURING SHORT-TERM CHANGES IN ENERGY INTAKE IN MAN

M.J. DAUNCEY
Agricultural Research Council Institute of Animal Physiology and formerly of the Medical Research Council Dunn Calorimetry Group, Babraham, Cambridge CB2 4AT, England

Summary

The metabolic effects of over-eating and under-eating were studied in eight volunteers (four men and four women) who each occupied a whole-body calorimeter at 26°C on three occasions. Heat production (HP) and total heat loss were measured continuously during the 28 h inside the calorimeter. The energy intake was designated high, medium or low and was of identical nutrient composition, in percentage terms, at each level of intake. The medium intake was based on the mean daily intake of the subject during the week preceding entry into the calorimeter. The mean 24 h intakes in the calorimeter were 13 800 (high), 8400 (medium) and 3700 (low) kJ, with a large s.d. Over-eating increased the 24 h HP by an average of 10% while under-eating decreased it by 6%. At night (2230–0930 h) there was a marked contrast in the relative effects of the high and low intakes; HP increased by 14% during over-eating compared with a decrease of only 1% for under-eating.

Introduction

In man, the metabolic effects of over-eating and under-eating are not clearly established. This is partly due to a lack of suitable equipment for measuring energy expenditure continuously over long periods. Also, studies with man have often altered the percentage nutrient composition and the total number of meals eaten per day, both of which may affect the energy expenditure. Some workers (Miller, Mumford and Stock, 1967; Glick *et al.*, 1977) report no effect of over-feeding on resting metabolic rate (RMR) at least 12 h after the last meal, while others find an increase (Durnin and Norgan, 1969; Apfelbaum, Bostsarron and Lacatis, 1971). Interest also centres on the extent to which some people may adapt to over-feeding by increasing their energy expenditure so that there is little or no gain in weight or change in body composition.

The metabolic effects in man of altering the energy intake for only 24 h are not known. This paper reports a study with eight adult volunteers who over-ate or under-ate during a 28 h period in a whole-body calorimeter, while both HP and total heat loss were measured continuously.

Materials and methods

Four men and four women took part in the study. Their mean (± s.e.) age, height and body weight were 35 ± 2.7 years, 1.69 ± 0.01 m and 60 ± 2.8 kg respectively. Each subject lived in the calorimeter at 26°C on three occasions of high, medium or low energy intake. The order was randomized and the sessions were separated by monthly intervals. A session lasted from 0930 h on Day 1 to 1400 h on Day 2. Evaporative and sensible heat losses and HP were measured continuously in the calorimeter, which measured 2.08 m long, 1.22 m wide and 1.95 m high (Dauncey, Murgatroyd and Cole, 1978). The subject wore standard clothing and followed a set pattern of activity which is indicated in *Figure 54.1.*

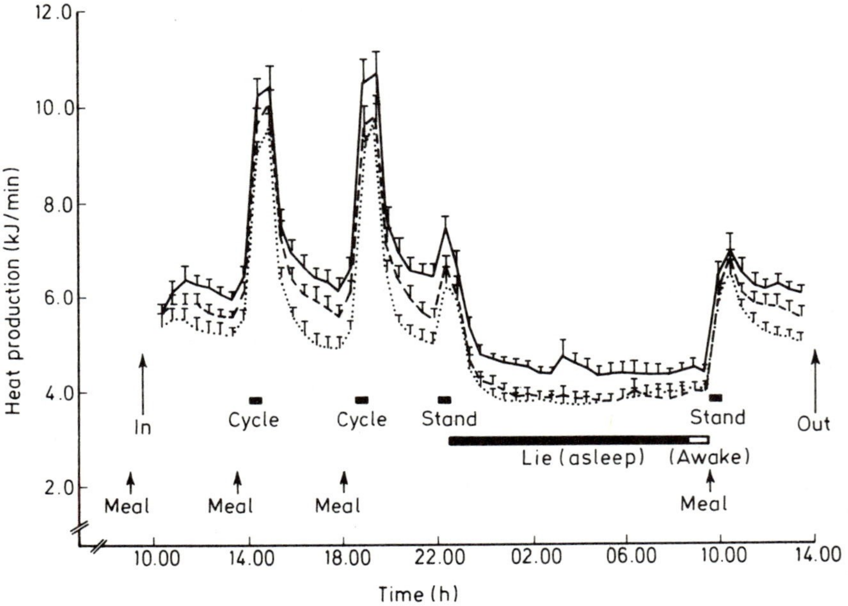

Figure 54.1 Heat production mean (± s.e., n = 8) *during three sessions of 28 h in the calorimeter at 26°C. Energy intakes were high* (——), *medium* (– – – –) *or low* (··········) *and of identical percentage protein, fat and carbohydrate. The sessions were at monthly intervals and the order was randomized. Unless otherwise stated the subject was sitting and able to read, write, listen to the radio or watch television. Cycling was at the low work load of 0.5 N and 50 rev/min. Mean values (kJ/min) for each 30 minute period are plotted.*

In a 24 h period in the calorimeter three identical meals were eaten. These were based on a unit system with one unit equivalent to a corned beef and crispbread sandwich, a tomato, two squares of chocolate and a digestive biscuit. This provided 1467 kJ metabolizable energy, 14 g protein, 20 g fat and 31 g carbohydrate. The energy intakes in the calorimeter are given in *Table 54.1.* The medium intake was based on the energy intake during everyday life, which was estimated from a week's weighed intake by the subject. The aim was for the subject to double or halve this intake for the high and low sessions. In practice all the subjects could not eat this amount of extra food and the intakes were reduced accordingly. This highlighted the importance of physical activity, the type of food available and the times at which it is eaten in determining energy intake in man. Further details of the methods are available (Dauncey, 1980).

Table 54.1 METABOLIZABLE ENERGY INTAKE AND HEAT PRODUCTION IN THE CALORIMETER ($\bar{x}$ ± s.e., n = 8)

Session in calorimeter	*Energy* (kJ/24 h)	
	Intake	*Expenditure*
High	13 829 ±475	8767*** ±288
Medium	8397 ±510	7986 ±297
Low	3703 ±359	7495** ±253

Paired t-tests for energy expenditure
*** significantly greater than Medium ($P<0.001$)
** significantly less than Medium ($P<0.01$)

Results

During the 24 h from 1330 h on Day 1 the difference between HP and total heat loss was only 1.2 ± 0.14%. Values for 24 h HP are given in *Table 54.1.* Despite the variation in the response of the subjects to over-feeding and under-feeding, the energy expenditure on the high intake was significantly greater ($P<0.001$) than that on the medium intake, while that on the low intake was significantly reduced ($P<0.01$). Comparisons were made with the paired t-test.

The relationship between the 24 h energy intake and energy expenditure could be summarized by a linear regression coefficient. The mean (± s.e.) linear coefficient of 0.125 ± 0.034 indicated that on average there was an increase in HP of 0.125 kJ for every 1 kJ increase in energy intake. Under the conditions of this study, the average energy intake needed to maintain energy balance was calculated as 7983 kJ/24 h.

Figure 54.1 compares the metabolic rates for the group during each level of energy intake. It shows that despite the wide range in energy intakes between the subjects during any one session, and the wide variation in energy intake during the week before entering the calorimeter, the range of energy expenditures during a particular session was narrow. There was a marked difference in the relative responses to over-feeding and under-feeding during the day and night. Although the effects of over-eating were greatest at night, those due to under-eating were most apparent during the day-time. The resting metabolic rate (RMR) 14 h after the last meal was significantly increased on the high intake ($P<0.01$), but there was no difference between that on the medium when compared with that on the low intake.

Discussion

This study shows that during only one day of altered energy intake the 24 h energy expenditure in man is dependent on the level of energy intake. The result was obtained even under conditions where the physical activity and energy intake were not standardized for several days before the measurements were made. During over-eating an average of 5400 kJ of additional metabolizable energy, the energy expenditure increased by 10% while the remainder was stored.

This finding agrees with work on other animals in which the 24 h energy expenditure within the zone of thermal neutrality is dependent on the level of energy intake (Graham *et al.*, 1959; Close, Mount and Start, 1971). In the present study the magnitude of the increase in the 24 h metabolic rate was of the same order as that calculated by Durnin and Norgan (1969). These workers found an increase of 10–12% after feeding an excess of 7100 kJ/d for six weeks. Apfelbaum, Bostsarron and Lacatis (1971) overfed a daily excess of 6300 kJ for two weeks and also found a similar increase in metabolic rate, with the RMR throughout the 24 h being increased by 10%. An increase in body weight which occurs after several weeks of over-feeding would be expected to affect the metabolic rate. However, it seems that the magnitude of the increase in energy expenditure after only one day of over-eating is similar to that found after several weeks.

The present work also showed that the RMR 14 h after the last meal could be affected by the previous day's energy intake. This finding is in agreement with work on growing animals (Dauncey and Ingram, 1979), although it did not agree with the conclusions of some workers (*see* Introduction). Since under-feeding did not alter the RMR 14 h after the last meal, the reduced RMR that has been observed by others during several days or weeks of under-feeding must take some time to become established (Keys *et al.*, 1950; Grande, Anderson and Keys, 1958). The increased metabolic rate during over-feeding may have been caused partly by the continued digestion, absorption and metabolism of the extra food. The possibility that there may also be an increase in thyroid hormone metabolism has been indicated from work with growing pigs (Ingram and Dauncey, *see* pages 411–415).

Acknowledgements

The author thanks Mr P.R. Murgatroyd for his engineering skills, Mr T.J. Cole for computing advice, Mr D.E. Walters for statistical advice, all the volunteers and Dr B.A. Cross, FRS, for allowing the work to be carried out at the Institute.

References

APFELBAUM, M., BOSTSARRON, J. and LACATIS, D. (1971). *Am. J. clin. Nutr.*, **24**, 1405–1409

CLOSE, W.H., MOUNT, L.E. and START, I.B. (1971). *Anim. Prod.*, **13**, 285–294

DAUNCEY, M.J. (1980). *Br. J. Nutr.*, **43**, 257–269

DAUNCEY, M.J. and INGRAM, D.L. (1979). *Br. J. Nutr.*, **41**, 361–370

DAUNCEY, M.J., MURGATROYD, P.R. and COLE, T.J. (1978). *Br. J. Nutr.*, **39**, 557–566

DURNIN, J.V.G.A. and NORGAN, N. (1969). *J. Physiol., Lond.*, **202**, 106P

GLICK, Z., SCHVARTZ, E., MAGAZANIK, A. and MODAN, M. (1977). *Am. J. clin. Nutr.*, **30**, 1026–1035

GRAHAM, N.McC., WAINMAN, F.W., BLAXTER, K.L. and ARMSTRONG, D.G. (1959). *J. agric. Sci., Camb.*, **52**, 13–24

GRANDE, F., ANDERSON, J.T. and KEYS, A. (1958). *J. appl. Physiol.*, **12**, 230–23

KEYS, A., BROZEK, J., HENSCHEL, A., MICKELSEN, O. and TAYLOR, H.L. (1950). *The Biology of Human Starvation.* Minneapolis: University of Minnesota Pre

MILLER, D.S., MUMFORD, P. and STOCK, M.J. (1967). *Am. J. clin. Nutr.*, **20**, 1223–1229

55

PLASMA THYROID HORMONES AND OXYGEN CONSUMPTION IN 14 HEALTHY VOLUNTEERS ON LOW-CALORIE AND WEIGHT-MAINTAINING DIETS. A PRELIMINARY STUDY.

P. SEROG, M. APFELBAUM, D. AUTISSIER, L. BRIGANT and F. BAIGTS
Laboratoire de Nutrition Humaine, Faculté Xavier Bichat, 16 Rue Huchard, 75018 Paris, France

Introduction

In man, restricted diets result in the lowering of oxygen consumption. This well-documented phenomenon (Benedict *et al.*, 1919 and Keys, 1951; reviewed by Apfelbaum, 1973) is particularly prominent even during very restricted diets when the nitrogen balance is equilibrated by a large protein load (Apfelbaum, Batsarron and Lacatis, 1971). During such a diet the energy expenditure adaptation is a day-to-day phenomenon. It seems chronologically unrelated to any hormonal changes previously studied. The serum cortisol and the serum glucagon show only transient peaks which return after a few days to the pre-diet levels. The serum insulin level falls rapidly and levels off (Apfelbaum, 1975).

Plasma levels of thyroid hormones during restricted dieting have been extensively investigated. The serum thyroxin levels do not change. Those of reverse T_3 (rT_3) were first studied by Danforth *et al.* (1975) who showed that the levels fell with over-eating. Spaulding *et al.* (1976) noted that rT_3 increased on fasting but could not demonstrate any changes in rT_3 on re-feeding. Leslie *et al.* (1978) showed that rT_3 is initially high in patients with anorexia nervosa and that it falls within the normal range after a 25% body-weight gain.

The aim of this experiment was to study the relationship between oxygen consumption ($\dot{V}O_2$) and thyroid hormones in subjects on two successive diets. The first was a carbohydrate-restricted diet which resulted in marked weight loss, and the second was an equilibrated low-calorie diet which maintained the subject's weight stable at the lower level.

Materials and methods

A group of 40 healthy young volunteers were studied before, during and after two weeks of each experimental diet. The composition of the very restricted diet is shown in *Table 55.1*. The low-calorie weight-maintenance diet was individually prescribed and the subjects' compliance assessed by daily questionnaires. (The diet contained an average of 78 g protein, 110 g of carbohydrates and 70 g of fat.)

Table 55.1 HIGH PROTEIN DIET (2.34 MJ/d)

Distribution					
Proteins	50%	70 g	$\frac{2}{3}$ Casein	14.0 g	
			$\frac{1}{3}$ Soya	22.0 g	
			Skimmed milk	34 g	
Lipids	25%	16 g	Linoleic acid	9.0 g	
Carbohydrate	25%	36 g	Lactose	7.2 g	
			Glucose	14.4 g	
			Galactose	14.4 g	
Minerals			Sodium	0.5 g	
			Potassium	2.0 g	
			Calcium	1.0 g	
			Magnesium	0.4 g	
			Iron	0.018 g	
			Zinc	0.015 g	
			Chloride	1.0 g	
			Phosphates	1.0 g	
Vitamins			A		5000 iu
			C		60 mg
			B_1		1.5 mg
			B_2		1.7 mg
			B_5		10 mg
			B_6		2 mg
			B_{12}		6 μg
			PP		20 mg
			Folic acid		400 μg
			Biotine		300 μg
			DL Tocopherolacetate		45 mg

Of the original 40 subjects 14 were eventually selected for their consistency of performance in closed-circuit oxygen consumption (the Cara model of the Benedict apparatus was used). Each subject was submitted to three consecutive measurements of $\dot{V}O_2$ at the end of each experimental period when blood was also taken for thyroid hormone determination.

Results

As expected, the subjects' weight fell during the first diet and thereafter remained stable (*Table 55.2*). The VO_2 decreased ($P<0.005$) during the restricted diet and increased again on refeeding (*Table 55.3*). It did not, however, reach the pre-diet levels.

Table 55.2 CHANGES IN WEIGHT AFTER A TWO-WEEK PROTEIN DIET AND THEN A TWO-WEEK REFEEDING DIET: 14 SUBJECTS

	Spontaneous diet		*High protein diet*		*Refeeding diet*
Weight (kg)	63.46± 10.78		59.00 ± 10.22		59.07 ± 9.57
Mean difference		4.46 ± 0.24		0.07 ± 0.34	
P		<0.001		NS	

Table 55.3 OXYGEN CONSUMPTION (ml O_2 STPD/min)

	Spontaneous diet	*High protein diet*	*Refeeding diet*
1	131	122	117
2	161	125	139
3	157	137	154
4	141	84	105
5	130	86	110
6	139	84	115
7	150	96	143
8	145	119	130
9	135	74	125
10	141	140	142
11	142	103	121
12	125	93	118
13	142	86	107
14	127	92	141
Mean	140.33	102.83	126.16
s.d.	10.50	21.50	15.33

Changes in the serum thyroxin were unremarkable (*Table 55.4*). Serum T_3 levels fell during the restricted diet and rose again on refeeding. Those of rT_3, however, increased dramatically on the restricted diet to nearly three times the initial values and then diminished on refeeding.

Discussion

It is obvious (*Figure 55.1*) that the changes in $\dot{V}O_2$ during the two experimental periods are exactly parallel to those of serum T_3, the changes of which mirror those of serum rT_3. This relationship strongly suggests that the specific deiodination of T_4 to rT_3 is the first step in the process of adaptation of energ expenditure to energy intake.

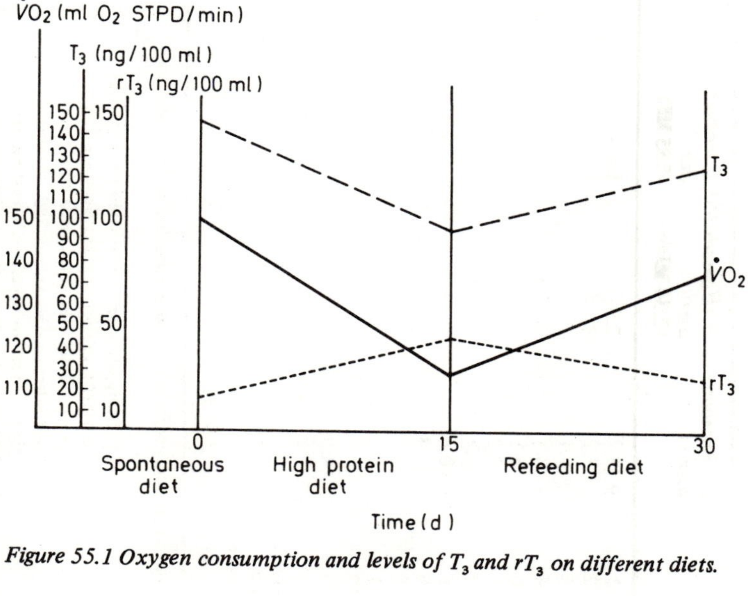

Figure 55.1 Oxygen consumption and levels of T_3 and rT_3 on different diets.

Table 55.4

Subject	T_4 (μg/100 ml)			T_3 (ng/100 ml)			rT_3 (ng/100 ml)		
	Spontaneous diet	*High protein diet* (2.34 MJ)	*Refeeding diet* (5.85 MJ)	*Spontaneous diet*	*High protein diet* (2.34 MJ)	*Refeeding diet* (5.85 MJ)	*Spontaneous diet*	*High protein diet* (2.34 MJ)	*Refeeding diet* (5.85 MJ)
1	9.2	7.9	8.0	150	96	115	16	59	37
2	7.2	8.8	8.0	135	115	150	17	45	24
3	6.1	7.9	8.0	130	135	158	16	25	27
4	8.2	9.1	7.0	170	110	145	11	45	17
5	8.8	7.4	6.2	145	92	115	20	50	29
6	7.2	6.3	5.6	120	77	130	13	45	21
7	6.3	6.6	6.9	135	105	180	20	43	30
8	10.8	8.7	7.3	150	115	145	14	27	13
9	7.4	7.3	7.3	140	80	100	14	47	20
10	5.4	5.8	4.3	135	79	100	14	47	16
11	8.2	7.3	6.0	170	84	130	18	30	19
12	7.6	8.0	5.8	160	100	120	20	57	23
13	4.9	6.0	4.4	130	70	86	12	31	15
14	7.9	7.0	5.8	150	92	118	14	34	20
Mean	7.51	7.43	6.47	144.28	96.42	128.00	15.64	41.78	22.21
s.d.	1.55	1.03	1.23	15.04	18.01	25.58	3.00	10.7	6.65
Normal range	7.69–9.77			141.30–165.80			14.21–20.00		

Statistical analyses were performed using the t-paired test

T_4	Spontaneous diet versus high protein diet	:	NS
	High protein diet versus refeeding diet	:	$P<0.01$
	Spontaneous diet versus refeeding diet	:	$P<0.02$
T_3	Spontaneous diet versus high protein diet	:	$P<0.001$
	High protein diet versus refeeding diet	:	$P<0.001$
	Spontaneous diet versus refeeding diet	:	$P<0.001$
rT_3	Spontaneous diet versus high protein diet	:	$P<0.001$
	High protein diet versus refeeding diet	:	$P<0.001$
		:	$P<0.001$

The factors that control this process are largely unknown; however, they are related in some way to changes in body weight and composition of the diet. The mechanisms that trigger the de-iodination of T_4 to rT_3 during restricted dieting have not been studied.

Acknowledgements

We gratefully acknowledge the kind assistance of Dr R. Cook, Department of Medicine, University of Cambridge, for his help with the manuscript. Grants were provided by DGRST (7871080), Institute de Recherche Servier, Sopharga.

References

APFELBAUM, M. (1973). In *Energy Balance in Man.* pp. 71–81. Ed. M. Apfelbaum, Paris: Masson Publications

APFELBAUM, M. (1975). In *Obesity in Perspective*, Vol. 2, Part 2, pp. 145–157. Ed. G. Bray. Washington, D.C.: US Government Printing Office

APFELBAUM, M., BOTSARRON, J. and de LACATIS, D. (1971). *Am. J. clin. Nutr.*, 24, 1045

BENEDICT, F.G. MILES, W.R., ROTH, P. and SMITH, H.M. (1919). *Am. J. Physiol.*, 30, 192

DANFORTH, E., DESILETS, E.J., HORTON, E.S., SINS, E.A.H., BURGER, A.G., BRAVERMAN, L.E., VAGENAKIS, A.G. and INGBAR, S.D. (1975). *Clin. Res.*, 23, 573A (Abstract)

KEYS, A. (1951). *Biology of Human Starvation.* Minneapolis University Press

LESLIE, R.D.G., ISAACS, A.J., GOMEZ, J., RAGGETT, P.R. and BAYLISS, R. (1978). *Br. med. J.*, 2, 526–528

SPAULDING, S.W., CHOPRA, I.J., SHERWIN, R.S. and LYALL, S.S. (1976). *J. clin. Endocr. Metab.*, 42, 197–200

56

THE REDUCTION IN METABOLIC RATE AND HEART RATE OF MAN DURING MEDITATION

D.J. FARRELL
Department of Biochemistry and Nutrition, University of New England, Armidale, New South Wales 2351, Australia

Summary

Measurements of metabolic rate by indirect calorimetry and of heart rate were made on 19 subjects for 30-minute periods before, during and after meditation. The metabolic rate was significantly reduced by 10% during meditation. Heart rate declined by 5% during meditation and was still significantly less 30 minutes after meditation.

Introduction

There are many factors known to influence the basal metabolic rate of man (Garrow, 1974). However, basal metabolism differs from resting metabolism in that the subject is in the postabsorptive state (Brody, 1945). During sleep there is a decrease of about 7% in the metabolic rate (Passmore and Durnin, 1955), but it has been suggested that a reduction occurs during transcendental meditation (Dhanaraj and Singh, 1977) which is practised in Western society.

The purpose of this chapter is to report the measurements made of both resting metabolism and heart rate of subjects before, during and after meditation.

Subjects and methods

Nineteen adults (eight females and 11 males) aged 18–67 years and weighing from 50 kg to 85 kg, who had been practising transcendental meditation for a few months to 13 years, volunteered for the study.

Measurement of energy metabolism was made by indirect calorimetry in a respiration chamber using the confinement method (Lulaine, 1884; cited by Brody, 1945). Volunteers sat in a comfortable chair within the chamber either between 1900 h and 2100 h, or between 1400 h and 1600 h, according to their preference. They were given details of the protocol to be followed and allowed to become familiar with the experimental conditions during a trial run. During the actual period of measurement, 15 minutes were allowed for the subject to become accustomed to the respiration chamber.

The respiration chamber (192 cm long × 152 cm high × 72 cm wide) with a small double-glazed window was of 2228 ℓ capacity and made of stainless steel sheets on a rigid-steel framework. Polyurethane (3 cm thick) was enclosed between the outer and inner walls to insulate the chamber. The hinged door when closed was sealed by inflating a bicycle tyre inner tube placed between the door and chamber perimeter. Temperature could be controlled by means of a bimetal contact thermometer that activated cooling and heating coils. The integrated electrical output of eight copper-constantan thermocouples sited at various points in the chamber and monitored on a chart recorder was used to record dry bulb temperature; an additional thermocouple measured wet bulb temperature.

The initial charge of air was circulated continuously, using a Thomas diaphragm pump (model 2727), at 150 ℓ/min through a cold trap that was located outside the chamber and held the relative humidity within the chamber at about 70%.

After 15 min, the chamber was sealed and an initial sample of chamber air was taken in a gas-tight metal syringe (500 ml) and held for subsequent analysis for carbon dioxide and oxygen concentrations using a Haldane analyser and a Beckman E_2 oxygen analyser. Temperature, humidity and barometric and chamber pressures were recorded initially and at intervals thereafter. Gas samples were again taken after 30, 60 and 90 minutes. After 30 minutes the subject was asked through a telephone system to commence meditating and 60 minutes later to stop. Checks were occasionally made to ensure that the subject was not sleeping.

Heat production for each of the three 30-minute periods (i.e. before (1), during (2) and after meditation (3)) was calculated from the change in volumes of oxygen and carbon dioxide in chamber air. These volumes were corrected for the volumes of the equipment in the chamber and the subject and reduced to STP. The equation used to calculate heat production (kJ) was:

$$16.2\ \ell\ O_2 + 5.0\ \ell\ CO_2$$

Heart rate was displayed on a meter outside the chamber from ECG waves sensed through wrist-plate electrodes. A panel light showed individual pulses. These were counted at 5-minute intervals and the mean value recorded. The means of the six observations for each 30-minute period are reported.

Data for each subject during the three 30-minute periods were analysed using a paired 't test'.

Results

Subjects did not find that the experimental conditions interfered with their meditation, and at no time during the experiment did the carbon dioxide concentration in expired chamber air exceed 1.2%.

Mean respiratory quotient, heat production and heart rate are given in *Table 56.1*. During meditation, heat production and heart rate declined ($P<0.01$) by 10% and 5% respectively, when compared with the period before meditation. Although heat production was not different ($P<0.05$) between periods 1 and 3, the heart rate was lower ($P<0.025$) during period 3, i.e. 30 minutes after meditation. Respiratory quotient remained the same throughout the three periods.

Table 56.1 MEAN RESPIRATORY QUOTIENT, HEAT PRODUCTION AND HEART RATE OF 19 SUBJECTS MEASURED FOR THREE 30-MINUTE PERIODS BEFORE, DURING AND AFTER MEDITATION

	Meditation				
	Before		*During*		*After*
Respiratory quotient	0.82^{a}[1]		0.82^{a}		0.81^{a}
s.e.d.[2]		0.01		0.01	
Heat production (kJ/kg per d)	109.4^{a}		98.9^{b}		116.0^{a}
s.e.d.		3.3		4.2	
Heart rate (beats/min)	77.2^{a}		73.3^{b}		75.5^{c}
s.e.d.		0.8		0.5	

[1] Values with different superscripts (a, b and c) are significantly different ($P<0.025$)
[2] Standard error of the difference between means

Discussion

It is clear that man can reduce his resting metabolism during meditation and this is accompanied by a decrease in heart rate. It appears from the continued decrease in heart rate following meditation that the effects of transcendental meditation persist for some time.

The respiratory quotient of 0.82 indicated that subjects were not in the post-absorptive state. No direction was given regarding food consumed prior to calorimetric measurements, although it is common practice not to consume food for several hours prior to meditating. However, values shown in *Table 56.1* are sufficiently close to a predicted mean basal metabolic rate of 100 kJ/kg per d using Table 14.5 compiled by Brody (1945) for our subjects whose average age was 37 years.

Heart rates of subjects appeared to be higher than frequently observed. Routt (1977) reported a mean value of 67 beats/min for 12 subjects during meditation in comparison with 78 beats/min for 12 controls, and observed a non-significant reduction in heart rate during meditation when compared with before and after. The results presented here support previous observations reported by Dhanaraj and Singh (1977) that during meditation there is a decline in oxygen uptake.

Acknowledgements

I thank the 19 subjects who participated in this study, Mr E. Thomson for his technical assistance and Mr D. Sharp for constructing the heart-rate counter.

References

BRODY, S. (1945). *Bioenergetics and Growth.* New York: Reinhold

DHANARAJ, V.H. and SINGH, M. (1977). In *Scientific Research on the Transcendental Meditation Programme*, Vol. 1, pp. 256–260. Eds. Orme-Johnson and J.T. Farrow. Maharishi European Research University Press

GARROW, J.S. (1974). *Energy Balance and Obesity in Man.* London: North-Holland Publishing Co.

PASSMORE, R. and DURNIN, J.V.G.A. (1955). *Physiol. Rev.,* **35**, 801–840

ROUTT, T.J. (1977). In *Scientific Research on the Transcendental Meditation Programme,* Vol. 1, pp. 137–139. Eds. Orme-Johnson and J.T. Farrow. Maharishi European Research University Press

57

ENERGY METABOLISM IN ADULT MINK IN RELATION TO PROTEIN-ENERGY LEVELS AND ENVIRONMENTAL TEMPERATURE

A. CHWALIBOG, N. GLEM-HANSEN, S. HENCKEL* and G. THORBEK
National Institute of Animal Science, Rolighedsvej 25, DK-1958 Copenhagen V, Denmark

Summary

Energy metabolism in eight adult mink, fed different protein-energy proportions (18, 34, 61%), feeding levels (150, 200, 250 g feed/d) and kept at different temperatures (20, 10, 0°C), was measured in 62 respiration experiments. Carbon dioxide production and oxygen consumption in relation to metabolic weight ($kg^{0.75}$) increased by about 40% when the temperature decreased from 20 to 0°C. Metabolizable energy (ME) was about 80% of gross energy (GE) for mixtures A (18% protein-energy) and B (34%) and was found to be independent of feeding levels. ME decreased to 72% for mixture C (61%) which was caused by a lower digestibility of the energy. ME was found to be independent of temperature for mixture B (34%). Retained energy (RE) was dependent on protein level, being highest for mixture B (34%) on all feeding levels, and on temperature, decreasing from 311 kJ at 20°C to −324 kJ at 0°C for mixture B (34%). The maintenance requirement based on 33 measurements with positive energy balance in the thermoneutral zone (20°C) was estimated as:

$$ME_m \text{ (kJ)} = 527 \times kg^{0.75}.$$

In this group (n = 33) the net efficiency of utilization of ME for production ($\frac{RE}{ME-ME_m}$) was found to be 0.67 ± 0.09.

Introduction

Some results concerning protein metabolism in mink can be found in the literature, but there are rather few on energy metabolism (Farrell and Wood, 1968). As fur production is an important part of Danish agriculture, the respiration unit for small animals at the Institute was rebuilt (Chwalibog, Lind and Thorbek, 1979) and used for trials with mink in order to measure energy metabolism at different levels of protein and energy intake. The data were used to estimate the maintenance requirement. Mink are often kept at temperatures below 0°C, and for that reason another experiment with adult mink kept at temperatures from 20 to 0°C was carried out in order to measure the influence of temperature on heat production (HP) and RE.

* Regnecentralen A/S, Copenhagen

Materials and methods

In the first experiment (I), eight adult male mink were measured individually in eight balance periods, consisting of seven days of collection and 24 hours of respiration measurements placed in the middle of each collection period. Three different feed compounds, mixtures A, B and C, with 18, 34 or 61% of protein-energy in relation to GE, respectively, were applied. The protein-energy (kJ) was calculated as crude protein (g) × 23.86. Three different feeding levels (150, 200 and 250 g/d: L, M and H) were applied and the temperature was 20°C with a humidity of 50–60%.

In the second experiment (II) the influence of a lower temperature (10 or 0°C, RH 50–60%) was investigated on animals fed 200 g of mixture B. Thereby the measurements carried out in series I on mink, which were fed with 200 g of mixture B at a temperature of 20°C, were used for comparison with the 10°C and 0°C experiments.

In series II a group of two mink was measured in four balance periods. The animals were kept for three weeks at each temperature and the measurements were carried out in the second and third week. The gas exchange was measured for the group of two and divided equally.

Results

The allocation of animals in different series and treatments, together with live weight and energy intake, is presented in *Table 57.1*. From the experiments where the mink were kept at different temperatures, the gas exchange was measured. The results have been regressed on metabolic weight as shown in *Table 57.2*. Carbon dioxide production and oxygen consumption in relation to metabolic weight increased by about 40% when the temperature decreased from 20 to 0°C.

The results from the balance and respiration measurements in the experiment on different protein-energy levels and at different temperatures have been used

Table 57.1 ALLOCATION OF ANIMALS, LIVE WEIGHT AND INTAKE OF ENERGY

	Ser. I (20°C)											
Feed (g)	*Mixture A*				*Mixture B*				*Mixture C*			
	Gross energy (kJ)	*Live weight*			*Gross energy* (kJ)	*Live weight*			*Gross energy* (kJ)	*Live weight*		
		n	*g*	*s.d.*		*n*	*g*	*s.d.*		*n*	*g*	*s.d.*
L.150	818	8	1218	273	927	12	1281	139	789	4	1214	26
M.200	1087	10	1313	157	1308	4	1417	80	1052	4	1450	143
H.250	1320	2	1172	–	1547	8	1442	138	1315	4	1332	81
Mixture B	*Ser. I* 20°C				*Ser. II* 10°C				*Ser. III* 0°C			
M. 200	1308	4	1417	80	1144	2	1496	–	1139	2	1319	–

Table 57.2 CARBON DIOXIDE PRODUCTION AND OXYGEN CONSUMPTION AT DIFFERENT TEMPERATURES IN RELATION TO METABOLIC WEIGHT ($kg^{0.75}$)

	20°C (n = 12)			10°C (n = 8)			0°C (n = 8)		
	Regr. coeff.	s_b	*s.d. of residuals*	*Regr. coeff.*	s_b	*s.d. of residuals*	*Regr. coeff.*	s_b	*s.d. of residuals*
Carbon dioxide (ℓ)	27.4	1.2	5.3	31.6	0.9	3.7	36.1	0.9	3.4
Oxygen (ℓ)	31.9	1.5	7.0	39.4	1.1	4.3	46.1	1.4	5.3

Table 57.3 ENERGY METABOLISM IN ADULT MINK IN RELATION TO PROTEIN-ENERGY LEVELS AND TEMPERATURES

	Ser. I (20°C)								
Feed (g)	*Mixture A*			*Mixture B*			*Mixture C*		
	ME (kJ)	*HE* (kJ)	*RE* (kJ)	*ME* (kJ)	*HE* (kJ)	*RE* (kJ)	*ME* (kJ)	*HE* (kJ)	*RE* (kJ)
L.150	652	777	−125	719	758	−39	561	678	−117
M.200	848	804	44	1063	752	311	755	744	11
H.250	1016	719	297	1238	893	344	944	775	169
Mixture B	*Ser. I* 20°C			*Ser. II* 10°C			*Ser. II* 0°C		
M. 200	1063	752	311	920	993	−73	901	1224	−324

Table 57.4 ENERGY METABOLISM EXPRESSED IN PER CENT OF GROSS ENERGY

	Ser. I (20°C)								
Feed (g)	*Mixture A* (n = 20)			*Mixture B* (n = 24)			*Mixture C* (n = 12)		
	ME (%)	*HE* (%)	*RE* (%)	*ME* (%)	*HE* (%)	*RE* (%)	*ME* (%)	*HE* (%)	*RE* (%)
L.150	80	95	−15	78	82	−4	71	86	−15
M.200	78	74	4	82	57	24	72	71	1
H.250	77	55	23	80	58	22	72	59	13
Mixture B	*Ser. I* 20°C (n = 4)			*Ser. II* 10°C (n = 2)			*Ser. II* 0°C (n = 2)		
M. 200	82	57	24	80	87	−6	79	107	−28

to calculate the ME, the total RE and the HP. The results obtained are shown in absolute values in *Table 57.3* and expressed as percentages of GE in *Table 57.4.*

The ME for each compound was found to be independent of feeding level. For mixtures A and B, the ME was about 80% of gross energy but only 72% in mixture C due to a lower digestibility of energy in this compound. By comparing ME values of mixture B at different temperatures, it was found that ME was about 80% of GE independent of temperature.

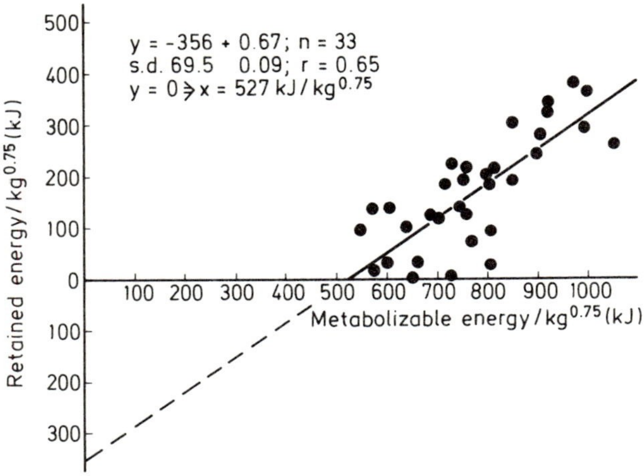

Figure 57.1 Retained energy in relation to metabolizable energy.

For each compound the RE increased with higher feeding levels. The RE was dependent on protein level, which was highest for mixture B as demonstrated in *Tables 57.3* and *57.4*. The RE fell rapidly with decreasing temperature together with increasing HP. An intake of 200 g resulted in an RE that was 311 kJ or 24% of GE at 20°C and –73 kJ (–6%) at 10°C, which decreased to a negative value of –324 kJ (–28%) at 0°C.

All measurements carried out in the thermoneutral zone (20°C), which gave a positive-energy balance (n = 33), were used for regression of RE/$kg^{0.75}$ on ME/$kg^{0.75}$ (*Figure 57.1*). This gave an estimate of the maintenance requirement as:

$$ME_m\ (kJ) = 527 \times kg^{0.75}.$$

In this group (n = 33), the overall efficiency of utilization of ME for production $(\frac{RE}{ME - ME_m})$ was found to be 0.67 ± 0.09.

References

CHWALIBOG, A., LIND, J. and THORBEK, G. (1979). *Z. Tierphysiol. Tierernähr. Futtermittelk.*, **41**, 154–162

FARRELL, D.J. and WOOD, A.J. (1968). *Can. J. Zool.*, **46**, 47–52

58

EFFECT OF WATER TEMPERATURE ON ENERGY UTILIZATION IN RAINBOW TROUT (*Salmo gairdneri*)

C.Y. CHO and S.J. SLINGER
Fisheries Branch, Ministry of Natural Resources and Department of Nutrition, University of Guelph, Guelph, Ontario, Canada N1G 2W1

Summary

The object of this study was to determine the heat increment of feeding and energy gain with resting rainbow trout fed a balanced diet at different water temperatures (7.5, 10, 15 and 20°C). Increasing water temperature resulted in increased digestible energy (DE) intake and gross heat increment. Heat production (HP; per kg BW/d) of fasting trout also increased as the temperature increased up to 15°C, but HP was decreased by 9% going from 15°C to 20°C. On the other hand, rainbow trout used a similar proportion (21–22% of GE) of their overall energy budget for maintenance except at 20°C when the requirement was 15%. Energy gain increased linearly from 44 to 53% of DE (r = 0.98) with increasing temperature and the heat increment was lowest at 1.9 kJ/g feed intake at 15°C and highest at 3.4 kJ at 7.5°C. This is in keeping with the knowledge that 15 °C is about the optimum temperature.

Introduction

Rainbow trout are poikilotherms and therefore environmental (water) temperature is the major factor regulating their metabolic rate. Since fish require little energy for thermoregulation their energy requirement for maintenance is much lower than in mammalian species. The standard environmental temperature (SET) for rainbow trout is 15°C (National Research Council, 1973), and this temperature seems to be close to the optimum for growth, feed utilization and other considerations. At lower temperatures the trout reduce their basal metabolic rate and consume less feed, and therefore the rate of gain is reduced in proportion to the reduction in temperature. At higher temperatures rainbow trout increase their feed intake, growth and feed : gain ratio. Smith, Rumsey and Scott (1978) studied the net energy maintenance requirement of salmonids at different water temperatures using direct calorimetry. They reported that the relationship between HP of fasting rainbow trout and water temperature (T) was described by the expression:

$$HP = 0.59 + 0.0525T \text{ kcal/kg per h}$$

which indicates a much higher rate of HP than previous reports.

Groups of hatchery-reared rainbow trout (50–140 g fish; 3 kg total biomass) were placed in an open-circuit indirect respirometer with an automated data processor, which was used for the study of Cho, Bayley and Slinger (1976). The HP was estimated by measuring the difference in the dissolved oxygen (DO) concentration of the inflowing and outflowing water, and this enabled the oxygen consumption to be calculated. The DO concentration in the influent was controlled at 9.5 ± 0.3 mg/ℓ and the DO in the effluent was never less than 5 mg/ℓ regardless of the water temperature. The oxygen consumption, water flow rate (4–6 ℓ/min) and temperature were monitored every minute by a mini-computer and the data were computed daily. The fish were resting on the bottom of a sealed tank (30 ℓ) with minimum voluntary activity and were offered three meals a day which lasted as long as the fish were actively feeding. Faeces were collected daily between 1700 and 0800 h to avoid possible contamination with feed residues. The apparent DE of the diet was measured using chromic oxide as a digestion indicator.

The diet used for all temperatures is shown in *Table 58.1.* The voluntary intake of DE was 67 kJ at 7.5°C and this increased to 304 kJ/kg BW per d at 20°C; however, live-weight gain was maximum at 17.5 g/kg BW per d at 15°C and decreased to 10.9 g/kg BW per d at 20°C. The feed : gain ratio was increased with rising temperatures and at 20°C it was highest (*Table 58.2*).

Table 58.1 COMPOSITION OF DIET

Ingredients (%)	%
Herring fish meal	35
Soybean meal	20
Wheat middlings	32
Vitamin premix	2
Mineral premix	1
Carboxymethylcellulose	3
Cr_2O_3	1
Herring oil	6
Digestive nutrients (% DM)	
Crude protein	42
Digestible protein	38
Digestible fat	9
Digestible energy	16 kJ/g

Table 58.2 FEED INTAKE AND GROWTH OF RAINBOW TROUT

Water temp (°C)	*7.5*	*10*	*15*	*20*
Average initial BW (g/fish)	93	136	47	107
DE intake/kg per d (kJ)	67	136	244	304
Live-weight gain/kg per d (g)	3.9	8.2	17.5	10.9
Feed:gain ratio	1.0	1.1	1.2	1.8

The oxygen consumption for the 7-day-feeding period was increased from 2.2 g to 7.1 g/kg BW per d as the water temperature increased from 7.5°C to 20°C. The oxygen consumption during the fasting period was increased from

1.3 g at 7.5°C to 4.5 g at 15°C and then decreased to 4.1 g at 20°C (*Table 58.3*). Gross heat increment of feeding was calculated to be 0.9, 1.2, 1.8 and 3.0 g oxygen at 7.5, 10, 15 and 20°C, respectively.

Table 58.3 OXYGEN CONSUMPTION (g/kg BW per d) IN FEEDING AND FASTING PERIOD

Water temp (°C)	*7.5*	*10*	*15*	*20*	
Feeding period (7 days)	2.2 ± 15*	3.9 ± 15	6.3 ± 11	7.1 ± 12	$P<0.05$
Fasting period (minimum after 3 days)	1.3	2.7	4.5	4.1	
Heat increment of feeding	0.9	1.2	1.8	3.0	

* Mean ± % CV

The partition of energy expressed as a percentage of gross energy (GE) intake is shown in *Table 58.4*. The loss as faecal energy ranged from 20 to 22% of dietary energy. The proportion ascribed to heat increment was lowest at 15°C and either lower or higher temperatures than 15°C gave higher heat increments. The energy required for maintenance as a percentage of GE intake was similar (21–22%) at 7.5, 10 and 15°C, but lower (15%) at 20°C.

Table 58.4 PARTITION OF ENERGY (% of GE) IN RAINBOW TROUT

Water temp (°C)	*7.5*	*10*	*15*	*20*
Energy				
Intake*	100	100	100	100
Digestible	78	78	80	80
Metabolizable	70	70	72	72
Net	55	60	63	61
Gain	34	38	42	46
Lost				
Faecal	22	22	20	20
Non-faecal (estimated)	(8)	(8)	(8)	(8)
Respired				
Heat increment	15	10	9	11
Maintenance	21	22	21	15

* Amount of energy intake per day (*See Table 58.2*).

The proportion of energy available for gain (% of DE) was increased from 44% to 58% as water temperature was raised from 7.5°C to 20°C (*Table 58.5*). The heat increment (kJ/g feed intake) was lowest at 15°C (1.9 kJ). The energy required for maintenance was increased from 19 to 66 kJ/kg BW per d when the water temperature was raised from 7.5°C to 15°C and was then reduced to 59 kJ/kg BW per d at 20°C.

Increasing water temperature resulted in increased DE intake accompanied by higher total oxygen consumption and gross heat increment of feeding (*Tables*

58.2 and *58.3*). Heat production of fasting trout (basal metabolism) also increased as the temperature increased up to 15°C, but heat production was decreased by 9% at 20°C which suggests that this water temperature is above the optimum. Fry (1971) stated in his review of the effect of temperature on 'active' metabolism that sockeye salmon decreased the 'active' metabolic rate

Table 58.5 UTILIZATION OF DIGESTED ENERGY BY RAINBOW TROUT FOR GAIN, MAINTENANCE AND HEAT INCREMENT

Water temp (°C)	*7.5*	*10*	*15*	*20*
Energy gain (% of DE)	44	49	53	58
Heat increment (kJ/g feed intake)	3.4	2.3	1.9	2.5
Maintenance (kJ/kg per d)	19	39	64	59

1 g of oxygen equivalent to 14.3 kJ

above an optimum temperature. Rainbow trout also used a similar proportion of their overall energy budget for maintenance except at 20°C (*Table 58.4*). Because of the time allotted to acclimatization and to faeces collection in these experiments the fish size increased from 47 to 136 g during the study; however, Brett and Glass (1973) indicated that the 'active metabolic rate of sockeye salmon was almost unaffected by fish weight for most temperatures (5–20°C). In spite of the facts that live-weight gain reached a plateau at 15°C, that there was 38% less gain at 20°C than 15°C and that there was poorer efficiency of converting feed to live weight at 20°C than at lower temperatures (*Table 58.2*), net energy gain (% of DE) increased linearly ($r = 0.98$) with increasing temperature (*Table 58.5*). While live-weight gain was lower at 20°C than at 15°C, the much higher fat accumulation in the carcass at 20°C than at 15°C (34 vs. 43% of DM) would explain this apparent anomaly. Maximum-weight gain and feed efficiency are of primary practical importance in fish culture. The excess fat accumulation in fish at higher water temperatures would appear to be undesirable for net-energy balance. Heat increment of feeding per unit of feed intake was highest at 7.5°C and lowest at 15°C which suggests that of the temperatures used 15°C was closest to the optimum for efficiency of feed utilization.

Most investigators of energy partitions have not defined their conditions at all clearly, have not employed balanced nutritional regimens and have employed forced swimming or 'excited' conditions in short-term experiments which usually last for less than 48 hours (*see* reviews by Beamish, Niimi and Lett, 1975 and Brett and Groves, 1979). Therefore, there are few meaningful quantitative estimations of energy partition of dietary intake in salmonids. The present results, particularly the maintenance requirement values, do not agree with those reported by others. The explanation of these differences needs resolution in future work. Also, further studies designed to partition the dietary-energy budget under practical culture conditions would be useful.

Acknowledgements

The authors thank the Ontario Ministry of Natural Resources, Fisheries Branch and Fisheries and Oceans Canada, Fisheries and Marine Services for financial support.

References

BEAMISH, F.W.H., NIIMI, A.J. and LETT, P.F.K.P. (1975). *Comparative Physiology – Functional Aspects of Structural Materials*, pp. 188–209. Eds. L. Bolis, H.P. Maddrell and K. Schmidt-Nielsen. Amsterdam: North-Holland Publishing

BRETT, J.R. and GLASS, N.R. (1973). *J. Fish. Res. Bd Can.*, **22**, 405–409

BRETT, J.R. and GROVES, D.D. (1979). *Fish Physiology*, Vol. VIII, pp. 279–352. Eds. W.S. Hoar, D.J. Randall and J.R. Brett. New York: Academic Press

CHO, C.Y., BAYLEY, H.S. and SLINGER, S.J. (1976). In *Energy Metabolism of Farm Animals*, pp. 299–302. Ed. M. Vermorel. EAAP Publication No. 19. Clermont-Ferrand: G. de Bussac

FRY, F.E.J. (1971). *Fish Physiology*, Vol. VI, pp. 1–98. Eds. W.S. Hoar and D.J. Randall. New York: Academic Press

NATIONAL RESEARCH COUNCIL (1973). *Nutrient Requirements of Domestic Animals*, No. 11, p. 57. Washington, DC: National Academy of Sciences

SMITH, R.R., RUMSEY, G.L. and SCOTT, M.L. (1978). *J. Nutr.*, **108**, 1017–1024

59

DISCUSSION PAPER: COMPARATIVE ASPECTS OF ENERGY METABOLISM

H.S. BAYLEY
Department of Nutrition, University of Guelph, Guelph, Ontario, Canada N1G 2W1

There are two approaches to the study of animal energetics: the empirical and the analytical. Animal scientists need to elucidate the quantitative relationship between dietary intake of energy and energy storage by the animal. This information for animals under given feeding and environmental regimens serves as a reliable predictor of performance for similar animals under the same conditions. Changes in the diet, in level of feeding or in the environment reduce the precision of the prediction because of the complexities of the physiological and biochemical processes involved in animal production. Analysis of these processes by which the chemical energy of the food components is made available to support biological work has allowed the definition of the energy changes associated with each individual step in the transformation to be made. However, the interrelationships of the processes *in vivo* have so far precluded the integration of this information into a reliable predictive model. The empirical description of feed-energy utilization is made by Harris (1966). A summary of the reactions by which chemical energy is made available for biological work has grouped the process of food catabolism into three phases: hydrolysis, simplification and oxidation (Krebs and Kornberg, 1957).

Current research on energy metabolism in the commercially important domestic animals must integrate both the empirical and analytical approaches in order to complete the description of the quantitative relationship between the animal and its food. Progress in biological research is often aided by reference to a wide range of species and so it is useful to devote a part of this book to a greater range of species than is usual.

Domestic animals provide some service to man in return for the resources expended in their husbandry, but in the process of domestication the animals have become specialized in relation to their product, rather than developing their ability to compete in a particular habitat. The full range of animal species is characterized by diversity as animals have adapted to exploit their particular niche in the biosphere. Although the empirical scheme for describing energy utilization in farm animals is well adapted to include both ruminant and simple-stomached animals, it does include some concepts that restrict its more general application. For instance, the energy cost of food catabolism is arbitrarily assigned to two components: energy lost in the urine and the general increase in heat production following the ingestion of a meal. It is clearly simpler to measure energy lost in the urine as nitrogen excretory products, but both losses do arise as a consequence of food catabolism. Beamish, Niimi and Lett (1975) found that this concept was difficult to incorporate into a scheme used to describe food energy utilization in ammonotelic fish.

The interrelationship of feeding level and the maintenance requirement is central to the development of feeding standards for domestic livestock. Although the maintenance requirement is simple to define algebraically using the relationship between energy intake and energy deposition, it is not a fixed biological entity for an animal of a given live weight but it may vary with both the current feeding level and the animal's dietary history. Feeding levels vary more widely in nature than among domestic animals. With the exception of the lactating cow, appetite limits the intake of farm animals to provide an energy intake of only modest multiples of their maintenance requirements (two to three fold). In contrast, hibernating mammals show a seasonal increase in dietary-energy intake during the autumn period, and rapidly store the extra energy as fat. This rapid accumulation of an energy reserve implies a marked elevation of feeding level. Neonatal aquatic mammals receive energy from their dams in amounts many times their maintenance requirement. This allows the rapid deposition of subcutaneous fat and provides the thermal insulation which is vital for their survival. At the other end of the range, adult man maintains constant weight (and thus energy content) for several decades. Much of the work considered in this section of this book has described the application of the techniques of animal calorimetry to human subjects in an attempt to understand the energy homeostasis achieved in adult man. To date, the mechanisms that relate energy needs to food intake have escaped precise definition, but they must be flexible enough to allow for the changes in food-energy intake, which occur at different stages of the life cycle and at different seasons of the year, as well as providing the stability characteristic of adult man.

Body temperature and metabolic rate of poikilotherms are influenced by environmental temperature. As a consequence the level of *ad lib.* food intake increases as temperature rises. Brett and Groves (1979) showed that for a given level of food intake the metabolic rate of Sockeye salmon was greater at 15°C than at 10°C (*Figure 59.1*). The rate of increase of the metabolic rate as the feeding level increased was greater at the higher temperature. This effect is shown very clearly for the carp, but these results show an anomalous plateau indicating that beyond a certain level of feeding further increases in food intake had no effect upon metabolic rate. Interest in culturing fish using commercial feeds has prompted research to quantitate the relationship between feed and the quantity and quality of fish produced. Smith, Rumsey and Scott (1978) measured heat increments of feeding for protein, fat and carbohydrate for salmonid fish. They found no difference between the heat increments for protein and carbohydrate, and ascribed this to the ammonotelism of the fish. They concluded that this may complicate the estimation of protein and energy requirements of fish determined in factorial experiments.

It is significant that three of the chapters in this section of the book have described calorimetric studies with man. The use of the type of chamber developed for experiments with animals allows observations of energy balance over periods of up to a day under conditions which may be closer to normal than those which require the subject to breathe directly into a Douglas bag. Thorbek,

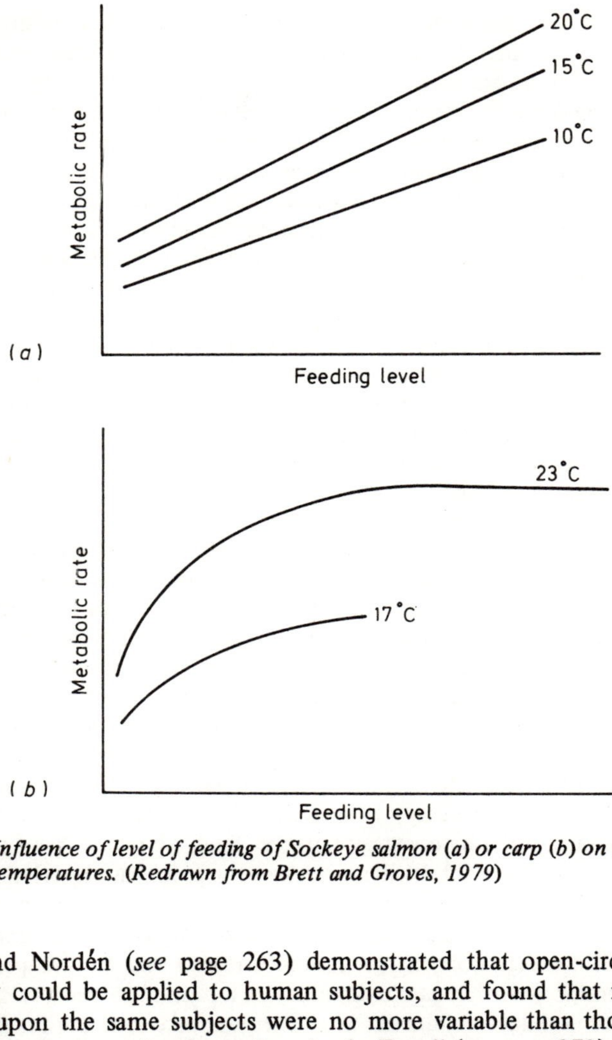

Figure 59.1 Influence of level of feeding of Sockeye salmon (a) or carp (b) on metabolic rate at different temperatures. (Redrawn from Brett and Groves, 1979)

Henckel and Nordén (*see* page 263) demonstrated that open-circuit indirect calorimetry could be applied to human subjects, and found that repeated observations upon the same subjects were no more variable than those obtained in the same apparatus using domestic animals. Farrell (*see* page 279) showed that there was substantial variation in metabolic rates measured in three successive 30-minute periods. Meditation significantly reduced both heart rate and heat production, and indicates the lability of short-term observations. Dauncey (*see* page 269) measured heat-production rates of human subjects confined within direct calorimeters for 24 hours, and showed that doubling food-energy intake increased the heat production and that the increase was equivalent to 10% of the extra dietary energy consumed at the higher level of food intake. This observation confirms that in man, as in other animals, an increasing energy intake increases the energy retention, and it discounts the idea that thermogenesis plays a substantial role in permitting changes in *man's* energy intake without affecting energy balance.

The subject of thermogenesis received attention from Müller and Kirchgeβner (*see* page 249), who found no difference in heat production of sows fed at the maintenance level upon either a mixed diet or a low-carbohydrate mixture of fat and protein. They discounted the idea that low-carbohydrate diets promote thermogenesis, and thus this work does not support the concept that particular dietary combinations are better suited to promote negative energy balance when fed at low intake levels. Rothwell *et al.* (*see* page 243) investigated the same phenomena using pigs growing from 6 kg to 20 kg, but their results were less conclusive. Tasaki and Kushima (*see* page 253) found an unusually large heat increment of feeding in response to a single meal of protein: more than half of the ingested energy was eliminated as heat by the cockerels. The corresponding heat losses when either carbohydrate or oil was fed were 5% and 0% of the ingested energy respectively. These results confirm that there is a greater heat increment of feeding for protein than for carbohydrate or fat, particularly for the uricotelic birds. The contradiction between the results of the investigations with the sows and the cockerels indicates the need for a systematic evaluation of the influence of dietary protein upon energy metabolism using a range of species and several levels of feeding. This uncertainty with regard to the effect of dietary protein upon heat production, and hence upon energy balance, is of great interest in human dietetics.

The investigation of the influence of environmental temperature upon heat production of young turkeys showed that the heat production rate did not reach a minimum value as temperature was increased from 20°C to 40°C (MacLeod, Tullett and Jewitt, *see* page 257). This is inconsistent with the concept of the thermoneutral zone of temperature'. However, the birds were allowed to feed *ad lib.*, so that the effect of environmental temperature may have been confounded with level of food intake, as shown for pigs by Close and Mount (1978). Chwalibog *et al.* (*see* page 283) gave a classic demonstration of the influence of environmental temperature upon energy balance: decreasing the temperature from 20°C to 10°C converted a positive energy balance (a third of metabolizable energy retained) to zero energy balance. Further reduction of the temperature to 0°C resulted in a negative energy balance. Such a large effect of temperature, at a fixed level of energy intake, demonstrates the practical importance of manipulating environmental temperatures in animal husbandry. The demonstration of the effect of water temperature upon fish productivity (Cho and Slinger, *see* page 287) showed that digestible-energy intake increased over the temperature range studied, although growth was maximum at an intermediate temperature. Energy lost as heat did not account for the extra energy intake so that a change in carcass composition would be expected, suggesting that it is possible to manipulate the type of carcass produced by alteration of either diet or water temperature.

Unfortunately the devotion of this part of the book to the comparative aspects of energy metabolism has achieved little in showing directly how information from a broad range of species could aid in the elucidation of the relationships between farm animals and their diets. Truly comparative studies with a variety of animal species may be of value in solving the persistent problems of farm animal energetics. A more rigorous understanding of the biological aspects of maintenance and the environmental, physiological and dietary factors which influence this would be a great achievment.

The demonstration that the classic techniques of animal calorimetry can be used to study energy balance in man is a great achievement, and further rigorous experimental work will ultimately clarify the ideas of the effects of dietary composition on energy balance, particularly in human adults.

References

BEAMISH, F.W.H., NIIMI, A.J. and LETT, P.F.K.P. (1975). In *Comparative Physiology –Functional Aspects of Structural Materials*, pp. 187–209. Eds. L. Bollis, H.P. Maddrell and K. Schmidt-Nielson. Amsterdam: North Holland

BRETT, J.R. and GROVES, T.D.D. (1979). In *Fish Physiology*, Vol. 8, 279–352. Eds. W.S. Hoav and P.J. Randall. New York: Academic Press

CLOSE, W.H. and MOUNT, L.E. (1978). *Br. J. Nutr.,* **40**, 413–421

HARRIS, L.E. (1966). *Biological Energy Interrelationships and Glossary of Energy Terms.* Washington, DC: National Academy of Science

KREBS, H.A. and KORNBERG, H.L. (1957). *Energy Transformation in Living Matter – A Survey.* Berlin: Springer-Verlag

SMITH, R.R., RUMSEY, G.L. and SCOTT, M.L. (1978). *J. Nutr.,* **108**, 1025–1032

60

GENERAL DISCUSSION

In discussing the metabolic rate of carp, it was pointed out that the increase in heat production caused by raising the ambient temperature for poikilotherms was the opposite of the effect in homeotherms. Doubt was expressed on whether heat production of carp plateaued as food consumption increased: such an effect implied a k_{pf} of 1.0. An alternative explanation was advanced that because of the short digestive tract of the carp, digestion may decline at a high feed intake so that metabolizable-energy intake would increase at a slower rate than food intake. It was noted that care should be exercised in reviewing reports of the effects of temperature on fish because the solubility of oxygen in water declines as temperature increases; thus the oxygen availability may limit metabolism at higher temperatures.

Effects of activity and excitement on heat production

There was wide interest in the effects on the variation in results of the procedures used to measure heat production. It was difficult to place mink in calorimeters without causing them to become so excited as to add a substantial activity or excitement component to any estimates of heat production. Estimations of the heat production of pigs were usually greater with unrestrained animals than with pigs restrained in cages. The maintenance requirements of unadapted water hens that were free to move inside a calorimeter chamber were 50–70% greater than those estimated for water hens adapted to confinement in small chambers.

Comparison of calorimetry on animals with that on man

There was general agreement on the need to use 24 h observations to measure heat production. Such studies gave reproducible results for repeated observations with the same individuals with standardized conditions of food and water intake. However, there is a far greater variation in measures of rates of heat production for man than for domestic animals. This larger degree of variability could be due to greater inherent individual differences in man than in other animals. However, the variation could also be due to the wider diversity of experience (diet, activity and physical environment) of human subjects than of animal subjects prior to their entry to the experiment. The latter explanation was supported by observations at the University of California that showed that long-term control (several months) of food intake, activity and environment reduced the variability of weight change in response to dietary manipulation between individuals so that the coefficient of variability was approximately 15%. This is similar to that observed in animal experiments and suggests that critical experimentation can be carried out with man provided that subjects are prepared to submit to long-term standardization of behaviour. This may allow objective tests of the hypothesis that thermogenic 'futile cycles' are responsible for between-individual variation.

Work at Babraham has shown that in normal daily living there are wide variations in daily energy intakes, and they may be as much as two-fold. In the calorimeter, subjects without a varied choice of foods or activity levels had great difficulty in consuming enough of the experimental diet to double their energy intake. Reservations were expressed concerning the applicability of measurements made under the controlled and confined circumstances in the calorimeter to normal living. However, such extrapolations were probably no greater than those made for domestic animals. The objective of such observations under restricted conditions was to isolate single experimental variables to allow orderly investigation.

Alternatives to calorimetry as a means of measuring energy balance have been discussed. Measurement of carbon dioxide entry rate using ^{13}C may hold promise, and the development of reliable and ethical means of estimating changes in body composition is proceeding.

It was concluded that calorimetry makes it possible to study energy balance in man with the precision attained in studies of domestic animals, but progress in this field of nutrition will be limited by the difficulties of making long-term studies.

PART 5

USE OF ENERGY IN REPRODUCTION AND LACTATION

61

PRODUCTS OF DIGESTION, HORMONES AND ENERGY UTILIZATION IN MILKING COWS GIVEN CONCENTRATES CONTAINING VARYING PROPORTIONS OF BARLEY OR MAIZE

J.D. SUTTON, J.D. OLDHAM and I.C. HART
National Institute for Research in Dairying, Shinfield, Reading RG2 9AT, England

Summary

Measurements were made in 16 intact cows and four cows with gastrointestinal fistulas. The difference in energy partition between diets containing 60% or 90% barley-based concentrates was related mainly to rumen volatile fatty acids (VFA). When ground maize replaced rolled barley, energy and starch digestion in the rumen were reduced and, with 90% concentrates, the yields of protein and lactose, but not fat, fell. Growth hormone tended to be lower and insulin higher on 90% concentrates.

The well established effects of diet composition on the utilization of dietary energy for milk production and on milk composition have been related to the molar proportions of VFA in the rumen but the relation between milk production and products of digestion remains poorly defined. The purpose of the present experiments was to measure the utilization of dietary energy (Experiment 1) and the products of digestion (Experiment 2) when diets were fed that were expected to result in wide differences in the proportions of VFA in the rumen and in the supply of starch in the duodenum.

Materials and methods

Four diets were fed. They consisted of 40% hay:60% concentrates (60) or 10% hay:90% concentrates (90) and the concentrates contained either rolled barley (B) or ground maize (M). They were designed to supply similar amounts of digestible energy (DE) and crude protein (about 1.8 kg/d). Diets 60 B and 60 M consisted of 6 kg hay and 9 kg concentrates daily (about 12.5 kg dry matter) and 90 B and 90 M consisted of 1.4 kg hay and 12.9 kg concentrates daily (about 12.2 kg dry matter). The cereal content of the concentrates was varied to try to equalize nitrogen intake and was 82% for 60 B, 78% for 60 M, 87% for 90 B and 83% for 90 M; the remainder of the concentrates consisted of soya bean meal and 2% vitamins and minerals.

EXPERIMENT 1

Eight weeks postpartum, 16 Friesian cows were allocated at random to one of the four diets for 12 weeks, of which the last eight were used for measurement of milk yield, milk composition and live weight. Faeces were collected for eight days and blood samples were taken hourly from an indwelling jugular catheter for 24 h during the last three weeks.

EXPERIMENT 2

Four Friesian cows were each fitted with a cannula in the rumen and a re-entrant cannula in the proximal duodenum anterior to the point of entry of the bile and pancreatic ducts. Each cow received each diet in a 4 X 4 Latin square design. Treatment periods began 6–8 weeks postpartum and lasted six weeks. In the fifth week of each period digesta flow was measured and sampled (2.5%) automatically for 72 h. No adjustments of flow rates were made for recovery of indigestible markers (Sutton, Youssef and Oldham, 1976). Faeces were collected during the last eight days of each period.

Results

EXPERIMENT 1

The four diets provided similar amounts of DE (*Table 61.1*). Milk yield on 90 B was over 30% greater than on 90 M with 60 B and 60 M intermediate. Differences between the diets in the content of protein and lactose in milk were generally small. However, wide differences in fat content were obtained not merely between the 60% and 90% concentrate diets but also between 90 B and 90 M; as a result, milk-fat yield was about 70% greater on the low-concentrate diets than on the high-concentrate diets. In contrast live-weight gain was greater on the high-concentrate diets.

Plasma prolactin was highest and growth hormone lowest on 90 M (*Table 61.1*); the high values for plasma insulin on this diet were due to one cow. Plasma growth hormone was lower on 90% concentrates than on 60% concentrates.

Table 61.1 EXPERIMENT 1: ENERGY DIGESTION, MILK YIELD AND COMPOSITION, LIVE-WEIGHT GAIN AND PLASMA HORMONES IN THE INTACT COWS

	60 B	*60 M*	*90 B*	*90 M*
Energy				
Digestibility (%)	70.8^{B}	69.1^{B}	75.0^{A}	72.5^{AB}
Digested (MJ/d)	157	158	164	158
Milk				
Yield (kg/d)	16.1^{B}	18.9^{AB}	20.6^{A}	15.6^{B}
Fat content (g/kg)	44.9^{A}	40.4^{A}	20.6^{C}	29.7^{B}
Protein content (g/kg)	31.5^{B}	30.0^{B}	30.3^{B}	34.3^{A}
Lactose content (g/kg)	45.2	45.2	46.2	45.5
Fat yield (g/d)	725^{A}	761^{A}	419^{B}	461^{B}
Protein yield (g/d)	506^{B}	562^{AB}	617^{A}	535^{B}
Lactose yield (g/d)	729^{B}	852^{AB}	954^{A}	714^{B}
Live-weight gain (kg/d)	0.04^{BC}	-0.11^{C}	0.34^{AB}	0.48^{A}
Hormones				
Prolactin (ng/ml)	8.0^{B}	10.8^{AB}	9.8^{AB}	15.3^{A}
Growth hormone (ng/ml)	7.8^{AB}	9.0^{A}	5.0^{BC}	2.0^{C}
Insulin (μu/ml)	7.4^{AB}	4.3^{B}	10.1^{AB}	18.5^{A}
Thyroxine (ng/ml)	43.5	26.9	30.6	32.7

ABC Means in the same line which do not share the same superscript differ significantly ($P<0.05$)

EXPERIMENT 2

Differences between the two barley diets in the digestion of energy and starch were generally small (*Table 61.2*). By far the largest difference between these two diets was in the molar proportions of VFA in the rumen.

Table 61.2 EXPERIMENT 2: ENERGY INTAKE AND DIGESTION, STARCH INTAKE AND FLOW, AND MOLAR PROPORTIONS OF RUMEN VFA IN THE FISTULATED COWS

	60 B	*60 M*	*90 B*	*90 M*
Energy				
Consumed (MJ/d)	232^{A}	227^{AB}	220^{B}	217^{B}
Total digestion (DE_T; %)	69.6^{B}	69.3^{B}	73.0^{A}	68.6^{B}
Rumen digestion (DE_R; %)	51.5^{A}	48.2^{A}	52.2^{A}	42.9^{B}
(DE_R/DE_T) x 100	74.1^{A}	69.5^{B}	71.6^{AB}	62.8^{C}
Starch (kg/d)				
Consumed	4.10^{A}	4.33^{A}	5.75^{B}	6.37^{C}
Duodenum	0.45^{A}	1.22^{B}	0.55^{A}	2.11^{C}
Faeces	0.05^{A}	0.35^{B}	0.08^{A}	0.68^{C}
Rumen VFA (molar %)				
Acetic	61.2^{A}	62.9^{A}	43.4^{C}	53.0^{B}
Propionic	21.9^{A}	21.0^{A}	41.0^{B}	28.0^{A}
Butyric	11.9	11.6	9.2	11.2

^{ABC}Means in the same line which do not share the same superscript differ significantly ($P<0.05$)

When maize replaced barley the contribution of rumen digestion to overall digestion of energy was considerably reduced. Starch flow to the duodenum and excretion in the faeces were much greater for the maize diets than for the barley diets. Proportions of VFA in the rumen on 60 M were similar to those on 60 B but the effect of increasing the proportion of cereal was less for the maize diets than for the barley diets.

Milk yields averaged about 15 kg/d and dietary effects on fat content were similar to those found in Experiment 1.

Discussion

The techniques used in the present experiment provide only approximate indications of the VFA supply from the rumen (Sutton, 1979); VFA production in the caecum, which could be considerable on the maize diets, is unknown. The amount of glucose derived from starch entering the duodenum is also very uncertain. Nevertheless it is apparent that the four diets did result in important differences in the supply of VFA and glucose to the animal.

The largest differences in energy utilization (*Table 61.3*) were associated with the proportion of concentrates in the diet, more energy being used for live-weight gain and less for milk production with the higher proportion of concentrates. In addition, important differences were established between the two 90% concentrate diets; whereas fat yield was similar, protein and lactose yields were significantly lower on 90 M than on 90 B (*Table 61.1*).

The lower milk fat yield and higher milk protein and lactose yield and live-weight gain on 90 B than 60 B were associated with only small changes in the site

Table 61.3 EXPERIMENT 1: NET ENERGY (MJ/d) FOR MILK[1], LIVE-WEIGHT GAIN (20 MJ/kg) and MAINTENANCE (0.30 MJ/kg$^{0.75}$)

	60B	*60M*	*90B*	*90M*
Milk	52.0^{AB}	56.7^{A}	46.7^{B}	42.1^{B}
Live weight	0.7^{BC}	-2.2^{C}	6.8^{AB}	9.4^{A}
Maintenance	31.1	32.7	31.1	31.8
NE/DE	0.53	0.55	0.52	0.53

[1] Equation 1 in Tyrrell and Reid (1965)
ABC Means in the same line which do not share the same superscript differ significantly ($P<0.05$)
NE = net energy; DE = digestible energy

of energy digestion and starch flow to the duodenum and it appears that the differences in energy partition between these two diets must be related primarily to the change in supply of acetic acid and propionic acid from the rumen.

Substitution of maize for barley increased postruminal digestion of starch and reduced energy digestion in the rumen. On the 60% concentrate diets these changes were small and had no detectable effect on energy utilization. However, on the 90% concentrate diets, the changes were larger and were accompanied by differences in rumen VFA proportions. Both diets resulted in reduced milk-fat secretion but the higher lactose and protein secretion on 90 B than on 90 M probably reflected the greater estimated production of propionic acid on 90 B, whereas the extensive postruminal digestion of starch on 90 M did not result in enhanced lactose synthesis.

The milk-fat content had the usual negative relationship to rumen propionic acid with no evidence of any effect from starch entering the duodenum. On the other hand the yield of milk fat and live-weight gain were similar on 90 B and 90 M despite the higher acetic:propionic acid ratio on 90 M and this could be interpreted as indirect evidence of an increased glucose supply from starch on this diet.

Despite the fact that all four hormones are implicated in the hormonal control of lactation and the secretion of each is known to be metabolically influenced, the relationship between the plasma hormone concentrations and milk yield, milk composition and the various metabolic parameters measured remains obscure. However the greater live-weight gain and lower milk-fat synthesis on the 90% concentrate diets may well be due to enhanced lipogenesis and reduced lipolysis in adipose tissue resulting from the tendency for higher insulin and lower growth hormone on those diets.

References

SUTTON, J.D. (1979). *Proc. Nutr. Soc.*, 38, 275
SUTTON, J.D., YOUSSEF, F. and OLDHAM, J.D. (1976). *Proc. Nutr. Soc.*, **35**, 100A
TYRRELL, H.F. and REID, J.T. (1965). *J. Dairy Sci.*, **48**, 1215–1223

ENERGY UTILIZATION BY JAPANESE BLACK CATTLE IN LATE PREGNANCY

K. IWASAKI, T. HARYU and K. KAMEOKA
Department of Animal Nutrition, National Institute of Animal Industry, Tsukuba 305, Japan

Summary

Effects of feeding levels and the advance of gestation on energy metabolism were studied wih six heifers and six cows of Japanese Black Cattle in late pregnancy. Energy metabolism experiments were conducted in an open-circuit respiration apparatus at three stages of pregnancy : 220, 250 and 275 days after conception. Heat production increased with raising of the feeding level and the advance of pregnancy. However, heat production per metabolic body size was not affected by the advance of pregnancy.

Introduction

Nutrient requirements for pregnancy in the *Japanese Feeding Standard for Beef Cattle* (National Research Council of Agriculture, Forestry and Fishery, 1975) have tentatively been inferred from the figures for dairy cattle obtained in this laboratory (Hashizume *et al.*, 1964b). Therefore, a series of experiments was commenced to establish the energy requirements for Japanese Black Cattle, which is the most popular beef breed in Japan. This chapter reports the results of energy balance trials in late pregnancy.

Materials and methods

Six heifers were divided into three groups and fed at the levels of (1) growing, (2) growing + pregnancy and (3) growing + twice pregnancy. Experiments with the same design were conducted with six cows using the nutrient level for maintenance instead of that for growing.

The rations consisted of rice straw wafer, ground barley and soya bean meal. Enhancement of energy in the ration for pregnancy was made by increasing the amount of barley. The amounts of the ration given were not altered according to the advance of pregnancy.

Energy metabolism trials were conducted in an open circuit respiration apparatus, described by Itoh *et al.* (1977), at three stages of pregnancy: 220, 250 and 275 days after conception. The respiration chamber was adjusted to

20°C and 60% relative humidity. Heat production was calculated by Brouwer's formula (1965).

Nitrogen balance and digestibility trials were performed for five days from the beginning of each stage and then respiration experiments were carried out for the succeeding 48 h. Fasting metabolism was estimated immediately after parturition, and the values of fasting-heat production proved to be very similar for heifers and cows, being 0.310 and 0.314 MJ/$W^{0.75}$, respectively. These values are almost the same as those for dairy cattle obtained previously in this laboratory (Hashizume *et al.*, 1964a).

Results

Results of the experiments are shown in *Tables 62.1* and *62.2* as the average figures for two animals.

Gross energy (GE) intake tended to decrease with the advance of pregnancy in the higher feeding levels. Refusal of rice straw was observed with the higher levels of feeding, but it was only 10% of the amount offered. Digestibility of energy increased with the higher energy levels owing to the higher intake of barley, whereas the advance of pregnancy seemed to have little effect on energy digestibility.

Table 62.1 RESULTS OF ENERGY BALANCE TRIALS FOR GROWING HEIFERS IN LATE PREGNANCY

Energy level	*Growing*			*Growing + pregnancy*			*Growing + 2 × pregnancy*		
Days after conception	*220*	*250*	*275*	*220*	*250*	*275*	*220*	*250*	*275*
GE intake (MJ)	110.4	109.3	106.9	118.2	121.3	117.7	134.6	127.4	128.1
Energy digestibility (%)	63.5	62.1	61.3	66.5	64.2	65.8	66.6	68.0	67.2
Energy loss in urine (%)	2.2	2.2	2.3	1.6	1.8	1.7	1.6	1.7	1.5
Energy loss as methane (%)	8.3	8.4	8.4	8.7	8.6	8.4	7.9	8.0	6.9
ME intake (MJ)	58.6	56.4	54.0	66.2	65.2	65.4	77.0	74.3	75.4
Heat production (MJ)	48.5	53.6	52.3	53.1	54.8	58.2	60.2	62.8	59.0
Energy balance (MJ)	10.1	2.8	1.7	13.1	10.4	7.2	16.8	11.5	16.4
Body weight (kg)	369.0	380.0	396.5	362.0	379.5	403.0	397.0	410.0	434.0
Body weight of calves right after birth (kg)		25.7			26.1			28.8	
Heat production (MJ/$W^{0.75}$)	0.580	0.623	0.588	0.638	0.639	0.645	0.679	0.691	0.619
Fasting heat production (MJ/$W^{0.75}$)		0.268			0.310			0.352	

Figures are shown as averages of two heifers
GE = gross energy; ME = metabolizable energy

Metabolizable energy (ME) intake was higher in the groups planned to be fed at higher energy levels, but the advance of pregnancy did not affect ME intake at each feeding level. Heat production became higher with higher energy levels as was expected, but, as for the effect of advance of pregnancy, no distinct difference was found between 250 days and 275 days after conception, although the values at 220 days after conception were smaller than those at the other

Table 62.2 RESULTS OF ENERGY BALANCE TRIALS FOR COWS IN LATE PREGNANCY

Energy level	*Maintenance*			*Maintenance + pregnancy*			*Maintenance + 2 × pregnancy*		
Days after conception	*220*	*250*	*275*	*220*	*250*	*275*	*220*	*250*	*275*
GE intake (MJ)	100.0	100.8	106.7	119.2	118.0	115.4	133.9	136.0	132.6
Energy digestibility (%)	53.0	52.7	50.7	56.4	56.2	55.6	58.0	58.2	57.6
Energy loss in urine (%)	1.9	2.1	1.9	1.5	1.8	1.5	1.4	1.3	1.6
Energy loss as methane (%)	7.5	8.0	7.9	8.0	8.2	8.0	7.7	8.2	7.8
ME intake (MJ)	43.5	42.9	43.6	55.5	55.1	53.1	65.5	65.3	64.3
Heat production (MJ)	41.3	45.7	49.5	48.8	50.7	50.7	54.5	57.7	58.2
Energy balance (MJ)	2.0	−2.8	−5.9	6.7	4.4	2.4	11.0	7.6	6.1
Body weight (kg)	398.0	404.5	(440.0)	389.5	405.0	430.0	403.5	419.0	439.5
Body weight of calves right after birth (kg)		25.2			27.6			29.7	
Heat production ($MJ/W^{0.75}$)	0.460	0.506	0.514	0.556	0.560	0.535	0.602	0.623	0.602
Fasting-heat production ($MJ/W^{0.75}$)		0.308			0.323			0.313	

Figures are shown as averages of two cows, except for the figure at 275 days after conception at the maintenance energy level due to parturition during the metabolism experiment

stages. Heat production per metabolic body size was not affected by the advance of pregnancy except for the low energy group at 220 days after conception.

Energy retention became smaller with the advance of pregnancy due to the increase in energy requirement for fetal development. Negative energy balance was found only in the cows fed on a maintenance ration 250 and 275 days after conception.

All the cattle showed increase in body weight, although the cows fed on a maintenance ration gained only 0.1 kg/d. Body weight of calves immediately after birth tended to be larger with the higher feeding levels though this was not statistically significant.

References

BROUWER, E. (1965). *Energy Metabolism*, p. 441. Ed. K.L. Blaxter. EAAP Publication No. 11. London and New York: Academic Press

HASHIZUME, T., MORIMOTO, H., HAMADA, T., MASUBUCHI, T., ABE, M., HORII, S., ZITZUKAWA, Y., YOKOTA, C. and ANBO, S. (1964a). *Spec. Rep. natn. Inst. Anim. Ind.*, No. 2, 1–107

HASHIZUME, T., MORIMOTO, H., MASUBUCHI, T., ABE, M., HAMADA, T., HORII, S., ZITSUKAWA, Y. and YOKOTA, C. (1964b). *Spec. Rep. natn. Inst. Anim. Ind.*, No. 5, 1–44

ITOH, M., MARYU, T., TANO, R. and IWASAKI, K. (1977). *Bull. natn. Inst. Anim. Ind.*, **32**, 45–61

National Research Council of Agriculture, Forestry and Fishery (1975). *Japanese Feeding Standard for Beef Cattle.* Tokyo: Central Association of Livestock Industry

63

EFFECT OF PROTEIN LEVEL AND BUFFERING CAPACITY ON ENERGY VALUE OF FEEDS FOR LACTATING DAIRY COWS

H.F. TYRRELL and P.W. MOE
Ruminant Nutrition Laboratory, Animal Science Institute, Agricultural Research Science and Education Administration, US Department of Agriculture, Beltsville, Maryland 20705, USA

Summary

Two experiments with lactating Holstein cows were carried out to determine the effect on the utilization of energy of varying crude protein from 11% to 20% of ration dry matter (DM). Digestible energy (DE) values improved with each increase in ration crude protein up to 20%, but the metabolizable energy (ME) value reached a maximum at 17% crude protein. Rations containing less than 14% crude protein resulted in a lower ME value when fed to lactating cows. Ration crude protein level had no effect on the partial efficiency of ME use for milk + body-tissue synthesis. The addition of 2.5% ground limestone to corn silage-based rations had no effect on any measurement of energy utilization.

Introduction

The apparent DE value in rations fed to lactating dairy cattle accounts for 85% of the variation in net energy (NE) value (Moe, Flatt and Tyrrell, 1972). The DE value of rations fed to lactating cows is lower than the DE value of the same ration fed to non-lactating cattle at maintenance intake (Tyrrell and Moe, 1975). Reducing the crude protein content of the ration fed to lactating cows from 16% to 11% caused lower digestive efficiency (Moe and Tyrrell, 1972). This observation, combined with the relatively small depression in digestibility at high intake of rations high in total protein content (Flatt *et al.*, 1969) led to two experiments designed to evaluate the effect of crude protein content on energy utilization by lactating Holstein cows.

Materials and methods

In Experiment 62, 12 cows in early lactation were assigned to 40% alfalfa hay: 60% gain rations containing 14%, 17% or 20% crude protein. Energy metabolism was measured using open-circuit respiration calorimetry on each cow at six-week intervals throughout the lactation and dry period. In Experiment 65, 24 cows representing early, middle and late stages of lactation plus pregnant-dry cows were assigned to 60% corn silage: 40% grain rations containing 11%, 14% or 17% crude protein. Rations of four cows at each protein level were supplemented with 2.5% ground limestone. Energy metabolism measurements were made at 12-week intervals on each cow throughout a lactation cycle. Within both experiments, crude protein content of rations was increased by removing ground maize and adding soya bean meal to the grain mix.

Results

The chemical composition of ration dry matter (*Table 63.1*) within each experiment was essentially constant except for the increasing crude protein associated with decreasing soluble residue owing to the substitution of soya bean meal for corn meal. Difference in chemical composition between experiments reflects the difference in forage source, corn silage being higher in hemicellulose and cellulose but lower in lignin than alfalfa hay.

Table 63.1 CHEMICAL COMPOSITION OF RATION DRY MATTER CONSUMED DURING METABOLISM TRIALS (EXPRESSED AS % DRY MATTER)

Ration	*Experiment 62*			*Experiment 65*		
	14	*17*	*20*	*11*	*14*	*17*
Organic matter	94.5	94.2	93.6	93.9	93.6	93.3
Crude protein	13.7	16.1	19.4	11.0	14.0	17.0
Ether extract	2.1	2.0	1.9	2.5	2.3	2.0
Hemicellulose[a]	12.9	11.9	11.4	16.3	16.2	16.7
Cellulose[a]	11.7	11.5	11.8	13.7	13.8	14.0
Lignin[a]	4.7	4.6	4.7	2.7	2.8	3.0
Soluble residue[b]	49.3	48.1	44.4	47.1	44.0	40.1

[a] Determined using acid and neutral detergent fibre extraction and oxidation of lignin by permanganate (Goering and Van Soest, 1970)
[b] Determined as the difference between organic matter and the sum of neutral detergent fibre, crude protein and ether extract. It is an estimate of carbohydrate soluble in neutral detergent

Within Experiment 62 intake and milk yield were lower for the 17% protein group than for either the 14% or 20% protein groups (*Table 63.2*). Inasmuch as there were only four cows per protein level in this experiment, this response can be attributable to chance assignment of cows to treatment. Within Experiment 65, cows fed the 11% crude protein diet consumed less than cows fed the 17% diet. DE and ME intakes were lower for the 11% than either of the two higher protein diets, resulting in lower milk yield. Tissue energy was not affected by dietary protein level within each experiment. Addition of limestone to corn silage-based rations had no effect on any energy partition measurement.

Both DE/intake energy (IE) and ME/IE were increased ($P<0.05$) within each experiment when the two high-protein diets were compared to the low-protein diet. Similarly, ME/DE was lower within each experiment for the high-protein diet as compared to the two low-protein diets. Although the effect of forage source and related factors requires further examination, one conclusion seems clear. Both DE/IE and ME/IE responded positively to increasing crude protein content up to 17% of the total ration dry matter fed to lactating dairy cows. Less than 14% crude protein does not appear adequate to support optimum

Table 63.2 PARTITION OF ENERGY BY LACTATING COWS

	Experiment 62			*Experiment 65*				
	Protein level			*Protein level*			*Limestone*	
Variable	*14*	*17*	*20*	*11*	*14*	*17*	–	+
IE (kJ/$W^{0.75}$)	2448^a	2104^b	2360^a	2623^a	2807^{ab}	2925^b	2791	2782
DE (kJ/$W^{0.75}$)	1552^a	1414^b	1615^a	1531^a	1761^b	1883^c	1728	1720
ME (kJ/$W^{0.75}$)	1360^a	1234^b	1397^a	1343^a	1540^b	1628^b	1506	1498
LE (kJ/$W^{0.75}$)	490^a	364^b	485^a	519^a	632^b	648^b	598	602
TE (kJ/$W^{0.75}$)	$+54^a$	$+92^a$	$+67^a$	-9^a	-23^a	$+18^a$	+1	−11
DE/IE	0.638^a	0.675^b	0.687^b	0.585^a	0.630^b	0.645^b	0.619	0.620
ME/IE	0.558^a	0.592^b	0.592^b	0.514^a	0.549^b	0.556^b	0.540	0.540
ME/DE	0.875^a	0.876^a	0.862^b	0.878^a	0.872^a	0.863^b	0.872	0.870
Body weight (kg)	602	638	664	543^a	611^b	633^b	590	602

[a,b] Values within an experiment which do not share the same superscript differ significantly ($P < 0.05$)

IE = intake energy; DE = digestible energy; ME = metabolizable energy; LE = lactation energy; TE = tissue energy

digestibility or metabolizability of energy. Crude protein levels above 17% may improve the digestibility of energy, but result in little improvement of the ME value. Crude protein levels did not affect the partial efficiency of milk + body tissue synthesis from ME.

References

FLATT, W.P., MOE, P.W., MOORE, L.A., HOOVEN, N.W. LEHMANN, R.P., ØRSKOV, E.R. and HEMKEN, R.W. (1969). In *Energy Metabolism of Farm Animals*, pp. 221–234. Eds. K.L. Blaxter, G. Thorbek and J. Kielanowski. London: Oriel Press

GOERING, H.K. and P.J. VAN SOEST (1970). *Forage Fiber Analyses.* Handbook #379. United States Department of Agriculture

MOE, P.W., FLATT, W.P. and TYRRELL, H.F. (1972). *J. Dairy Sci.*, **55**, 954–958

MOE, P.W. and TYRRELL, H.F. (1972). *J. Dairy Sci.*, **55**, 318–324

TYRRELL, H.F. and MOE, P.W. (1975). *J. Dairy Sci.*, **58**, 1151–1163

THE UTILIZATION BY HIGH-YIELDING COWS OF ENERGY FROM ANIMAL TALLOW OR SOYA BEAN OIL ADDED TO A DIET RICH IN CONCENTRATES

Y. VAN DER HONING
Research Institute for Livestock Feeding and Nutrition, Postbox 160, 8200AD Lelystad, The Netherlands

Summary

Digestibility and utilization of energy by dairy cows on a diet rich in concentrates were studied in a 3 × 3 Latin square design. The concentrates used contained either 5% or no animal tallow or soya bean oil. Fat supplementation increased digestibility of lipids, but did not alter the digestibility of other organic matter markedly. Methane losses were slightly lowered. Soya bean oil depressed the milk-fat percentage. Utilization of metabolizable energy (ME) did not differ significantly between treatments. Milk production tended to increase with the addition of fat to the diet, possibly owing to the higher intake of ME.

Introduction

Dietary fats are of particular interest as potential sources of energy for high-yielding dairy cows, whose intake capacity is often too small. Effects of different types of fats and oils on milk yield and composition have been given considerable attention, but few studies are available on the digestion and utilization of energy from fat-enriched diets.

Increasing the proportion of lipids in a ruminant ration resulted in variable depressions of digestibility of organic matter and energy. Crude fibre digestion is said by some to be especially depressed owing to a reduction of cellulolytic activity of the rumen flora (Dijkstra, 1960; Cottijn, Buysse and Boucqué, 1971), although others have not found such results with diets containing up to 10% lipids (Swift *et al.*, 1948; Palmquist, 1978). As most data on digestion come from sheep and most data on feeding value from feeding trials with cows, energy and nitrogen-balance trials with high-yielding dairy cows were performed to study the effect on digestion and utilization of energy of an addition of 5% animal fat or soya bean oil to a concentrates mixture in a low-roughage diet. For comparison, the digestibility and ME content of hay and each concentrates mixture were determined with sheep at maintenance.

Materials and methods

Grass hay and a mixture of concentrates (25% soya bean meal, solvent extracted; 10.8% coconut expeller; 10% linseed meal, solvent extracted; 16% maize gluten feed; 20% dried beet pulp; 5% cane molasses; 10% tapioca meal; and 3.2% of a mixture of minerals and vitamins, including extra magnesium oxide) served as a control ration. In the experimental rations (F and S) the concentrates were composed of 95% of the above mixture and either 5% animal tallow or 5% soya bean oil.

Six mature, high-yielding dairy cows in early lactation received a diet of 7 kg of grass hay and 13–14 kg of concentrates. Treatments C (control), F and S were arranged in a 3 X 3 Latin square design. A preliminary period of 4 weeks for adaptation to the diet was followed by a 12-day collection period of urine, faeces and milk. Gas exchange was also measured in this period 2 or 3 times during 48 h in an open-circuit system.

Except for the lipids the usual (Weende) analysis was done. Crude fat (XL) was extracted with hexane after boiling the samples of feed and faeces with diluted acid to prevent losses as soaps.

Four mature wethers were used to determine digestibility of the ration components at the maintenance level of feeding and to measure urine and methane losses.

Results

XL of concentrates in rations F and S was 55 and 44 g/kg higher than in ration C. Dry matter intake on all treatments was equal (17.5 kg), with slightly more than 30% from hay. Intake of XL was 449, 1099 and 964 g/d on C, F and S, respectively.

Digestibility and composition of the ration measured with cows are presented in *Table 64.1*, as well as the calculated digestibility for the same diet, using sheep digestibility coefficients. Compared to the sheep data, coefficients of organic matter, energy and fat of cows show a similar difference of 4–5 units for each diet and somewhat more for crude protein (XP). Between treatments neither organic matter and nitrogen-free extract (XX), nor crude fibre (XF) were significantly different. Digestibility of XL was higher for S and F than for C, owing partly to relatively lower endogenous fat excretion and partly to higher digestibility of the supplemented fat or oil. The fat supplementation seemed to depress only digestion of XP slightly. Digestibility of energy of the F ration was slightly higher.

Table 64.1 AVERAGE CONSUMPTION OF DRY MATTER OF DIETS (ℓ; g/kg AND MJ/kg) AND COEFFICIENTS OF DIGESTIBILITY AS MEASURED WITH COWS (2) AND AS CALCULATED WITH SHEEP DATA ON DIGESTIBILITY AT MAINTENANCE (3)

	C			*F*			*S*		
	1	*2*	*3*	*1*	*2*	*3*	*1*	*2*	*3*
Organic matter	910	74.4	79.3	913	74.6	78.9	914	74.1	78.8
XP	224	74.1	81.8	214	72.5	80.0	214	72.6	80.7
XL	26	60.6	64.6	63	79.3	83.3	55	72.7	77.4
XF	161	55.6	62.8	159	55.8	58.1	156	54.5	59.8
XX	500	81.4	84.3	478	81.1	85.2	488	81.1	83.9
Energy	17.87	72.2	76.9	18.59	73.2	77.5	18.45	72.0	77.0

Methane losses were significantly lower on F and S versus C: 5.3% and 5.0% versus 5.9% of gross energy (GE) respectively. Therefore metabolizability (q) of F and S was slightly higher than of C: 63.4, 62.6 and 61.7.

Milk yield on treatments C, F and S was 24.1, 26.4 and 26.5 kg with a percentage of fat of 3.84, 3.65 and 2.67 and of protein of 3.27, 3.04 and 3.14, respectively. A similar significant depression of milk fat due to soya bean oil was reported by MacLeod and Wood (1972).

Utilization of ME is presented in *Table 64.2.* Average nitrogen and energy balances, expressed per unit metabolic body weight ($W^{0.75}$), were positive on all diets. Feeding level relative to maintenance was 3.3–3.5. Milk energy per $W^{0.75}$ was lowest on diet S. However, utilization of ME (k_1) for milk + tissue energy and maintenance did not differ significantly between treatments; neither did the deviation from the general regression equation used in the Dutch feed evaluation system (van Es, 1978).

Table 64.2 METABOLIZABILITY (q), MILK ENERGY (LE) AND ENERGY BALANCE (RE) AND UTILIZATION OF ME (k_1) BY DAIRY COWS

	C	*F*	*S*
q	61.7	63.4	62.6
RE (kJ/$kg^{0.75}$)	63	117	133
LE (kJ/$kg^{0.75}$)	586	616	542
ME (kJ/$kg^{0.75}$)	1617	1710	1669
k_1	0.582	0.600	0.581
Difference * (kJ/$kg^{0.75}$)	–45.2	–25.1	–54.8

* Calculated as the difference (kJ/$kg^{0.75}$) between net energy measured and net energy predicted from (0.6 + 0.0024 (q–57)) ME (kJ/$kg^{0.75}$), with an assumed net energy for maintenance equal to 293 kJ/$kg^{0.75}$.

The overall efficiency, calculated by multiplying q × k_1, was for C, F and S 36.0%, 38.1% and 36.4% respectively, but it did not differ significantly between treatments, although for the F diet it tended to be higher.

Discussion

The lower methane losses with diet F and especially with diet S suggest changes in rumen fermentation due to the added fat, so does, again for S, the low milk-fat content. Obviously, these changes did not affect fibre digestibility, q or ME utilization markedly. The higher milk yield for F and S versus C and the higher quantity of milk energy for F versus C might be due to the higher ME intake and/or to a specific effect of fat. Such effects on milk production can only be proved to exist with long-term (feeding) trials.

Acknowledgements

The digestion trials with sheep were carried out under the supervision of ir. A. Steg. The contribution of R. Terluin and his staff in the execution of the animal trials and of A. Nijenhuis and his staff in carrying out the analysis is gratefully acknowledged.

References

COTTIJN, B.G., BUYSSE, F.X. and BOUCQUÉ, C.V. (1971). *Z. Tierphysiol. Tierernähr. Futtermittelk.*, **27**, 252–259

DIJKSTRA, N.A. (1969). *Neth. J. agric. Sci.*, **17**, 27–40

van ES, A.J.H. (1978). *Live Stk Prod. Sci.*, **5**, 331–345

MacLEOD, G.K. and WOOD, A.S. (1972). *J. Dairy Sci.*, **55**, 439–445

PALMQUIST, D.L. (1978). *Distillers Feed Res. Counc. Conf. Proc.*, **33**, 12–16

SWIFT, R.W., BRATZLER, J.W., JAMES, W.H., TILLMAN, A.D. and MEEK, D.C. (1948). *J. Anim. Sci.*, **7**, 475–485

65

PASTURE INTAKE AND HEAT PRODUCTION OF BREEDING EWES

J.L. CORBETT, E.P. FURNIVAL, M.W. INSKIP, F.S. PICKERING and J. PLAZA*
CSIRO, Division of Animal Production, Armidale, New South Wales 2350, Australia

Summary

The following measurements were made over five-day periods at three stages of pregnancy in 21 grazing Merino ewes and concurrently in 19 non-pregnant (NP) ewes of about 38 kg live weight: pasture intake expressed as metabolizable energy (ME), daily heat production determined by the carbon dioxide entry rate technique (CERT) and by difference of energy retention (ER). Estimates of ME for maintenance of ewes 54, 100 and 130 days pregnant, from regressions of ER on ME intake, were greater than those of NP ewes by 0.46, 0.66 and 1.95 MJ/d respectively. The estimated minimum ME requirement at day 130 of pregnancy for the average ewe was 0.23 MJ/kg or 0.58 MJ/$kg^{0.75}$.

Introduction

Systems that define the energy needs of livestock require definition of maintenance requirements (ME_m). Determinations of fasting metabolism (F) under standardized conditions (Blaxter, 1967) indicate directly the ME_m of some mature animals, for example those kept for fibre production or breeding females if non-lactating and in early pregnancy. If conditions preceding determinations have been non-standard, for example feeding levels have been above maintenance or there has been fasting during lactation (Holter, 1976), the values obtained are elevated. Webster, Brockway and Smith (1974) described F as 'irrelevant' in assessing the needs of growing animals, but an equation of Graham, Searle and Griffiths (1974) that adjusted F for age of animal and previous energy intake and gain gave ME_m for grazing lambs in good agreement with estimates from their ME intake and energy retention determined by slaughter (Perez *et al.*, 1976).

Estimates of 'practical maintenance' made by the factorial approach should be tested under practical conditions; techniques used should preferably yield measurements of energy balance and not just of animal live weight and production (van Es, 1972). The CERT for determining heat production (Corbett *et al.*, 1971; Engels, Inskip and Corbett, 1976) allows estimation of the energy balances of numerous animals housed in normal conditions or at pasture. Results in this chapter include such estimates made on 40 grazing sheep over five consecutive days at each of three stages of pregnancy.

* Present address: INIA, Apartado 22, Badajoz, Spain

Materials and methods

ANIMALS AND PASTURES

Oestrus was synchronized in 100 fine-wool Merino ewes 3.5 years old and 30–50 kg live weight, and lambings occurred in a six-day period, most within four days (2–5 September 1977). To promote variation in energy intake, one experimental flock of about 50 sheep grazed pastures of ***Phalaris aquatica*/*Trifolium repens*** that provided abundant herbage of 2000–3000 kg dry matter (DM) per ha, while a second 50 grazed similar pastures with about half those amounts; anthelmintic was administered regularly. Both flocks included 20 pregnant (P) and 20 non-pregnant (NP) ewes, all the latter having shown oestrus (one was later found to be pregnant), and initially all four groups of 20 were similar in live weight. Additional sheep in each flock included three with oesophageal fistulas (OF) and three NP ewes fitted with harness and faecal collection bags in three periods when these and experimental ewes were dosed twice daily with 1 g chromium oxide (Cr_2O_3). Concentrations of chromium oxide in the collected faeces were compared with those in accumulated 'grab' samples, taken twice daily from the rectum of the same sheep in the same periods, to measure bias in the grab sample values.

MEASUREMENTS

At days 21, 80, 114 and 140 of gestation all sheep were fasted without water for 24 h and weighed; tritiated water was injected for estimation of body composition but results are not reported because those for the P ewes were erratic, as found also by Houseman, Robinson and Fraser (1978), who used deuterium oxide. Rates of wool growth were determined from dyebands (Chapman and Wheeler, 1963) applied at each fast and fleece weights at the subsequent shearing. Fasted weights were adjusted to a fleece-free basis (FW) as were the live weights at pasture (W).

Measurements were made during the first five days of each of three periods (days 48–60, 94–105 and 125–135 of gestation) of the digestible organic matter intake (DOMI) and the daily heat production (CERT) of five P and five NP ewes in both flocks; during the second five days the same measurements were made on a further 20 ewes similarly distributed. DOMI only was measured on an additional five P and five NP in each flock over five days in periods II and III, while similar numbers of 'control' ewes were subjected to neither DOMI nor CERT procedures.

DOMI was calculated from faeces output, estimated from grab-sample chromium oxide concentration, adjusted for bias, and herbage digestibility determined on OF extrusa. ME (kJ/g DOMI), calculated as [14.23 + 5.86 digestible crude protein (DXP)] (van Es, 1978), varied between 15.1 and 15.35; DXP (g/g DOMI) was estimated, as by Holter and Reid (1959), from XP (nitrogen × 6.25) in extrusa, adjusted to XP in feed eaten, using a relationship established with OF sheep given cut herbage.

For CERT, infusions of carbon-14 sodium bicarbonate (about 1.1 MBq/d) were begun 16 h before parotid saliva, the source of body carbon dioxide for radioactivity assay, was sampled continuously for five days (about 140 ml/d). In period I, infusions were intravenous (jugular) but the catheters were found vulnerable to displacement. In periods II and III, infusions were intraperitoneal; catheters and infusate were sterile and there was never any infection. Studies in respiration chambers showed the change in route of infusion did not alter the relationship reported by Engels, Inskip and Corbett (1976) between carbon dioxide entry rate and heat production at respiratory quotient (RQ) 0.86.

Results

Infusions in half of period I were over four days, not five, and CERT (heat production, kJ/d) was determined for 122 sheep-days out of the possible 180, the principal problem being with jugular catheters. Overall, CERT was determined for 460 sheep-days out of 580, there being occasional faults in the parotid catheters or the infusion pumps. Regression analyses of the many sets of CERT over four or five consecutive days showed no significant change in the values with time from start of infusion; absence of a progressive reduction indicated there was no substantial recycling of labelled carbon. DOMI of sheep used for CERT was about 80% ($P<0.05$) of those whose intakes only were measured, but over the experiment these ewes and the 'control' ewes did not differ in mean W or FW or changes in these, or in wool growth, or in birth weights of their lambs. Five ewes (12.5%) bore twins.

Regressions of $\log_{10}$ CERT and of $\log_{10}$ DOMI on $\log_{10}$ W gave exponents for W of 0.95 ± 0.13 and 1.28 ± 0.33 respectively. ME_m from regressions of energy retention (ER = ME intake (MEI) – CERT) on ME intake and solving for ER = 0 are given per $kg^{0.75}$ as well as per kg and per sheep (*Table 65.1*). The term MEI^2 was significant in period I only when the equations calculated on a per sheep basis, but not those for unit W, differed ($P<0.05$) between ewes pregnant 54 days and those NP and gave ME_m of 6.37 and 5.91 MJ/d respectively. DOMI was

Table 65.1 ESTIMATES OF METABOLIZABLE ENERGY FOR MAINTENANCE (ME_m) OF GRAZING MERINO EWES PREGNANT (P) 54,100 and 130 DAYS AND OF SIMILAR EWES NON-PREGNANT (NP) FROM REGRESSIONS OF ME INTAKE (MEI) MINUS HEAT PRODUCTION (THAT IS, ENERGY RETENTION, ER) ON ME INTAKE. THE MODEL USED[1] WAS: $ER = a + b\,MEI + c\,MEI^2$ AND SOLVED FOR ER=0

	Period I (May)		*Period II (July)*		*Period III (Aug)*	
	P (54 days)	*NP*	*P (100 days)*	*NP*	*P (130 days)*	*NP*
No. of Ewes	20	16	21	19	21	19
Mean live weight fleece free (kg)	38.1	37.4	40.0	36.8	43.0	37.9
Mean MEI (MJ/d)	6.89	7.24	7.85	6.51	14.67	11.40
Mean ER (MJ/d)	0.73	1.26	2.40	1.70	5.23	3.80
ME_m: (kJ/kg per d)	← 153 →		← 123 →		222	202†
($kJ/kg^{0.75}$ per d)	← 377 →		← 304 →		564	496**
(MJ/d)	6.37	5.91*	4.91	4.25	9.25	7.30***

[1] Quadratic term significant only in period I; rectilinear equation used in periods II and III
† $P = 0.074$ * $P<0.05$ ** $P<0.01$ *** $P<0.001$

increased in ewes pregnant 100 days but ME_m/kg did not differ from that of NP ewes and the values were less than in period I. ME_m per sheep was greater for P than NP ewes by 0.66 MJ/d, a value considered significant biologically though not statistically ($P>0.05$). All values increased substantially in period III when the ME_m of ewes pregnant 130 days was greater than that of NP ewes by 1.95 MJ/d ($P<0.001$) or 65 kJ/kg$^{0.75}$ ($P<0.01$) or 20 kJ/kg ($P = 0.074$).

Regressions of DOMI on W and change (g/d) in FW gave ME_m/kg values greater than those in *Table 65.1* by 25–76% (Corbett, 1980). The greatest value was for NP ewes in period III, associated with a non-significant coefficient for change in FW; ME_m for P ewes (130 days) was less than for NP ewes by 21% but the coefficient for FW was 60 kJ ME/g.

Discussion

Adverse effects of CERT and any of DOMI procedures on the ewes were apparently confined to the days these were applied because longer term performances, such as W and wool, were unimpaired. About 40% of individual ER values were negative but period means were all positive so the ME_m may relate closely to true practical values. The erratic results from partitioning DOMI between W and change in FW are an example of the problems encountered with this approach (van Es, 1972).

In period II, ME_m values resembled those for housed ewes and were less than in period I when there were strong cold winds with rain and sleet. In periods I, II and III the mean maximum and minimum temperatures (°C) were respectively 10.1, 11.0, 13.2 and 0.6, −1.4, −2.8; mean wind runs (km/d) were 185, 124, 109; rainy days (and total mm of rain fall) were 5(15.5), 3(0.7), 1(25.0); herbage values (kg DM/ha) were 1440, 1550, 1760 in low-availability pastures and 2540, 3000, 3300 in high-availability pastures. The 'spring rise' in ME_m values during period III for NP as well as P ewes, and the low midwinter values, may however indicate a seasonal variation in metabolic rate such as that observed in white-tailed deer (Holter *et al.*, 1975).

Mean mass of lamb(s) per ewe at birth was 4.2 kg and fetal mass at 130 days' gestation was calculated (using a Gompertz equation derived from slaughter data from similar ewes) to be 2.4 kg, increasing by 77 g/d. From the results of Rattray *et al.* (1974) this corresponded to 320 kJ/d or, with all products of conception, 430 kJ/d. Graham (1964) calculated the heat increment of gestation as 376 kJ/kg fetal tissue, the tissue being defined as fetus, placenta and fluids adjusted to a common DM content, namely that of the fetus. The heat increment of our ewes at 130 days' gestation can be calculated as $[43 \times (222 - 202)] = 860$ kJ/d (*see Table 65.1*) and fetal tissue (from information in Rattray *et al.*, 1974) as 2.85 kg. On these bases the heat increment of gestation was 302 kJ/kg.

The gross increments per ewe pregnant 54, 100 and 130 days of, respectively, 0.46, 0.66 and 1.95 MJ/d, include the additional energy expenditures associated with their greater W than NP ewes. On average, the minimum ME for a grazing Merino ewe in a flock of 43 kg mean W, pregnant 130 days and in which about 15% bear twins, could be taken as ME_m plus energy accretion in conception products; that is $(9.25 + 0.43) = 9.7$ MJ/d, about 0.23 MJ/kg or 0.58 MJ/kg$^{0.75}$. The value 9.7 MJ is about 1.1 MJ greater than the estimate of MAFF (1975) which has earlier been questioned (Russell *et al.*, 1977).

References

BLAXTER, K.L. (1967). *The Energy Metabolism of Ruminants.* London: Hutchinson

CHAPMAN, R.E. and WHEELER, J.L. (1963). *Aust. J. Sci.,* **26**, 53–54

CORBETT, J.L., FARRELL, D.J., LENG, R.A., McCLYMONT, G.L. and YOUNG, B.A. (1971). *Br. J. Nutr.,* **26**, 277–291

CORBETT, J.L. (1980). *Proc. N.Z. Soc. Anim. Prod.* (in press)

ENGELS, E.A.N., INSKIP. M.W. and CORBETT, J.L. (1976). In *Energy Metabolism in Farm Animals,* pp. 339–342. Ed. M. Vermorel. EAAP Publication No. 19. Clermont-Ferrand: G. de Bussac

GRAHAM, N.McC. (1964). *Aust. J. agric. Res.,* **15**, 127–141

GRAHAM, N.McC., SEARLE, T.W. and GRIFFITHS, D.A. (1974). *Aust. J. agric. Res.,* **25**, 957–971

HOLTER, J.B. (1976). *J. Dairy Sci.,* **59**, 755–759

HOLTER, J.B. and REID, J.T. (1959). *J. Anim. Sci.,* **18**, 1339–1349

HOLTER, J.B., URBAN, W.E., HAYES, H.H., SILVER, H. and SKUTT, H.R. (1975). *Can. J. Zool.,* **53**, 679–685

HOUSEMAN, R.A., ROBINSON, J.J. and FRASER, C. (1978). *Proc. Nutr. Soc.,* **37**, 48A

MAFF (1975). *Energy Allowances and Feeding Systems for Ruminants.* London: HMSO

PEREZ, C.J., FURNIVAL, E.P., PICKERING, F.S. and CORBETT, J.L. (1976). In *Energy Metabolism in Farm Animals,* pp. 359–360. Ed. M. Vermorel. EAAP Publication No. 19. Clermont-Ferrand: G. de Bussac

RATTRAY, P.V., GARRETT, W.N., EAST, N.E. and HINMAN, N. (1974). *J. Anim. Sci.,* **38**, 613–626

RUSSELL, A.J.F., MAXWELL, T.J., SIBBALD, A.R. and McDONALD, D. (1977). *J. agric. Sci., Camb.,* **89**, 667–673

van ES, A.J.H. (1972). In *Handbuch der Tierernährung,* Vol. 2, pp. 1–54. Eds. W. Lenkeit and K. Breirem. Hamburg: Paul Parey

van ES, A.J.H. (1978). *Live Stk Prod. Sci.,* **5**, 331–345

WEBSTER, A.J.F., BROCKWAY, J.M. and SMITH, J.S. (1974). *Anim. Prod.,* **19** 127–139

66

ENERGY EXCHANGES IN TWIN-PREGNANT EWES

P.V. RATTRAY, T.E. TRIGG and C.F. URLICH
Ruakura Animal Research Station, Hamilton, New Zealand

Summary

Two comparative slaughter trials were conducted with ditocous ewes in late pregnancy. In the first, five increasing levels of feeding after day 95 of gestation resulted in weight changes of maternal tissues from –3.91 to 2.03 kg. All groups lost fat (–3.41 to –1.47 kg) and energy (–143 to –43 MJ) but gained water (0.15 to 3.32 kg). In the second experiment, where ewes were fed differentially from days 45 to 95 and after day 95 of gestation, level of feeding in late pregnancy had the largest effect on ewe body composition, while level of feeding during both periods influenced fetal weight. It appears that ewe body condition, as determined by the feeding level in mid-pregnancy, can influence the partitioning of nutrients to fetal growth in late pregnancy. Heavy ewes compensate for undernutrition by supporting fetal growth from maternal reserves, while thin ewes tend to replenish depleted tissues at the expense of the conceptus.

Introduction

In many parts of the world where sheep are grazed on pastoral or range land little or no supplement is fed during pregnancy, and during seasons of poor feed supply the ewe has to rely partly on body reserves to support the growth of the conceptus. The extent to which the ewe's maternal reserves can compensate for nutritional deprivation during pregnancy is not known. Two comparative slaughter trials were conducted to examine some of these aspects in ditocous ewes.

Materials and methods

EXPERIMENT 1

At day 95 of gestation, nine initial slaughter animals were taken from a group of synchronized six-year-old Coopworth ewes (mated to Suffolk rams), that had been diagnosed ditocous by X-ray. The remainder were offered a pelleted diet of 60% lucerne meal and 40% barley meal at one of five feeding levels (1.0, 1.1, 1.3, 1.5 and 1.7 times non-pregnant maintenance (M), where M was assumed to be 40 g dry matter (DM)/kg $W^{0.75}$ and W is the initial wool-free empty body weight). Feed was offered individually in pens indoors and was increased weekly in the following manner: 85%, 90%, 95%, 100%, 110% and 120% of the average nominal feeding level. The ewes were slaughtered at approximately 140 days of gestation, after being shorn and fasted for 18 h. Slaughter and analytical procedures were similar to those used by Rattray *et al.* (1974).

EXPERIMENT 2

Coopworth ewes, aged six years and averaging 57.7 kg live weight, that had been synchronized and mated to Suffolk rams, were randomized into two groups at about 45 days postmating and fed differentially until they achieved average weights of 51.4 kg (L) and 67.2 kg (H) by day 90 of gestation, when they were X-rayed to identify ditocous ewes. Two initial slaughter groups were killed on day 95 and the remainder from each mid-pregnancy feeding level were shorn and individually fed, outdoors, at average levels of either 1.0 (L) or 1.75 (H) X M until slaughter at day 135. The diet, feeding regimens, slaughter procedures and chemical analyses were similar to those in Experiment 1.

Results and discussion

EXPERIMENT 1

For the initial slaughter group (day 95) the components of the empty body (wool-free, conceptus-free and ingesta-free) were as follows: weight 39.4 kg, water 22.5 kg, fat 9.4 kg, protein 5.9 kg, ash 1.6 kg and energy 502 MJ. Average fetal weight was 0.64 kg (86.2% water, 1.4% fat, 9.6% protein, 2.8% ash) and total conceptus energy 5.9 MJ. The composition of these 95-day fetuses is similar to those of Rattray *et al.* (1974).

The data from the final slaughter group are summarized in *Table 66.1.* As level of feeding increased, live weight change, water and protein retention increased, while losses of fat, ash and energy decreased. Robinson, Smart and Pennie (1978)

Table 66.1 AVERAGE CHANGES IN EMPTY-BODY COMPONENTS FROM DAY 95 TO 140 OF GESTATION, AVERAGE FETAL WEIGHTS AND TOTAL CONCEPTA ENERGY (EXPERIMENT 1)

Feeding level × *M*	*No. of ewes*	*Change in empty body* *Weight* (kg)	*Water* (kg)	*Fat* (kg)	*Protein* (kg)	*Ash* (kg)	*Energy* (MJ)	*Fetal weight* (kg)	*Conceptus energy* (MJ)
1.0	8	−3.9	0.2	−3.4	−0.5	−0.15	−143	3.9	33.5
1.1	7	−1.2	1.2	−2.5	0.1	−0.07	−92	3.9	34.0
1.3	7	−0.2	2.1	−2.7	0.4	−0.05	−93	4.2	36.9
1.5	7	2.0	3.3	−2.0	0.7	−0.04	−60	4.5	41.0
1.7	8	2.0	3.0	−1.5	0.6	−0.03	−43	4.7	42.3
s.e. (diff.)		1.32	0.80	1.37	0.16	0.06	49.7	0.30	2.81
Significance among means		$P<0.001$	<0.01	>0.10	<0.001	<0.10	<0.10	<0.05	<0.05
Significance of trend									
linear		$P<0.05$	<0.05	<0.05	<0.01	>0.10	>0.10	<0.10	<0.05
quadratic		$P>0.10$	>0.10	>0.10	<0.10	>0.10	>0.10	>0.10	>0.10

report changes in fat and protein similar to those on the lower three feeding levels. The hydration and protein retention have been noted previously (Lodge, 1972). While in the work of Heaney and Lodge (1975) these changes were accounted for by increases in the mammary gland and uterus, this was true only for the lowest two feeding levels in this study.

Groups which gained weight were in negative energy balance, losing fat but gaining protein and water. In the light of these body composition changes during late pregnancy, ewe body-weight gains, even if corrected for conceptus, uterus and mammary increases, could be misleading in indicating nutritional status.

Fetal weights and total conceptus energy increased with level of feeding. Composition of the fetuses was not altered by the level of nutrition and averaged 81.5% water, 2.0% fat, 13.5% protein and 3.0% ash, values very close to those of fetuses of the same age reported by Rattray *et al.* (1974). Except for the highest feeding level there was considerable variation in fetal weight and total conceptus energy within groups, with total conceptus energy being related to predicted initial empty-body energy (r = 0.8–0.9) and loss of tissue energy (r = 0.6–0.9). This suggests that maternal energy reserves influence fetal growth and energy retention during late pregnancy.

EXPERIMENT 2

The data for the initial slaughter groups (day 95) are summarized in *Table 66.2.* Average fetal weight and total conceptus energy were significantly affected by feeding prior to day 95. Fetal composition did not differ and averaged 85.3% water, 1.3% fat, 11.2% protein and 2.4% ash, which except for the higher protein content is similar to Experiment 1 for fetuses of the same age.

The data for the final slaughter groups are summarized in *Table 65.3.* In contrast to Experiment 1 all groups lost weight, even at similar levels of feeding, presumably because of greater energy expenditure for thermoregulation caused by their shorn condition and outdoor housing situation.

Level of feeding after day 95 had a greater effect on the changes in ewe body weight and composition than prior to day 95; however, fetal weight and total conceptus energy were significantly influenced by level of feeding both before and after day 95. There was a significant interaction for protein ($P<0.01$) and ash ($P<0.05$) retention.

The HL (high weight, low feeding level) group had the same fetal weight and

Table 66.2 AVERAGE WEIGHTS OF EMPTY BODY AND COMPONENTS, AVERAGE FETAL WEIGHTS AND TOTAL ENERGY IN THE EMPTY BODIES AND CONCEPTA OF THE INITIAL SLAUGHTER GROUPS (EXPERIMENT 2)

Feeding level	*No. of ewes*	*Total empty body*						*Fetal weight* (kg)	*Conceptus energy* (MJ)
		Weight (kg)	*Water* (kg)	*Fat* (kg)	*Protein* (kg)	*Ash* (kg)	*Energy* (MJ)		
L	5	40.5	23.0	9.7	6.3	1.5	523	0.62	6.0
H	5	52.5	26.7	16.5	7.5	1.8	811	0.71	7.3
s.e. (diff.)		5.81	2.17	3.10	0.74	0.24	134.3	0.033	0.42
Significance		$P<0.10$	>0.10	<0.10	>0.10	>0.10	<0.10	<0.05	<0.05

Table 66.3 AVERAGE CHANGES IN EMPTY-BODY COMPONENTS FROM DAY 95 TO 135 OF GESTATION, AVERAGE FETAL WEIGHTS AND TOTAL CONCEPTA ENERGY (EXPERIMENT 2)

Feeding level	*No. of ewes*	*Total empty body*						*Fetal weight* (kg)	*Conceptus energy* (MJ)
		Weight (kg)	*Water* (kg)	*Fat* (kg)	*Protein* (kg)	*Ash* (kg)	*Energy* (MJ)		
LL	7	−7.8	−2.3	−4.4	−1.0	−0.09	−193	2.9	25.4
LH	9	−1.0	1.5	−2.8	0.3	0.15	−104	3.4	32.4
HL	6	−8.0	−2.4	−4.9	−0.4	−0.16	−201	3.4	32.3
HH	7	−3.5	0.3	−3.3	−0.2	−0.18	−136	3.7	35.0
s.e. (diff.)		1.04	0.77	0.81	0.26	0.089	30.0	0.11	2.65
Significance of feeding									
pre day 95		$P<0.05$	>0.10	>0.10	>0.10	<0.01	>0.10	<0.001	<0.05
post day 95		$P<0.001$	<0.001	<0.01	<0.001	<0.05	<0.001	<0.001	<0.05

conceptus energy as the LH group but lost more body weight ($P<0.01$), fat ($P<0.05$) and energy ($P<0.05$), showing that heavy ewes, in contrast to light ewes, have greater maternal reserves that can be mobilized to support fetal growth during long-term undernutrition in late pregnancy.

The LH (low weight, high feeding level) group gained protein and ash, while the HH (high weight, high feeding level) group lost these components, suggesting that for ewes in poor condition at this time, homeostatic mechanisms partition dietary nutrients toward replenishment of maternal reserves rather than fetal growth.

These findings have important practical implications in pastoral conditions, where in years of pasture shortage it is commonly advocated that ewes should be fed at submaintenance after mating (Coop and Clarke, 1969). While this practice has its merits in conserving pasture *in situ*, it could lead to depletion of maternal reserves, and it is therefore critical to feed ewes in poor condition at liberal levels during late pregnancy if satisfactory birth weights, lamb survival and mammary development are to be achieved. In contrast, for ewes which are well fed in mid-pregnancy, level of feeding in the last trimester is of less importance.

Acknowledgements

The authors are grateful to the staff of the Nutrition Section, Ruakura Animal Research Station for conducting feeding trials and carrying out slaughter work and chemical analyses.

References

COOP, I.E. and CLARKE, V.R. (1969). *J. agric. Sci., Camb.*, **73**, 387–394

HEANEY, D.P. and LODGE, G.A. (1975). *Can. J. Anim. Sci.*, **55**, 545–555

LODGE, G.A. (1972). *Handbuch der Tierernährung.* Eds. W. Lenkeit, K. Breirem and E. Craseman, pp. 157–189. Hamburg and Berlin: Verlag Paul Parey

RATTRAY, P.V., GARRETT, W.N., EAST, N.E. and HINMAN, N. (1974). *J. Anim. Sci.*, **38**, 613–626

ROBINSON, J.J., SMART, R.I. and PENNIE, K. (1978). *Anim. Prod.*, **26**, 390–391

67

ENERGY UTILIZATION BY EWES DURING PREGNANCY AND LACTATION

M. TISSIER, M. THERIEZ, A. PURROY and A. BRELURUT
I.N.R.A.–C.R.Z.V. de Theix, 63110 Beaumont, France

Summary

In two feeding experiments, the *in vivo* measurement of ewe body energy by the deuterium oxide technique made possible calculations of energy utilization. The ewes received various levels of metabolizable energy (ME) in late pregnancy (Experiment 1) or in lactation (Experiment 2). The ME efficiencies calculated for energy gain in the maternal body during pregnancy, or for milk production during a six-week lactation, together with efficiency of body energy losses for milk production, are in general agreement with the published data. The technique of measuring body composition *in vivo* with water isotopes can be of great interest in studying energy utilization of the ewe.

Introduction

During the last 15 years the efficiency of ME utilization in pregnant ewes has been the subject of many studies made either by indirect calorimetry or more often by successive slaughter (Langlands and Sutherland, 1968; Lodge and Heaney, 1973; Rattray *et al.*, 1974). Estimates of efficiency for conceptus growth (fetus, membranes and fluids) are closely related and small (on average 13%), the fetus(es) using energy for maintenance and growth. There are very few estimates for milk production in the ewe (Gardner and Hogue, 1964; Hadjipieris *et al.*, 1966) but they appear to be higher than in the dairy cow: on average 73% compared with 60% (van Es, 1975). Little is known about the contribution of body reserves to the energy balance in pregnant or lactating ewes while many studies have been made on lactating dairy cows (Moe, Tyrrell and Flatt, 1971; Journet, Verite and Remond, 1976). Two feeding trials have been carried out in which measuring body energy variations made it possible to calculate the efficiency of ME utilization during pregnancy and lactation.

Materials and methods

Experimental conditions are shown in *Table 67.1*. In each trial, 24 adult ewes were penned individually and given nitrogenous supplies according to current recommendations and various levels of energy: 146% and 168% of maintenance requirement (M) during late pregnancy (Trial 1), 231%, 209% and 184% of M during lactation (Trial 2) (*Table 67.1*).

Table 67.1 EXPERIMENTAL DESIGN OF THE TWO EXPERIMENTS. THERE WERE 24 EWES IN EACH EXPERIMENT – LIMOUSINE BREED OR CROSSBREED ROMANOV X LIMOUSINE. THEY WERE FED MEADOW HAY (7.6 MJ OF ME/kg DRY MATTER) AND CONCENTRATES (MAIZE + SOYA BEAN IN VARIOUS PROPORTIONS)

	Experiment 1 (*Pregnancy: weeks* −*6 to* −*1*)	*Experiment 2* (*Lactation: weeks* +*1 to* +*6*)	
Litter size	1–4 fetuses (2 or 3 in 18 ewes)	mainly twins suckling (15 ewes)	
Levels of ME supplies ($kJ/kg\ BW^{0.75}$ at week + 1)	High 580 (1.46 M)*	High	916 (2.31 M)
		Medium	828 (2.09 M)
	Low 666 (1.68 M)	Low	730 (1.84 M)

* The maintenance requirement (M) is estimated at 397 kJ of ME/kg $BW^{0.75}$ week +1.

ME supplies were calculated from measurements of diet digestibility. Total body energy and its changes were determined by measuring live weight and body water (using deuterium oxide (D_2O) dilution technique) several times and by applying equations obtained previously from slaughtered sheep (*Table 67.2*). The quantities of energy retained in the gravid uterus were obtained from the birth weight of the lamb with relationships calculated from published data (*Table 67.2*) (Tissier and Theriez, 1978). The ewes' milk production was calculated from the weight and growth rate of the lambs, using the relationships of Robinson, Foster and Forbes (1969).

The efficiency of ME utilization for body energy gain and for milk production as well as that of body energy for milk production were calculated by multiple linear regression.

Table 67.2 ESTIMATING EQUATIONS FOR BODY ENERGY AND ENERGY CONTENTS OF GRAVID UTERUS

Body components	*Estimating equations*	
BE Pregnancy (MJ)	$BE = 26.539\ BW - 22.284\ D_2O - 114.110$	RSD = 6.1%
BE Lactation (MJ)	$BE = 31.003\ BW - 27.743\ D_2O - 173.570$	RSD = 8.9%
FW* (kg)	$FW = LBW \exp[1 - \exp(-0.0103\ (t - 146))]$	
FW† as % of gravid uterus weight	$FW\% = 3.218\ t - 0.011\ t^2 - 173\ 141$	
Energy content of gravid uterus† (kJ/kg)	$E_{GU} = 30.9\ t - 880.7$	

BE = body energy; BW = body weight (kg) at D_2O infusion; D_2O = body water (kg) estimated by D_2O dilution technique; FW = fetal weight; LBW = lamb birth weight (kg); t = days of pregnancy

* From data of Wallace (1948); Joubert (1956); Lodge and Heaney (1973); Rattray *et al.* (1974)

† From data of Langlands and Sutherland (1968); Lodge and Heaney (1973); Rattray *et al.* (1974); Robinson *et al.* (1977)

Results and discussion

EVOLUTION OF COMPONENT WEIGHT IN THE WHOLE BODY

The mean characteristics of the ewes in the two trials at the time of body energy measurements are summarized in *Table 67.3.*

Table 67.3 MEAN WEIGHTS OF THE EWES AND THEIR ESTIMATED BODY COMPONENTS

	Experiment 1: Pregnancy				*Experiment 2: Lactation*			
Weeks	−6		−1		+1		+6	
Days from lambing	−42.1	(2.9)*	−6.5	(2.0)	+2.3	(0.5)	+40.1	(3.3)
Body weight (kg)	71.3	(5.0)	80.2	(5.2)	61.2	(5.0)	55.6	(4.9)
Estimated body components								
Water (kg)	44.4	(3.5)	51.2	(4.9)	41.2	(4.0)	38.8	(3.2)
Lipids (kg)	15.4	(1.8)	17.1	(1.9)	10.7	(2.8)	8.2	(2.3)
Proteins (kg)	7.8	(0.4)	8.7	(0.6)	7.0	(0.6)	6.5	(0.5)
Energy (MJ)	789	(75)	873	(74)	583	(110)	474	(97)

* Figures in brackets are standard deviations

In late pregnancy (−42 to −7 days) whole body energy, including the gravid uterus, increases by 15% approximately, mainly owing to the rise in body lipids in the ewe. In fact, if the quantities retained in the gravid uterus, calculated with the relationships in *Table 67.2* are subtracted, the actual weight of the mother rises by 2.6 kg, 1.6 kg of which is lipids: this corresponds to a weight-gain energy value of approximately 20.1 MJ/kg.

At the beginning of lactation (+ 2 to + 40 days), the ewes on average lose large amounts of body energy, even at the high level of energy supply, which could not however meet all the estimated maintenance and milk-production requirements (Tissier and Theriez, 1978). The weight losses (5.7 ± 2.7 kg) are largely lipids: their energy value is comparable to that of the weight gain (19.4 MJ/kg).

ME UTILIZATION IN LATE PREGNANCY

The data of 20 ewes in positive energy balance for the maternal tissue were used. Because of the small variability in energy retained in the gravid uterus (26.81 ± 4.33 MJ), it was not possible to consider this as a separate variable in the regression model. Different values for efficiency of ME utilization for uterus growth (k_p) were therefore tested, and the relationship between on the one hand ME available for maintenance and energy gain in maternal tissues (X) and on the other hand energy actually retained in these tissues (Y) was calculated. The two variables were expressed per kg of metabolic weight after lambing (week + 1). The best fit to the data was found with k_p values of between 0.13 and 0.16. For example, when $k_p = 0.16$, the relationship calculated is:

$$Y = 0.39\ (\pm 0.21)\ X - 74.12 \qquad r = 0.40$$

The ranges of variation are 3.7–208.4 for Y, and 316.5–526.3 for X, Y and X being expressed as kJ/kg $BW^{0.75}$ at week + 1.

The efficiency of ME for maternal tissue gain would therefore be 39%, a value lower than those in the relationships put forward by Garrett, Ferrell and Rattray (1976) of 49–52%.

ME UTILIZATION IN LACTATION

Equations were calculated similarly to those proposed by Moe, Tyrrell and Flatt (1971) associating produced milk energy (Y) with ingested ME (X_1) and energy from body reserves (X_2), each of these factors being expressed as kJ/kg metabolic weight after lambing (week + 1). Only 21 ewes in negative energy balance were used.

For the first six weeks of lactation the equation calculated is:

$$Y = 0.69 (\pm 0.19) X_1 - 0.70 (\pm 0.27) X_2 - 177.10 \qquad r = 0.68$$

The ranges of variation are 258.0–595.6 for Y, 749.1–1116.5 for X_1 and 4.3–222.0 for X_2.

The efficiencies of transforming ingested ME and body energy into milk energy are 0.69 and 0.70 respectively. These estimations agree well with those already published for the ewe (Gardner and Hogue, 1964; Hadjipieris *et al.*, 1966).

Although the coefficients calculated are all significantly different from zero, the associated standard deviations are even higher. In addition extrapolation at zero body energy changes and/or zero milk energy would lead to abnormally low maintenance requirements. This inaccuracy is most probably due to the small number of animals used in the calculations and to the fact that the levels of energy were not sufficiently different. As well as this, it was not possible to measure directly the milk production of the ewes.

These results show that the technique of measuring body composition *in vivo* with water isotopes could be of great interest in studies on energy utilization by the ewe. In particular, it enables data to be obtained quickly from a large number of animals kept in conditions close to those of usual husbandry. Such results can complement the information obtained with more accurate and sophisticated techniques applied to a small number of animals.

References

GARDNER, R.W. and HOGUE, D.E. (1964). *J. Anim. Sci.*, **23**, 935–942

GARRETT, W.N., FERRELL, C.L., RATTRAY, P.V. (1976). In *Energy Metabolism in Farm Animals*, pp. 315–318. Ed. M. Vermorel. EAAP Publication No. 19. Clermont-Ferrand: G. de Bussac

HADJIPIERIS, G., JONES, J.G.W., WIMBLE, R.H. and HOLMES, W. (1966). *J. agric. Sci., Camb.*, **66**, 341–349

JOUBERT, D.M. (1956). *J. agric. Sci., Camb.*, **47**, 382–427

JOURNET, M., VERITE, R. and REMOND, B. (1976). In *Energy Metabolism in Farm Animals*, pp. 323–326. Ed. M. Vermorel. EAAP Publication No. 19. Clermont-Ferrand: G. de Bussac

LANGLANDS, J.P. and SUTHERLAND, H.A.P. (1968). *Br. J. Nutr.*, **22**, 217–227
LODGE, G.A. and HEANEY, D.P. (1973). *Can. J. Anim. Sci.*, **53**, 479–489
MOE, P.W., TYRRELL, H.F. and FLATT, W.P. (1971). *J. Dairy Sci.*, **54**, 548–553
RATTRAY, P.V., GARRETT, W.N., EAST, N.E. and HINMAN, N. (1974). *J. Anim. Sci.*, **38**, 383–393
ROBINSON, J.J., FOSTER, W.H. and FORBES, T.J. (1969). *J. agric. Sci., Camb.*, **72**, 103–107
ROBINSON, J.J., MacDONALD, I., FRASER, C. and CROFTS, R.M.J. (1977). *J. agric. Sci., Camb.*, **88**, 539–552
TISSIER, M. and THERIEZ, M. (1978). In *I.N.R.A. Alimentation des Ruminants*, 403–448. Versailles: I.N.R.A. Publications
van ES, A.J.H. (1975). *Live Stk prod. Sci.*, **2**, 95–107
WALLACE, L.R. (1948). *J. agric. Sci., Camb.*, **38**, 243–401

68

THE ENERGY COSTS OF PREGNANCY IN THE SOW

J. NOBLET* and W.H. CLOSE
ARC Institute of Animal Physiology, Babraham, Cambridge CB2 4AT, England

Summary

The partition of the metabolizable energy (ME) intake of gilts has been investigated at two levels of feeding, 1.8 and 2.5 kg/d, at several stages of pregnancy and compared with that of non-pregnant animals of similar age. Protein retention of the pregnant animals was higher than that of the control animals, especially during the mid and late periods of pregnancy. Fat retention for both groups of animals decreased with duration of pregnancy. The mean energy cost of maintenance (kJ/kg$^{0.75}$ per d) and efficiency of protein and fat synthesis for the pregnant animals were 407, 0.56 and 0.93, respectively; corresponding values for the controls were 409, 0.40 and 0.96, respectively.

Introduction

Pregnancy stimulates weight gain in animals in excess of that accounted for by uterine and mammary tissue development (Salmon-Legagneur and Rerat, 1962). This additional gain may be attributed to changes in the rate and composition of maternal tissue deposition. This suggests that the efficiency with which energy is utilized during pregnancy is greater than it is in non-pregnant animals of similar age and body size. On the other hand, it has been shown (Brody, 1938; Verstegen, van Es and Nijkamp, 1971) that heat production increases markedly in the later stages of gestation, suggesting a decrease in the efficiency of energy utilization or an increase in the animal's maintenance energy requirement. There is little available information to show whether the maintenance energy requirement or the efficiency of energy utilization of animals changes during pregnancy and the experiments reported here were designed in an attempt to provide this information in the pig.

Materials and methods

The experiments were designed in a 2 × 3 factorial arrangement involving two levels of feeding with animals at three stages of gestation. The feeding levels,

* Present address: I.N.R.A., Station de Recherches sur l'Élevage des Porcs, 78350 Jouy-en-Josas, France

which remained constant throughout pregnancy, were 1.8 (low) and 2.5 (high) kg food d. The food contained 13.3 (± 0.8) MJ ME and 165 (± 1.0) g crude protein (CP)/kg dry matter (DM). The three stages of gestation at which the animals were investigated were 40–60 days (early), 60–80 days (middle) and 90–110 days (late). Four gilts were investigated at each treatment. A similar number of non-pregnant litter sisters (controls) of comparable age at each stage of pregnancy were also investigated. The experiment therefore involved a total of 48 animals.

Following mating, both pregnant and control animals were kept in batches of four in a straw-bedded sow house. Four weeks prior to the appropriate stage of gestation, the animals were moved into a farrowing house and accommodated in individual farrowing pens, similar in design to that constructed within the calorimeter. Following habituation the animals were removed to a heat-sink calorimeter (Close *et al.*, 1978) maintained at 20 (±0.5) °C. Each animal remained in the calorimeter for a seven-day period during which its heat loss (H) and energy and nitrogen balances were determined. The ME intake, H and protein retention (P) of the animals were determined directly while their energy retention (ER) and fat retention (F) were determined by difference. At the end of the balance period the pregnant animals were slaughtered and their uterine and mammary tissue removed for subsequent chemical analysis.

Results and discussion

The body-weight gain of the pregnant animals was significantly greater ($P<0.05$) at each stage of pregnancy than that of the control animals (*Table 68.1*). By partitioning total gain during pregnancy into maternal and reproductive components it was established that maternal growth in early and middle pregnancy occurred at a rate equal to that of the corresponding control animal. In late pregnancy it was less than that of the control animals.

Table 68.1 MEAN BODY WEIGHT, BODY-WEIGHT GAIN (ΔW), METABOLIZABLE ENERGY INTAKE (MEI), HEAT LOSS (H), ENERGY RETENTION (ER), PROTEIN (P) AND FAT (F) DEPOSITION OF GILTS AT DIFFERENT STAGES OF PREGNANCY COMPARED TO CONTROL ANIMALS OF SIMILAR AGE

	Stage of pregnancy (days)	*Body weight* (kg)	*ΔW** (g/d)	*ME*	*H*	*ER* (MJ/d)	*P*	*F*
Pregnant	47	116	404	20.7	15.3	5.4	0.8	4.6
	49	139	576	29.2	17.2	12.0	1.1	10.9
Control	–	114	259	20.2	14.8	5.4	0.9	4.5
	–	140	410	30.1	17.5	12.6	0.4	12.2
Pregnant	75	144	380	20.2	16.7	3.5	1.5	2.0
	68	147	605	28.2	20.8	7.4	1.7	5.7
Control	–	129	134	20.7	16.7	4.0	0.9	3.1
	–	133	364	28.0	19.7	8.3	1.6	6.7
Pregnant	98	154	368	19.6	19.6	0.0	1.3	–1.3
	98	200	623	29.3	23.6	5.7	1.9	3.8
Control	–	132	212	20.4	17.5	2.9	1.0	1.9
	–	154	382	28.4	22.0	6.4	1.4	5.0

* Calculated from the first day of pregnancy or an equivalent period for the control animals

The energy retained by the animals is also given in *Table 68.1*. ER, for both pregnant and control animals, decreased with advancement of pregnancy, that is, as body weight increased. There was little difference in ER between pregnant and control animals in early gestation. In middle and late gestation, however, ER of the pregnant animals was less than that of the controls. The linear regression equations relating ER (y; kJ/kg$^{0.75}$ per d) to ME intake (x; kJ/kg$^{0.75}$ per d) were for pregnant animals:

$$y = 0.91\ (\pm 0.06)\ x - 398\ (\pm 37) \qquad r = 0.95 \qquad (1)$$

and for control animals:

$$y = 0.83\ (\pm 0.07)\ x - 349\ (\pm 47) \qquad r = 0.93 \qquad (2)$$

The data used for the compilation of equation (1) included three sets of data where ME intake was below the level needed to maintain zero ER, that is, from the late pregnant animals at the lowest level of feeding (*Figure 68.1*). If these data sets are excluded and only values associated with positive ER are used in the linear regression analysis, the relation becomes, for pregnant animals:

$$y = 0.84\ (\pm 0.07)\ x - 345\ (\pm 43) \qquad r = 0.93 \qquad (3)$$

Comparison of equations (2) and (3) suggests that there is little difference between pregnant and control animals in the efficiency with which energy is utilized for productive purposes (k).

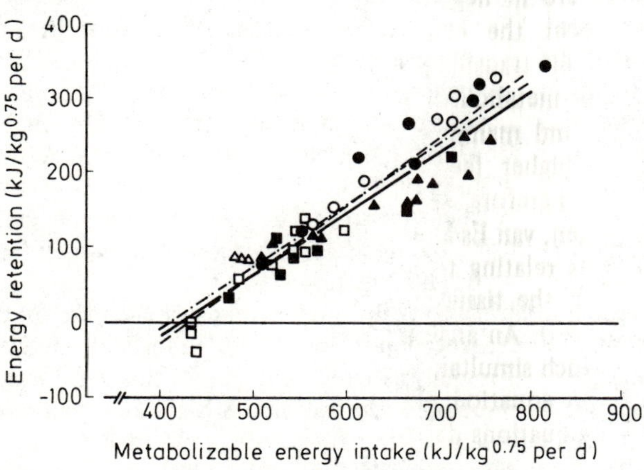

Figure 68.1 The relation between metabolizable energy intake and energy retention for pregnant (open symbols) and control (closed symbols) animals during early (○, 40–60 days), middle (△, 60–80 days) and late (□, 90–110 days) pregnancy. The regression lines relating ER to ME intake are given for both the pregnant (– – – –, all data sets included; –·–·–·–, only data sets above ER = 0 included) and control (——) animals

The differences in ER reflect variations in the protein and fat gains of the animals. With the exception of one group of animals, P decreased with decrease in ME intake at each period of gestation investigated. In both the pregnant and control animals, P appeared to be maximum during the middle and late stages of gestation. At each of these stages, however, the retention of the pregnant animals was greater than that of the controls. F also decreased with decrease in ME intake at each stage of gestation investigated. However, unlike P, F decreased with advancement of pregnancy; the decrease in F per unit decrease in ME intake was greater for the pregnant animals, 0.91 compared with 0.81 kJ/kJ ME.

From the linear regression equations relating ER to ME intake (equations (1), (2) and (3)) it was possible to calculate the animal's maintenance energy (ME_m) requirement as that intake at which ER = 0. The estimates of ME_m derived from these equations are 432 (1), 420 (2) and 411 (3) kJ/kg$^{0.75}$ per d, and suggest that there is little difference in ME_m between pregnant and control animals. The difference in ME_m and the partial efficiency of ER (k) calculated for the pregnant animals, that is, from equations (1) and (3) suggests that these entities are dependent on the data from which they are derived. At low ME intakes, however, particularly those associated with negative ER, it is possible that the animals were below their critical temperature and at these intakes both ME_m and k are high relative to those at higher intakes where the animals were within the zone of thermal neutrality (Close, 1978). If all data sets are included in the linear regression analysis then the overall values of ME_m and k would be higher than if only data relating to higher ME intakes are included.

Figure 68.1 suggests that during late pregnancy the heat production of the animals at the low level of feeding is higher relative to that at earlier stages of gestation or at higher food intakes. There are a number of possibilities which could explain this increase in metabolism. (1) Following prolonged application of low feed intakes the body weight of the animals is lower and their carcasses are leaner relative to younger animals or animals receiving higher food intakes. The energy cost of the maintenance of this tissue may be higher and may result in an increased metabolism. (2) At the late stage of gestation the animals on the low feed intakes are in negative energy balance and maternal tissue is being catabolized to meet the energetic demands of the developing uterine and mammary tissue. This transfer of energy necessitates a cost to the animal which would increase its metabolism. (3) Relative to their body weight, the total weight of uterine and mammary tissue is greater compared to that of heavier animals receiving higher feed intakes. The heat production associated with pregnancy would, therefore, be higher. Similar conclusions may be drawn from the data of Verstegen, van Es and Nijkamp (1971).

The calculations relating to ME_m and k take no account of the change in the composition of the tissue deposition of the animals at energy equilibrium, that is, where ER = 0. An analysis which overcomes this deficiency should construct equations which simultaneously partition ME intake into a requirement for ME_m, P and F. An equation of the form: ME = aP + bF + c would fulfil this requirement. The equations derived from such a relationship were for pregnant animals:

$$ME = 1.78\ (\pm\ 1.18)\ P + 1.07\ (\pm\ 0.15)\ F + 407\ (\pm\ 43) \qquad r = 0.90 \qquad (4)$$

and for control animals:

$$ME = 2.47\ (\pm\ 0.46)\ P + 1.04\ (\pm\ 0.07)\ F + 409\ (\pm\ 20) \qquad r = 0.84 \qquad (5)$$

where ME, P and F are in kJ/kg$^{0.75}$ per d. Equation (4) has been derived from the same series of data sets as equation (3), that is, above ER = 0.

The reciprocals of the coefficients a and b represent estimates of the partial efficiencies of protein (k_p) and fat (k_f) deposition, while the intercept represents the estimates of ME_m. ME_m (kJ/kg$^{0.75}$ per d), k_p and k_f for the pregnant animals were 407, 0.56 and 0.93, respectively, compared to 409, 0.40 and 0.96 for the control animals. Thus pregnant animals appear to have a similar ME_m requirement and energetic efficiency for fat synthesis as control animals but a higher energetic efficiency for protein synthesis.

References

BRODY, S. (1938). *Mo. agric., expl. Sta. res. Bull.* No. 283

CLOSE, W.H. (1978). *Br. J. Nutr.*, **40**, 433–438

CLOSE, W.H., HEAVENS, R.P., NOBLET, J. and START, I.B. (1978). *J. Physiol. (Lond.)*, **284**, 11–12

SALMON-LEGAGNEUR, E. and RERAT, A. (1962). In *Nutrition of Pigs and Poultry.* pp. 207–233. Eds. J.T. Morgan and D. Lewis. London: Butterworths

VERSTEGEN, M.W.A., van ES, A.J.H. and NIJKAMP, H.J. (1971). *Anim. Prod.*, **13**, 677–683

69

EFFECT OF PROTEIN CONTENT AND METABOLIZABILITY OF THE DIET ON THE EFFICIENCY OF UTILIZATION OF METABOLIZABLE ENERGY IN DAIRY COWS

W. SCHNEIDER, D. FRITZ, J.R. REICHL and K.H. MENKE
Institute of Animal Nutrition, University Hohenheim, 7000 Stuttgart 70, PO Box 106-06100, Federal Republic of Germany

Summary

Forty respiration experiments with lactating cows were evaluated by multiple regression analysis. Protein content of the diet had a significant negative effect on the efficiency of utilization of metabolizable energy (ME), but metabolizability (q) had no significant effect in these experiments with diets of rather high metabolizability. Inclusion of results from other laboratories lead to square functions of q or M (MJ ME/kg) which showed maxima in the range of q = 0.6–0.7 and M = 11–12. The best fit was achieved with the function:

$$k_1 = 0.1189\,M - 0.00502\,M^2 - 0.0082 \text{ g crude protein (XP)/MJ ME};\ n = 248;\ r = 0.61$$

Introduction

The quantitative description of the effect of diet composition on the efficiency of utilization of ME in lactating cows is still a matter of discussion. Frequently the results of respiration experiments have been described by linear functions of q. The following experiments have been evaluated by non-linear multiple regression analysis in order to find an explanation for the discrepancies in the literature.

Materials and methods

In 52 respiration experiments with lactating cows the effects of feeding level (IME^X, MJ/kg$^{0.75}$ per d), protein–energy ratio (g XP/MJ ME), metabolizability (q = ME/GE) and ME content (M = MJ ME/kg dry matter (DM)) on energy retention (LE^X+RE^X) and efficiency of ME utilization (k_1) have been studied in an open-circuit respiration chamber. The means and the variations of the variables are given in *Table 69.1.* Experimental details will be published by Fritz (1980). Twelve animals with changes in body energy of more than 0.2 MJ RFE^X/d were excluded from this evaluation, since reliable correction factors are not available. The abbreviations are similar to those used by Berschauer, Gaus and Menke (*see* page 101).

Table 69.1 MEANS AND VARIATIONS OF THE VARIABLES

Laboratory *No. of cows*		IME^x (MJ/W^x)	LE^x+RE^x (MJ/W^x)	k_1 (ME_m^x=0.39)	*q*	*M* (MJ ME/kg T)	*XP/ME* (g/MJ)
Ho	$\bar{x}$	1.246	0.523	0.612	0.631	11.62	13.5
40	s.d.	0.283	0.178	0.065	0.053	1.03	2.5
Ro	$\bar{x}$	1.320	0.521	0.560	0.612	11.58	14.9
33	s.d.	0.157	0.094	0.039	0.030	0.81	2.0
Wa	$\bar{x}$	1.354	0.517	0.536	0.533	9.71	17.0
147	s.d.	0.158	0.095	0.044	0.046	0.86	2.4
Be	$\bar{x}$	1.266	0.479	0.545	0.584	10.75	17.1
28	s.d.	0.129	0.092	0.056	0.058	1.12	4.1

Ho = Hohenheim (these experiments); Ro = Rostock (Schiemann *et al.*, 1970–1972); Wa = Wageningen (Van der Honing, 1975); Be = Beltsville (Moe, Flatt and Tyrrell, 1972)

Results

A linear regression between energy retention (LE^x+RE^x, x) and energy intake (IME^x, y) resulted in a maintenance turnover of 0.46 MJ ME/W^x and a k_1 value of 0.66 ($1/b_1$, equation 1 in *Table 69.2*). Inclusion of q as a second independent variable did not improve the correlation, but protein content (XP/ME) had a significant effect on ME-requirement (equation 3). Regression coefficients of q and q^2 were not significant, but did tend to cause a minimum ME-requirement at q = 0.70. Analogously the regression of k_1 (y) on q, q^2 and XP/ME (equation 3) showed a significant negative effect of protein content on k_1, whereas the

Table 69.2 REGRESSION EQUATIONS FROM THE HOHENHEIM EXPERIMENTS; N = 40; s_b IN BRANCHES

Eq. No.	b_0	b_1 (LE^x+RE^x)	b_2 (q)	b_3 (q^2)	b_4 (XP/ME)	*y'=0 at q=*	*r* *RSD*
	y = IME^x						
1	0.460	+1.502** (0.083)					0.95** 0.092
2	0.634	+1.525** (0.086)	−0.294 (0.286)				0.95** 0.092
3	0.989	+1.482** (0.084)	−2.142 (5.65)	+1.527 (4.75)	+0.0164** (0.0061)	0.70	0.96** 0.086
	y = k_1 (ME_m^x = 0.39 MJ)						
4	−0.353		+3.27 (3.85)	−2.41 (3.25)	−0.00977** (0.00423)	0.68	0.46* 0.060
	b_0		b_2 (M)	b_3 (M^2)	b_4 (XP/ME)	*y'=0 at M=*	*r* *RSD*
5	−0.219		+0.1565 (0.1837)	−0.00625 (0.00848)	−0.01026** (0.00432)	12.5	0.45*

* $P<0.05$
** $P<0.01$

(non-significant) regression coefficients of q and q^2 tended to a maximum of k_1 at $q = 0.68$. The use of M instead of q as the independent variable leads to equation 5 with the corresponding maximum at 12.5 MJ ME/kg T. The multiple regression analysis with k_1 as the dependent variable showed the best correlations when the maintenance requirement was adopted to be 0.39 MJ ME/W^x.

Tables 69.3 and *69.4* show the results of similar calculations with data from other laboratories (animals with changes in body energy of more than 0.2 MJ RFE^x again were excluded). The effect of protein content on k_1 is significant in all experiments, with only the exception of those from Rostock (probably because of the lower variation in protein content, as is shown in *Table 69.1*). Inclusion of the square of XP/ME or of the level of feeding (IME^x) as further variables did not improve the correlation. The effect of M (*Table 69.4*) on k_1 seems to be more pronounced than that of q (*Table 69.3*). Regression coefficients of M and M^2 are significant in the Rostock experiments and in equations 23–28 where data from different stations have been evaluated together. The same is true – with few exceptions – in equations 11–17 with regard to the regression coefficients of q and q^2. The maxima of the functions of M (*Table 69.4*, $y'=0$) frequently are in the range of M of 11–12 MJ/kg T. Functions of q show a greater variation, but tend to maxima at corresponding values of q in the range of 0.6–0.7. The best correlation is found with equation 27 (n=248, r=0.61, RSD=0.44).

Table 69.3 REGRESSION EQUATIONS OF DATA FROM OTHER LABORATORIES: k_1 (y) AS FUNCTION OF METABOLIZABILITY (q) AND PROTEIN–ENERGY RATIO (XP/ME) OF THE DIET; s_b IN BRANCHES

Eq. No. Lab.	b_0	b_1 (q)	b_2 (q^2)	b_3 (XP/ME)	b_4 ($(XP/ME)^2$)	*y'=0 at q=*	*r RSD*
6	1.033	−1.489	+1.306	−0.00351			0.18
Ro		(7.55)	(6.30)	(0.0037)			0.040
7	0.235	+1.934	−1.599	−0.02942	+0.0008	0.60	0.20
Ro		(10.14)	(8.52)	(0.050)	(0.0016)		0.041
8	0.355	+0.715	−0.314	−0.0065**		1.14	0.56**
Wa		(1.59)	(1.48)	(0.0014)			0.037
9	0.515	+0.836	−0.399	−0.0309	+0.00073	1.05	0.57**
Wa		(1.59)	(1.48)	(0.0233)	(0.00070)		0.037
10	0.687	+0.351	−0.547	−0.00927**		0.32	0.64**
Be		(2.65)	(2.20)	(0.0023)			0.454
11	0.036	+1.933*	−1.498	−0.00606**		0.65	0.54**
Wa+Ro		(1.14)	(1.04)	(0.0012)			0.038
12	0.374	+0.3015**					0.33**
Wa+Be		(0.065)					0.044
13	−0.290	+3.267**	−2.748**	−0.00758**		0.59	0.55**
Wa+Be		(0.902)	(0.813)	(0.00114)			0.039
14	0.379	+0.294**					0.35**
Wa+Ro+Be		(0.054)					0.043
15	−0.226	+2.996**	−2.507**	−0.00694**		0.60	0.54**
Wa+Ro+Be		(0.853)	(0.764)	(0.00106)			0.039
16	0.249	+1.224	−0.814	−0.00775**		0.75	0.61**
Wa+Ro+Be+Ho		(0.831)	(0.731)	(0.00108)			0.045
17	0.270	+1.393*	−0.964	−0.01606*	+0.00025	0.72	0.61**
Wa+Ro+Be+Ho		(0.849)	(0.748)	(0.0088)	(0.00026)		0.045

* $P<0.05$
** $P<0.01$

Table 69.4 REGRESSION EQUATIONS OF DATA FROM OTHER LABORATORIES: k_1 (y) AS FUNCTION OF ME-CONTENT (M) AND PROTEIN–ENERGY RATIO (XP/ME) OF THE DIET

Eq. No. Lab.	b_0	b_1 (M)	b_2 (M^2)	b_3 (XP/ME)	b_4 ($(XP/ME)^2$)	$y'=0$ *at M=*	*r* *RSD*
18	−1.160	+0.309*	−0.0135*	−0.00255		11.4	0.39
Ro		(0.166)	(0.0071)	(0.0034)			0.038
19	−0.928	+0.303*	−0.0133*	−0.0267	+0.00073	11.4	0.41
Ro		(0.167)	(0.0072)	(0.0342)	(0.00103)		0.038
20	0.344	+0.0425	−0.00117	−0.00646**		18.2	0.56**
Wa		(0.084)	(0.0043)	(0.00137)			0.037
21	0.504	+0.0533	−0.00162	−0.0334	+0.00081	16.5	0.56**
Wa		(0.085)	(0.0043)	(0.0237)	(0.00071)		0.037
22	0.207	+0.1005	−0.00506	−0.0089**		9.9	0.62**
Be		(0.140)	(0.0063)	(0.0023)			0.046
23	−0.149	+0.1445**	−0.00645**	−0.00623**		11.2	0.55**
Wa+Ro		(0.0374)	(0.00181)	(0.00121)			0.037
24	−0.228	+0.1661**	−0.00759**	−0.00746**		10.9	0.56**
Wa+Be		(0.0434)	(0.00213)	(0.00113)			0.039
25	0.406	+0.0133**					0.33**
Wa+Ro+Be		(0.0026)					0.043
26	−0.206	+0.1594**	−0.00724**	−0.00689**		11.0	0.56**
Wa+Ro+Be		(0.0339)	(0.00162)	(0.00103)			0.038
27	−0.001	+0.1189**	−0.00502**	−0.00820**		11.8	0.61**
Wa+Ro+Be+Ho		(0.0367)	(0.00174)	(0.00106)			0.044
28	0.045	+0.1271**	−0.00541**	−0.0190**	+0.00032	11.7	0.61**
Wa+Ro+Be+Ho		(0.0372)	(0.00177)	(0.0087)	(0.00026)		0.044

* $P<0.05$
** $P<0.01$

Linear functions of q and M (equations 12, 14, 25) have correlation coefficients of about 0.30–0.35).

Surprisingly these results with dairy cows show similarities with those published by Schiemann *et al.* (1971), who found that efficiency of energy utilization in adult oxen with positive energy balance followed a curvilinear function with a maximum at 74% digestibility of energy (corresponding to q = 0.63).

References

FRITZ, D. (1980). *Untersuchungen über die Verwertung der umsetzbaren Energie bei der Milchkuh.* Dissertation Hohenheim (in preparation)

MOE, P.W., FLATT, W.P. and TYRRELL, F.H. (1972). *J. Dairy Sci.,* **55**, 945–958

SCHIEMANN, R., JENTSCH, W., HOFFMANN, L. and WITTENBERG, H. (1970–1972).
1. *Mitt. Arch. Tierernähr.,* **20**, 227–251
2. *Mitt. Arch. Tierernähr.,* **20**, 423–440
3. *Mitt. Arch. Tierernähr.,* **22**, 675–695
4. *Mitt. Arch. Tierernähr.,* **22**, 697–720

SCHIEMANN, R., NEHRING, K., HOFFMANN, L. JENTSCH, W. and CHUDY, A. (1971). *Energetische Futterbewertung und Energienormen,* Figure 5.1, Equation 5.96, p. 141. Berlin: Verl. VEB Dt. Landw. Verl.

VAN DER HONING, Y. (1975). Intake and Utilization of Energy of Rations with Pelleted Forages by Dairy Cows. Thesis Wageningen, Agric.Res.Rep. 836, Centre f. Agric. Publ. a. Docum

THE ENERGY METABOLISM OF DAIRY COWS UNDERFED IN EARLY LACTATION

T.E. TRIGG, K.E. JURY, A.M. BRYANT and C.R. PARR
Ruakura Animal Research Station, Hamilton, New Zealand

Summary

Immediate and carry-over effects of underfeeding pasture-fed dairy cattle for six weeks after calving have been studied using respiration calorimetry. Results from 72 energy-balance trials have been subjected to regression analysis on a within-cow basis. This type of analysis is a departure from that generally used, and small but statistically significant differences in energy partitioning were distinguished between well-fed cows and their underfed co-twins in early lactation.

Introduction

In New Zealand the dairy industry is almost completely dependent on improved grassland for the supply of nutrients. Owing to the seasonal nature of pasture growth, it is almost inevitable that animals will be subjected to periods of under-nutrition, often about the time of calving. The effects of this on production, both current and subsequent to the underfeeding, are therefore of considerable local significance. Most early work summarized by Broster (1972, 1974) emphasized the importance of level of feeding in early lactation but some recent work (e.g. Gordon, 1977; Bryant and Trigg, 1979) has demonstrated that the immediate effects of underfeeding can be small and carry-over effects almost non-existent.

The effects of underfeeding in early lactation on consequent energy partitioning and production of cows offered pasture throughout lactation have been studied; and data from 72 energy balances using respiration calorimetry, collected during the first two years of a three-year programme, are presented here.

Materials and methods

Eight sets of monozygous twins (three in year 1, five in year 2) in their third or later lactation were used. Four days after calving (August–September) one animal from each set was offered *ad lib.* (H) grazing; grazing of the co-twin (L) was such that dry matter intake (DMI) during six weeks was 50–60% of expected *ad lib.* intake. All animals were offered generous grazing for the remainder of the trial except during balance measurements which were at two stages of lactation: in year 1, weeks 10–16 (early) and 22–28 (mid) and in year 2, weeks 8–18 (early) and 22–32 (mid). At these times animals were housed in individual stalls and offered only fresh pasture, cut twice daily. For each animal and stage of lactation, complete energy balances (Bryant *et al.* 1977) were established at two feeding levels: *ad lib.* (A) and 75% National Research Council (1966) (R) requirements. Balance periods of 10 days included three (R) or four (A) consecutive days of gas collection and followed a 14-day preliminary feeding period.

BIOMETRICAL ANALYSIS

Data from years 1 and 2 were pooled within stages of lactation. Analysis of nutrient partition data was on a within-cow basis; a regression slope for each cow was calculated by:

$$\frac{Y_A - Y_R}{X_A - X_R}$$

where Y and X are the dependent and independent variables respectively and A and R represent *ad lib.* and low levels of feeding during energy-balance measurements. The combined estimate of the slope for each treatment (H or L) was calculated as the average of the individual slope, each being weighted first by $^1/(X_A - X_R)^2$. Comparison of the H and L coefficients allowed for the correlation within twin sets by modifications of the method of Yates (1939).

Results

During the six weeks following calving, for groups H and L respectively, mean (±s.e.) DMI was 13.1 and 6.9±0.3 kg DM per head per d and average live weight (LW) change during this period was 0.1 and –1.2 kg/d. Subsequent intakes were measured only during indoor feeding periods and averages (±s.e.) for animals while at the *ad lib.* feeding level (A) were 3.76±0.12 and 3.27±0.13 kg/100 kg LW respectively for early and mid lactation; equivalent data while restricted (R) were 2.49±0.13 and 1.92±0.08 kg/100 kg LW. Mean digestibility coefficients (±s.e.) for organic gross energy matter (OM)/(GE) and nitrogen were 0.722±0.003, 0.738±0.003 and 0.709±0.004 respectively in early lactation and 0.723±0.006,

Table 70.1 TOTAL FULL-CREAM MILK (FCM) AND FAT PRODUCTION

Weeks of Lactation	*FCM* (kg)				*Fat* (kg)			
	H	*L*	*s.e.*(d)	*P*	*H*	*L*	*s.e.*(d)	*P*
1–6	869	622	36	***	36.7	26.0	2.5	**
7–12	811	689	15	***	35.0	29.3	1.0	***
13–18	659	605	12	**	27.8	25.2	0.6	**
19–24	575	617	39	NS	24.6	25.1	1.7	NS
25–36	722	756	22	NS	34.6	32.5	1.0	NS

0.674±0.007 and 0.701±0.008 respectively for mid lactation. Total nitrogen levels of pasture were on average (±s.d.) 2.75±0.03 and 2.95±0.03 g/100 g DM for early and mid lactation respectively.

Underfeeding during weeks 1–6 reduced current yields of fat-corrected milk (FCM) and fat to about 70% (*Table 70.1*) but these differences disappeared after 12 weeks.

The fates of digestible energy (DE) and metabolizable energy (ME) during weeks 8–18 of lactation are shown in *Tables 70.2* and *70.3* respectively. For each partition, regression coefficients (b) were not significantly affected by previous nutritional level and only the pooled coefficient is given. In contrast, the intercept (c) differs in all cases between data from previously well fed (c_H) and underfed (c_L) animals, indicating the presence of a carry-over effect, albeit small. Similar unpresented data from mid lactation did not show significant carry-over effects on energy partitioning except for methane: group L partitioned 4 kJ/kg$^{0.75}$ more DE ($P<0.001$) to methane than group H.

The average values for composition of tissue change calculated from carbon: nitrogen balance data for early lactation (*Table 70.4*) show that considerably more fat was catabolized from H cows than from L cows. Although considerable

Table 70.2 PARTITION OF DIGESTED ENERGY (X) IN EARLY LACTATION (MJ)*; Y = bX + c

Y	*b±s.e.*†	C_H	C_H–C_L±*s.e.*(d)	*P*
Urine E	0.058±0.011	0.042	–0.0047±0.0019	*
Methane E	0.055±0.001	0.054	–0.0046±0.0016	*
ME	0.888±0.016	–0.101	+0.0092±0.0026	**

* All values divided by kg W$^{0.75}$
† All regression coefficients significant ($P<0.01$)

Table 70.3 PARTITION OF METABOLIZABLE ENERGY (X) IN EARLY LACTATION (MJ)*; Y = bX + c

Y	*b±s.e.*†	C_H	C_H–C_L±*s.e.*(d)	*P*
Heat E	0.429±0.054	0.336	0.018±0.005	**
Retention E	0.571±0.054	–0.335	–0.018±0.005	**
Milk E	0.192±0.056	0.368	0.080±0.018	**
Tissue E	0.378±0.086	–0.704	–0.098±0.021	**

* All values divided by kg W$^{0.75}$
† All regression coefficients significant ($P<0.01$)

Table 70.4 COMPOSITION OF TISSUE CHANGE IN EARLY AND MID LACTATION (MEAN ± s.e.)[1]

	Early			*Mid*		
	H	*L*	*P*	*H*	*L*	*P*
Fat (g/d)	–168±84	+11±77	***	–4±55	30+54	NS
Protein (g/d)	–85±27	–94+26	NS	–27±33	–20+33	NS

[1] Mean of data obtained during (A) and (R) periods

tissue protein was lost from both groups, no difference was apparent between treatments. These data correspond to a calculated mean net energy change from tissue of –9.56 and –1.73 MJ/d from H and L cows respectively. Similar data for mid lactation were –0.78 and 0.72 MJ/d. Observed LW change was –1.3 and –0.9 kg/d for the H and L cows respectively in early lactation, reducing to –0.6 and –0.5 kg/d in mid lactation.

Discussion

Relative DMI and LW change between groups H and L during underfeeding indicate that the L nutritional treatment was severe. It is reasonable to expect that underfed cows were also at a disadvantage in terms of quality of feed ingested. Despite this, only small effects were noted on subsequent production. For example, 71.2% of FCM production losses of the L group occurred during the six weeks of undernutrition. The effects are similar to those described by Broster (1974) but are considerably smaller in magnitude. Reasons for this are uncertain, but differences between New Zealand cows and those of European origin, together with feed type, quality and allocation, may be involved.

The experimental design of this trial allowed within-set comparisons to be made by regression analysis. It is of considerable interest that the relatively small differences in production were confirmed by small but significant differences between H and L cows in partitioning of energy (*Tables 70.3* and *70.4*), indicating a high level of precision of the experimental design. The design and analysis contrast with previous energy metabolism studies in lactating dairy cows. Generally, data have been obtained from cows offered the one feeding level and regression analysis was done using data obtained from different animals. Variations in energy intake have been achieved by using cows at different stages of lactation, using cows at the same stage of lactation but with different intakes or by using the same animals at different stages of lactation. The within-cow approach used in this work facilitated the evaluation of the effects of variations in feeding level for the same cow at any one time.

Another feature of this work which contrasts with similar dairy cow studies is that pasture was the sole ration.

A number of important points emerge from the data presented. The higher proportion of DE partitioned to urine possibly reflects a greater protein oxidation in underfed animals, more than is required for milk production. The greater energy loss as methane of the L group has not been reported elsewhere, and no explanation can be deduced from data available from this experiment. It is of interest that a similar small effect was noted in mid lactation and the explanation for this long-term carry-over effect can only be speculative.

The greater proportion of ME partitioned to heat by the H group possibly reflects the marginally higher milk production of these animals. Cows well fed in early lactation partitioned 80 $kJ/kg^{0.75}$ ME more to milk than L cows in early lactation. For an average-sized New Zealand cow this amounts to approximately 7 MJ/d additional ME available for milk production. The ME advantage of H animals comes largely from tissue reserves and reflects their magnitude and ability for mobilization.

Data in *Table 70.4* indicate that animals underfed in early lactation mobilize less fat than well-fed cows but similar amounts of protein. An alteration of energy partitioning and depletion of reserves during underfeeding may be responsible; the authors' data confirm the first alternative and do not exclude the second.

In addition to tissue fat, well-fed cows catabolized protein in early lactation, reflecting a demand for specific amino acids either for milk protein synthesis or as lactose precursors. These effects on tissue protein were found in mid lactation though they were smaller in magnitude than in early lactation; however, declining milk yields at this time would account for this.

Acknowledgements

The skilled technical assistance of M.A.S. Cook and R.P. Newth and many staff at the Nutrition Centre is acknowledged.

References

BROSTER, W.H. (1972). *Dairy Sci. Abstr.*, **34**, 265–288

BROSTER, W.H. (1974). Biennial Review of the National Institute of Research in Dairying, 14–34

BRYANT, A.M., HUGHES, J.W., HUTTON, J.B., NEWTH, R.P., PARR, C.R., TRIGG, T.E. and YOUNG, J. (1977). *Proc. N.Z. Soc. Anim. Prod.*, **37**, 158–162

BRYANT, A.M. and TRIGG, T.E. (1979). *Proc. N.Z. Soc. Anim. Prod.*, **39**. In press

GORDON, F.J. (1977). *Animal Production*, **25**, 373–379

NATIONAL RESEARCH COUNCIL (1966). *Nutrient Requirements of Dairy Cattle.* Washington: National Academy of Sciences

YATES, F. (1939). *Proc. R. Soc. of Edin.*, **59**, 184–198

71

ENERGY AND NITROGEN METABOLISM IN LACTATING COWS FED GRADED QUANTITIES OF AMMONIUM LACTATE

L.S. BULL*, R.W. HEMKEN, R.A. ERDMAN and J. O'LEARY
University of Kentucky, Lexington 40546, Kentucky, USA

Summary

The inclusion of an ammonium lactate product of whey fermentation in the rations of lactating cows results in a reduction in intake at levels of total lactate greater than 6.5% of the ration. However, the energy consumed is utilized with equal efficiency in all diets studied. A trend toward more tissue deposition and elevated urinary nitrogen output suggests that the limitation on use for milk production is the ability to use the ammonia in the rumen.

Introduction

The feeding of fermented feeds of various types is popular in ruminant production. This has resulted in concern about the ability of ruminants to utilize a large quantity of lactic acid and the metabolic consequences (Satter and Esdale, 1968; Byers and Goodall, 1979). Compared with other volatile fatty acids (VFA) (pk 4.7–4.9) the value for lactate (3.1) is of concern. In lactating cows, the suspected pathway of lactate through propionate could have an adverse effect on milk-fat production and the partition of energy between milk and tissue. This experiment was conducted to determine the influence of added lactate (as the ammonium salt) on energy metabolism and related parameters in lactating cows. Owing to the use of the ammonium salt of lactate, nitrogen metabolism was also investigated.

Materials and methods

Four lactating Jersey cows were used in a 4 × 4 Latin square arrangement of treatments. All cows were fed a completely mixed ration with dry matter (DM) consisting of 60% corn silage (6.7% lactate) and 40% concentrate (50% corn, 45% soya, 4% minerals). Soya bean meal (44% crude protein (CP)) was replaced

* Current address: Department of Animal and Veterinary Sciences, Hitchner Hall, University of Maine, Orono, Maine 04469, USA

by ammonium lactate (44% CP)* at 0%, 33%, 66% or 100% to produce diets with 4%, 6.5%, 8.5% and 11% lactate and 15% CP. Cows were fed to achieve a slight positive energy and nitrogen balance according to the National Research Council (1978). Periods were of 42 days' duration. Days 0–21 were preliminary (adjustment) and on days 22–28 feed, orts, faeces, urine and milk were collected. Respiratory exchange was determined on each cow for three consecutive 24-h periods during days 29–42. Hourly blood (jugular) and urine samples were taken for 12 h, starting at feeding on day 42 for pH, $P\text{CO}_2$, $P\text{O}_2$, $[HCO_3^-]$, lactate and glucose (enzymatic).

Respiratory exchange was measured in an open-circuit calorimeter, with entry airlock and milking device. Gas flow was measured by calibrated dry test meters. Gas analysis of residual and spirometer composite samples was by paramagnetic (oxygen) and infrared (carbon dioxide, methane) methods. Heat production and balance calculations were based on equations of Brouwer (1965) and Flatt and Tabler (1961). Net energy (lactation) value of the rations was computed using the method of Moe, Flatt and Tyrrell (1973), which includes factors for pregnancy, excess nitrogen intake and tissue gain or loss. Blood pH and gases were measured by electrometric techniques (Instrumentation Laboratories Model 213–05 blood-gas and pH analyser). Lactate in feed was measured by the method of Barker and Summerson (1941) and blood L(+) lactate by an enzymatic procedure (Sigma Technical Bulletin 726–UV).

Statistical analysis of the data was by methods of Steel and Torrie (1960). Computer formulae in Fortran logic were used to compute all data.

Results and discussion

The results of the energy-balance measurements are given in *Table 71.1.* Respiration data were taken as the average of the three consecutive 24-h measurements and used as a single value. On two occasions, mechanical difficulties resulted in the averaging of only two consecutive periods.

There was a gradual decline in intake as the amount of lactate in the diet increased. The cause of this inhibition cannot be attributed solely to lactate because there was also an increase in ammonia consumption with the lactate. Blood data offer some possible explanations. Plasma glucose (mg/100 ml) was lower (69 versus 81) for the two higher lactate diets than for the lower two ($P<0.05$). Plasma L(+) lactate (mg/100 ml) was slightly higher (12.1 versus 9.4) for the 11% diet than for the others, which were equal (NS). Acid–base parameters (pH, $P\text{CO}_2$, $P\text{O}_2$, $[HCO_3^-]$) showed no treatment effects. The possible role of ammonia might be indicated by the fact that urine ammonia (g/ℓ, % of total nitrogen, and g/day) increased with all levels of added ammonia.

Although digestible energy (DE, %) increased with increased lactate, the reduced intake can explain much of the variation. The regression of DE (%, Y) on gross energy (GE) intake (kJ/kg$^{0.75}$) produced:

$$Y = 80.4 - 0.0047X, \qquad r = 0.52$$

There was no evidence that any treatment group should not be treated with the

* Provided by Calor Agricultural Research, Okemos, Michigan. An ammonia-buffered by-product of whey fermentation and condensation with 38% lactate, 7% nitrogen, 39% water as fed

others. This regression suggests about a 3.9 percentage units decline in DE per incremental intake of a multiple of maintenance.

Methane energy losses in this study were very low, representing only 2–2.5% of consumed energy. However, the nature of the ration and the level of consumption may have been the cause. Recovery of standard gas which included methane indicated that detection should not have been a problem (Kane, 1978).

Energy balance and partition of energy between milk and tissue reflect level of intake and also suggest a trend (NS) toward more tissue deposition at a higher intake of lactate. Milk-fat percentage was not affected by treatment and was slightly higher when the lactate content of the ration was increased, probably owing to the reduced yield. A confounding factor might be the increased intake of ammonia and the reduced use of nitrogen at high intake, as suggested above. Such a situation could reduce the amino-acid supply to the mammary gland and push energy into tissue. There was no evidence that the ration influenced the utilization of metabolizable energy (ME) for production above maintenance (assumed to be 485 kJ ME/$g^{0.75}$ per d). All of the cows were pregnant during most of the trial, and average days pregnant were: 126, 138, 128 and 125 for diets with 4%, 6.5%, 8.5% and 11% lactate, respectively.

A major objective of this study was to determine the net energy (NE, lactation) value of the rations with varying amounts of lactate. These calculations are shown in *Table 71.1.* Although there was a tendency for the diets with added lactate to have higher values, the differences were not significant. One would conclude that the NE (lactation) value of the ammonium lactate product would be, then, equal to the soya bean meal which it replaced, on a dry basis. On an as-fed basis the whey product would be 68% equivalent to soya. The validity of these measurements would seem to rest in the fact that at three times maintenance intake, the predicted NE (lactation) value of the basal (4% lactate) ration

Table 71.1 PARTITION OF DIETARY ENERGY IN LACTATING COWS FED DIETS WITH INCREASING QUANTITIES OF LACTATE

Parameter[1]	*Dietary lactate (%)*			
	4	*6.5*	*8.5*	*11*
Intake	2874^{A}	$2795^{A,B}$	2695^{B}	2293^{C}
Faeces	950^{A}	$904^{A,B}$	837^{B}	657^{C}
Methane	59	59	54	50
Urine	109	113	84	105
Heat	941	954	853	858
Balance	815	766	867	623
Milk[2]	678^{A}	644^{A}	$573^{A,B}$	485^{B}
Tissue	137	121	294	138
DE (%)	67.0^{A}	$67.7^{A,B}$	68.9^{B}	71.4^{C}
ME (%)	61.1^{A}	61.5^{A}	$63.8^{A,B}$	64.6^{B}
EB/ME (net) (%)	63.6	62.8	65.6	64.1
NE(1) (MJ/kg DM)	7.03	7.20	7.53	7.32

A,B,C Means on same row with different superscripts differ significantly ($P<0.05$)

DE = digestible energy; ME = metabolizable energy; EB = energy balance; NE = net energy; DM = dry matter

[1] All units as kJ/$kg^{0.75}$ per d unless otherwise stated.

[2] Average yield (kg) and fat content (%) were: 14.3, 5.18; 13.2, 5.23; 11.4, 5.51; 10.2, 5.53 for the diets in the order shown

is 7.07 MJ/kg DM (National Research Council, 1978). The value of 7.03 derived in this study (3.6 times maintenance intake) is in excellent agreement.

Nitrogen metabolism data are presented in *Table 71.2.* Intake reflects dietary consumption. Absorption is proportionally increased with added non-protein nitrogen (NPN), a fact demonstrated by others (Poos, 1977). The reduction in utilization of nitrogen, as shown by the increased urine excretion and reduced milk output probably reflects an excess of ruminally available ammonia relative to energy for microbial amino-acid synthesis (Roy *et al.*, 1977). This factor could be a driving force in the trend toward a greater partition of energy into body

Table 71.2 PARTITION OF DIETARY NITROGEN IN LACTATING COWS FED DIETS WITH INCREASING QUANTITY OF LACTATE

Parameter[1]	*Dietary lactate (%)*			
	4	*6.5*	*8.5*	*11*
Intake	345^{A}	319^{B}	306^{B}	270^{C}
Faeces	134^{A}	$114^{A,B}$	$89^{A,B}$	74^{B}
Urine	111	127	132	131
Milk	89^{A}	85^{A}	76^{A}	58^{B}
Retained	11	−7	9	7
DP (%)	61.2^{A}	64.3^{A}	70.9^{B}	72.6^{B}
Urine (% DP)	54	62	61	67
Used (% DP)	47	38	39	33

A,B,C Means on same row with different superscripts differ significantly ($P<0.05$)
DP = digestible protein
[1] All units as g/d unless otherwise stated

tissue (*Table 71.1*). There were no signs of ammonia toxicity among these cows, even though more than 62% of the total diet nitrogen was as NPN (including the silage) in the 11% lactate diet. The ammonia toxicity index of ammonium lactate is only about 25% that of urea (Henderson, H.E., unpublished data, Michigan State University).

References

BARKER, S.B. and SUMMERSON, W.H. (1974). *J. Biol. Chem.*, **138**, 535–554

BROUWER, E. (1965). In *Energy Metabolism in Farm Animals*, pp. 441–443. Ed. K.L. Blaxter. EAAP Publication No. 11. New York and London: Academic Press

BYERS, F.M. and GOODALL, S.R. (1979). *J. Anim. Sci.*, **48**, 624–632

FLATT, W.P. and TABLER, K.A. (1961). In *Energy Metabolism in Farm Animals*, pp. 39–47. Eds. E. Brouwer and A.J.H. van Es. EAAP Publication No. 10. The Netherlands: Wageningen

KANE, E. (1978). PhD Dissertation, University of Kentucky, Lexington

MOE, P.W., FLATT, W.P. and TYRRELL, H.F. (1972). *J. Dairy Sci.*, **55**, 945–958

NATIONAL RESEARCH COUNCIL (1978). *Nutrient Requirements of Domestic Animals*, No. 3. Washington, DC: National Academy of Science-National Research Council

POOS, M.I. (1977). PhD Dissertation, University of Kentucky, Lexington
ROY, J.B.H., BALCH, C.C., MILLER, E.L., ØRSKOV, E.R. and SMITH, R.H. (1977). *Proc. 2nd Int. Symp. on Protein Metabolism and Nutrition*, pp. 126–129. Wageningen, The Netherlands: Centre for Agricultural Publishing and Documentation
SATTER, L.D. and ESDALE, W.J. (1968). *J. appl. Microbiol*, **16**, 680
STEEL, R.G.D. and TORRIE, J.H. (1960). *Principles and Procedures of Statistics*. New York: McGraw-Hill

SOME ASPECTS OF ENERGY AND PROTEIN UTILIZATION IN RUMINANTS GIVEN SILAGE DIETS

P.C. THOMAS, N.C. KELLY, D.G. CHAMBERLAIN and J.S. CHALMERS
The Hannah Research Institute, Ayr KA6 5HL, Scotland

The importance of ensiled forages for the winter feeding of farm livestock in the United Kingdom has led to a recent surge of interest in the utilization of silage diets by ruminants. This chapter summarizes investigations that have been made into aspects of energy balance and energy and protein metabolism in sheep and cows given mainly grass silages prepared using formic acid as an additive. Topics considered are: regulation of energy intake; digestion of silage energy and protein constituents; utilization of silage energy; and dietary energy and protein interactions.

REGULATION OF ENERGY INTAKE

With silage diets, *ad lib.* intake can be impaired by amino-acid degradation products and organic acids formed during fermentation in the silo (Wilkins, 1974). However, with well-preserved silages as used by the authors fermentation products are not regulatory factors of primary importance. Concentrations of ammonia and butyric acid, an index of clostridial activity, in the silages are low, and there is little evidence of substantial amino-acid breakdown. The silages have a high lactic-acid content and a low pH but the lactic acid is rapidly fermented in the rumen (*see below*) and partial neutralization of silages with sodium bicarbonate has shown that intake is not restricted by silage acidity (Farhan and Thomas, 1978). In sheep given a chopped silage containing 13.8 MJ digestible energy (DE) kg dry matter (DM), DE intake was increased by approximately 12% when the silage was finely minced (Thomas, Kelly and Wait, 1976), and in cows given a silage containing 13.2 MJ DE/kg DM intake was reduced by approximately 16% by the insertion of water-filled bladders into the rumen via a rumen fistula (Farhan and Thomas, 1978). These effects indicate that, even for silages containing more than 10.5–11.3 MJ DE/kg DM (the values suggested to typify the point of change from physical to metabolic mechanisms of regulation of intake with hay and concentrate diets; Baumgardt, 1970), energy intake is regulated through a physical, rumen-fill mechanism. Metabolic regulatory mechanisms seem unlikely to be involved with diets containing a substantial proportion of silage.

DIGESTION OF SILAGE ENERGY AND PROTEIN CONSTITUENTS

In sheep given diets of silage alone all the dietary water-soluble carbohydrates and 0.90–0.99 of the digestible cellulose in the diet were fermented in the rumen. Supplementation of a silage (644 g DM/d) with barley (151 g DM/d) reduced the proportion of digestible cellulose fermented from 0.90 to 0.77; the passage of starch to the duodenum with the barley diet was only 10.7 g/d. For three silage diets and one silage and barley diet examined the mean (with standard error) proportions of DE apparently digested in the rumen, small intestine and caecum and colon were 0.643 ± 0.01, 0.240 ± 0.02 and 0.118 ± 0.01. These values indicated that 0.621 ± 0.01 of DE was absorbed from the rumen and caecum and colon as short-chain fatty acids and that 0.525 ± 0.02 of DE was absorbed as fatty acids from the rumen alone.

Rumen short-chain fatty acid proportions were studied in a number of experiments with both sheep and cows. Typically, with diets of silage alone molar proportions were acetate 0.61–0.71, propionate 0.16–0.26 and butyrate 0.07–0.12. A consistent finding was that the inclusion of cereal concentrates in the diet reduced the proportion of acetate, increased the proportion of butyrate and produced little increase or occasionally a reduction in the proportion of propionate. Significantly, in sheep given silage alone (784 g DM/d) protozoal counts were 2.78×10^5/ml but in those given barley (420 g DM/d) in addition, the counts were increased ($P<0.01$) to 16.48×10^5/ml.

Nitrogen balance studies have indicated that silage nitrogen is utilized inefficiently in the rumen (Wilkins, 1974). However, studies of the digestion of silage and silage and concentrate diets in sheep (Thomas, Chamberlain and Alwash, 1976 and later unpublished work) showed that the duodenal passage of crude protein (CP; Y, g/g CP intake) was related to the intake of digestible organic matter (X, g/g CP intake) by the equation:

$$Y = 0.12 + 0.2\,X \qquad r = 0.87,\ P<0.01,\ n = 7$$

This equation provided values only 10–12% lower than the corresponding equation obtained by Hogan (1975) in studies with dried forage diets, and in both studies the digestibility of CP in the intestines was similar. With the silage diets, however, the rate of bacterial protein synthesis in the rumen was low and the duodenal CP contained a high proportion of non-bacterial protein, which was assumed to derive mainly from the low-degradability plant proteins which had survived microbial attack in the silo. As a consequence of its protein make-up, the duodenal CP had a lower content of lysine and a much lower content of methionine than had been observed in animals receiving a variety of hay and hay and concentrate diets (Chamberlain and Thomas, 1979).

UTILIZATION OF SILAGE ENERGY

Efficiency of energy utilization

Calorimetric studies of the efficiency of utilization of metabolizable energy (ME) for maintenance and fattening in sheep were reported by Kelly and Thomas (1978). In most cases the efficiencies observed were in broad agreement with those predicted from equations derived from studies with dried conserved forages (Agricultural Research Council, 1965). For one silage the efficiency of utilization of ME for fattening (k_{f-}) was especially low and it was initially speculated that this might be related to the high D(−) lactic-acid content of the silage. This proposal has since been refuted. Both L(+) and D(−) lactic acids were rapidly fermented in the rumen and D(−) lactic acid did not appear in important concentrations in the peripheral blood. Moreover, normal k_{f-} values were obtained with a further high-lactate silage.

A recent observation was that in sheep, the calorimetric efficiency of utilization for diets of silage and groundnut meal (72:28) was higher than that for diets of silage and barley (68:32) although both diets were theoretically adequate in protein. A similar effect of protein supplements on the efficiency of utilization of ME for lactation ($k_{\ell o}$) was pointed out by Thomas and Castle (1978) in calculations made from the results of closely controlled feeding experiments. Calculated $k_{\ell o}$ values were 46%, 39%, 54% and 47% for diets of silage, silage and barley, silage with groundnut cake and silage with barley and groundnut cake, respectively.

Effects of silage diets on milk composition

Low milk protein and solids-not-fat contents, not associated with inadequate energy intake, in cows given highly digestible silage diets were first described by Murdoch and Rook (1963) and have since been reported in a number of studies. In this context, the effects of the proportion and type of concentrate in the diet on milk composition have been investigated. In four experiments, silage diets containing a high proportion of flaked maize concentrate reduced milk-fat content in a way similar to that reported with diets of hay and flaked maize (Rook, 1976), but there was little or no associated increase in milk-protein content. This poor response in protein was consistent with the effects of diet on the molar proportion of propionic acid in the rumen (*see above*) but the small changes in propionate do not provide the full explanation. In two experiments with cows given silage diets, intraruminal infusions of propionic acid (500 ml/d or 650 ml/d) failed to produce a significant ($P<0.05$) increase in milk-protein content.

ENERGY AND PROTEIN INTERACTIONS

In feeding experiments with formic-acid silage diets providing adequate amounts of digestible CP, supplementation of the diet with a protein-rich food in addition to or instead of barley has given improvement in growth rate (Griffiths, Spillane and Bath, 1973) and milk yield (Thomas and Castle, 1978) often, although not invariably, associated with improvements in silage intake. The effects on intake are partly accounted for by the physical regulation of silage intake and by the influence of barley on the rate of digestion of silage structural carbohydrates in the rumen (*see above*), but the effects of supplementary protein on the efficiency of utilization of energy for growth and lactation have also to be explained. Calculations from studies of silage digestion indicate that in animals given silage and barley diets the supply of total amino acids from the small intestine is adequate for high rates of animal production but that the supply of methionine

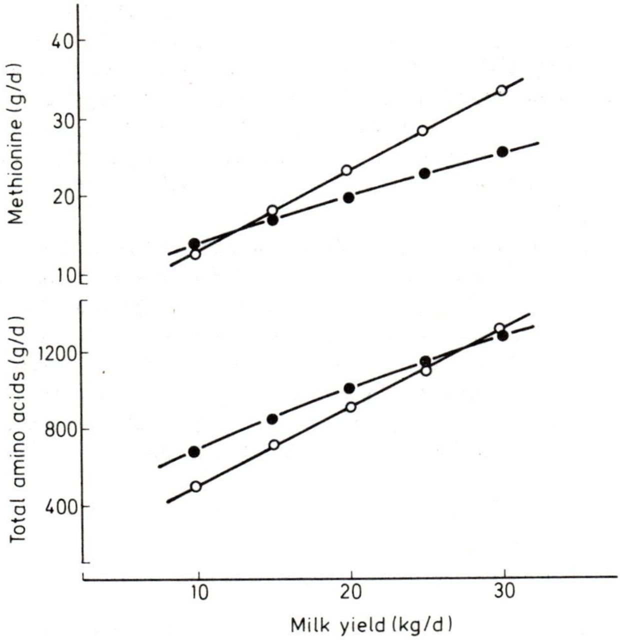

Figure 72.1 The relationship between the requirement (○) and the supply (●) of total amino acids and methionine absorbed from the small intestine and milk yield. The figures are for a 600 kg cow giving milk containing 30 g protein/kg and receiving its requirement of ME for maintenance and milk production in a diet of silage (11.22 MJ ME/kg DM, 140 g CP/kg DM) and barley (13.7 MJ ME/kg DM, 108 g CP/kg DM) in proportions 65:35. Values have been calculated using data from the Agricultural Research Council (1979), Gill and Ulyatt (1979), Paul and Southgate (1978) and Thomas, Chamberlain and Alwash (1976)

is limiting (*Figure 72.1*). On this basis, the effects of supplementary dietary proteins on energy utilization are probably due to the influence of the proteins on methionine supply. There is direct evidence that methionine is limiting for growth in sheep (Kelly and Thomas, 1975), but the occurrence of similar limitations on growth in cattle, on milk yield and possibly on milk-protein content needs to be examined.

References

AGRICULTURAL RESEARCH COUNCIL (1965). *The Nutrient Requirements of Farm Livestock. No. 2 Ruminants.* London: HMSO

AGRICULTURAL RESEARCH COUNCIL (1979). *The Nutrient Requirements of Farm Livestock, No. 2 Ruminants.* London: HMSO

BAUMGARDT, B.R. (1970). In *Physiology of Digestion and Metabolism in the Ruminant*, pp. 235–253. Ed. A.T. Phillipson. Newcastle upon Tyne: Oriel Press

CHAMBERLAIN, D.G. and THOMAS, P.C. (1979). *J. Sci. Fd Agric.*, **30**, 677–686

FARHAN, S.M.A. and THOMAS, P.C. (1978). *J. Br. Grassld Soc.*, **33**, 151–158

GILL, M. and ULYATT, M.J. (1979). *Br. J. Nutr.*, **41**, 605–609

GRIFFITHS, T.W., SPILLANE, T.A. and BATH, I.H. (1973). *J. agric. Sci., Camb.*, **80**, 75–88

HOGAN, J.P. (1975). *J. Dairy Sci.*, **58**, 1164–1177

KELLY, N.C. and THOMAS, P.C. (1975). *J. Br. Grassld Soc.*, **30**, 237–239

KELLY, N.C. and THOMAS, P.C. (1978). *Br. J. Nutr.*, **40**, 205–219

MURDOCH, J.C. and ROOK, J.A.F. (1963). *J. Dairy Res.*, **30**, 391–397

PAUL, A.A. and SOUTHGATE, D.A.T. (1978). *The Composition of Foods*, pp. 280–281. London: HMSO

ROOK, J.A.F. (1976). In *Principles of Cattle Production*, pp. 221–236. Eds. H. Swan and W.H. Broster. London: Butterworths

THOMAS, P.C. and CASTLE, M.E. (1978). *The Hannah Research Institute Report 1978*, pp. 108–117. Ayr: The Hannah Research Institute

THOMAS, P.C., CHAMBERLAIN, D.G. and ALWASH, A.H. (1976). *J. Br. Grassld Soc.*, **31**, 123–128

THOMAS, P.C., KELLY, N.C. and WAIT, M.K. (1976). *J. Br. Grassld Soc.*, **31**, 19–22

WILKINS, R.J. (1974). In *University of Nottingham Nutrition Conference for Feed Manufacturers: 8*, pp. 167–189. Eds. H. Swan and D. Lewis. London: Butterworths

EFFICIENCY OF UTILIZATION OF CORN-SILAGE DIETS BY PREGNANT AND LACTATING SHEEP

A. CRISTIAN, E.E. MENESES*, G. BURLACU and J. STAVRI
Institute of Animal Nutrition, Baloteşti, Ilfov, Romania

Summary

The efficiency of energy utilization of diets based on corn silage in pregnant and lactating sheep with an average body weight at lambing of 62 ± 0.8 kg has been studied. An efficiency of dietary utilization of 67.1–69.2% as digestible energy (DE) and of 60–61% as metabolizable energy (ME) was found for pregnant sheep and of 66–68% as DE and of 57–60% as ME for lactating sheep, depending on the content of corn silage in the diet.

The energy requirement for maintenance and pregnancy (Y) is shown by the equation:

$$Y = 401 + 0.0265e^{0.0611t}$$

where t = days of pregnancy. The efficiency of ME utilization for lactation was 72.7%, the maintenance requirement being 468 kJ/kg$^{0.75}$.

Introduction

Only a few studies are available on energy and protein requirements for breeding sheep, insufficient for a correct estimation of energy and protein requirements for both pregnancy (Langlands and Sutherland, 1968; Lodge and Heaney, 1973; Tissier and Theriez, 1978; Robinson *et al.* (quoted by Tissier and Theriez, 1978) and lactation (Coop (quoted by Tissier and Theriez, 1978); Treacher 1974).

Materials and methods

Studies were made of eight Palas Merino sheep of average weight of 62 ± 0.8 kg, inseminated in June–July 1977 so that lambs were obtained in November and December of the same year. Sheep were fed with corn-silage diets (34% dry matter (DM)) supplemented with compounds of 70% sunflower meal, 15% alfalfa meal, 6% maize, 2% chalk, 3% bone meal, 2.5% salt and 1.5% premix. Energy-balance trials were carried out on 203 occasions, 74 on dry ewes or those in

* Current address: Faculty of Veterinary Medicine, National University of Cahamarca, Peru

the first three months of pregnancy, 64 on ewes in the last two months of pregnancy, and 65 on lactating ewes. The balance trials were made in six-day periods, determining food intake, urine, faeces and respiratory exchanges in 24-h periods. The energy-balance method was shown in a previous paper (Burlacu *et al.*, in press). The total quantity of energy used for fetal growth was determined, using the calculated values of fetal composition during pregnancy as shown by Jarrige, Petit and Tissier (1978). Fetal energy so calculated was subtracted from the total energy determined to estimate the maternal tissue balance. The ME intake was set to zero maternal tissue balance, taking into account the efficiency of ME above and below the maintenance level of 70% and 84%, respectively.

The milk production was estimated by 48-h period milking weekly, and chemical composition (DM, protein and fat) and heat of combustion were determined.

Results

The dry sheep or those in the first three months of pregnancy, with an average weight of 62 ± 0.8 kg, ate a quantity of food consisting of 57.2% DM corn silage and 42.8% compound equivalent to 1042 MJ/kg$^{0.75}$ gross energy (GE) of which they utilized 67.1% as DE. The gross energy lost as methane was 5.2% and that as urine was 2%. Therefore, the ME was 59.9% of the GE intake; 43.4% of this is lost as heat production, so that the average gain is 16.5%. The net efficiency of ME computed by regression is 66.7 ± 6% and the maintenance requirement is 401 kJ/kg$^{0.75}$ per d.

In the last two months the pregnant sheep, weighing on average 71.0 ± 0.5 kg, ate a diet made up of 49.2% DM corn silage and 50.8% DM compound, equivalent to 804 kJ/kg$^{0.75}$ GE, of which 69.2% was used as DE and 60.8% as ME. The heat production of advanced pregnant sheep was also greater (51.3% lost from the GE intake) and was due to maintenance requirement and to fetal growth. The energy requirement for maintenance and pregnancy (computed at zero maternal tissue balance) is given by the following equation (at the average lambing weight of 4.5 kg):

$$Y = 401 + 0.0265e^{0.0611t} \qquad s_b = 0.05$$

where Y = energy requirement and t = days of pregnancy. It can also be seen that ME requirement for pregnancy increases from 401 kJ in the first months to 585 kJ/kg$^{0.75}$ in the last days of pregnancy. If the ME needed by pregnancy according to the equation is 85.8 MJ for a sheep of 60 kg and single lambing (lamb of 4.5 kg) and the lamb's energy value at lambing is 20.6 MJ (25.8 MJ totally fixed in uterus × 0.80 (% of fetus); Jarrige, Petit and Tissier, 1978), then the efficiency of ME utilization for pregnancy is:

$$\frac{296 \times 100}{85.8} = 24\%$$

The lactating sheep in the first months of lactation, weighing 64.0 ± 0.4 kg, ate a diet with 61.1% DM corn silage and 38.9% DM compound equivalent to 1225 MJ GE/kg$^{0.75}$, which they utilized as DE at 64.2% of GE and as ME at 56.5%

of GE. The energy losses as methane, urine and heat were, as a percentage, similar to those recorded in the sheep in the first period of pregnancy. The milk energy is 17.5% of GE intake, so that in the first month of lactation the energy balance is negative (–3.2% of GE intake). In the second month, lactating sheep, mean weight 61.7 kg, ate 1285 MJ/kg$^{0.75}$ of a diet consisting of 60.7% DM corn silage and 39.3% DM compound, which was utilized as DE at 67.6% DE intake and as ME at 59.6% of GE intake. The losses as methane, urine and heat are as percentages similar to those recorded in non-pregnant sheep or in the first months of lactation. The energy secreted as milk decreases (12.6% GE) in comparison with the first month, the total energy balance being positive (+6.9% GE).

The efficiency of utilization of ME for milk is 72.7% and the maintenance requirement is 468 kJ ME/kg$^{0.75}$, that is greater than for dry sheep or those in the first months of pregnancy.

Discussion

The results concerning food consumption for breeding sheep in different physiological states, compared with those obtained by Clancy *et al.* (1977) and Tissier, Theriez and Molenat (1977) show similar ratios between intake energy values (per kg$^{0.75}$) in dry, pregnant and lactating sheep. The energy requirement expressed in ME/kg$^{0.75}$ in non-pregnant sheep is the same as that noted by Tissier and Theriez (1978), calculated from the average values indicated in the literature for different sheep breeds. For the last two months of pregnancy an efficiency of ME utilization for fetal growth emerged greater than that accepted (13%) in the new French system of estimation of food requirement for ruminants. For lactating sheep, Gardner and Hogue (1964, 1966) found, as here, a net efficiency of ME utilization for milk production greater (65–85%) than for dairy cows, whose general efficiency is 60%. Coop (quoted by Tissier and Theriez, 1978) and Treacher (1975) also assume an increase of maintenance requirement for lactating sheep of 10–15%.

References

BURLACU, G., BAIA, G., CHRISTIAN A., RUS, V. and STAVRI, J. (1978). *Arch. Tierernähr.*, **28**, 235

BURLACU, G., BAIA, G., RUS, V. and STAVRI, J. 'Energy utilization of corn silage diets by growing sheep.' In press.

CLANCY, M., BULL, L.S., WANGSNESS, P.J. and BONMGARDT, B.R. (1977). *J. agric. Sci.*, **42**,4,960

GARDNER, R.W. and HOGUE, D.E. (1964). *J. Anim. Sci.*, **23**, 935

GARDNER, R.W. and HOGUE, D.E. (1966). *J. Anim. Sci.*, **25**, 789

JARRIGE, R., PETIT, M. and TISSIER, M. (1978). 'Reproduction, gestation et lactation.' In *Alimentation des Ruminants*, p. 235. Versailles: I.N.R.A.

LANGLAND, J.R. and SUTHERLAND, H.A.M. (1968). *Br. J. Nutr.*, **22**, 217

LODGE, G.A. and HEANEY, D.P. (1973). *Can. J. Anim. Sci.*, **53**, 479

TISSIER, M., THERIEZ, M. and MOLENAT, G. (1977). *Annls Zootech.*, **26**, (2), 149

TISSIER, M. and THERIEZ, M. (1978). 'Ovins.' In *Alimentation des Ruminants*, pp. 404–408. Versailles: I.N.R.A.

TREACHER, T.T. (1974). *Management and Diseases of Sheep.* British Council Course No. 324, Edinburgh, April

74

EFFICIENCY OF UTILIZATION OF FOOD ENERGY BY PREGNANT AND LACTATING SOWS

G. BURLACU, M. ILIESCU, P. CĂRĂMIDĂ and J. STAVRI
Institute of Animal Nutrition, Baloteşti, Ilfov, Romania

Summary

The efficiency of dietary energy (DE) utilization based on maize, soya bean meal and starch for pregnant sows and on maize and soya bean meal for lactating sows, with an average weight at farrowing ranging between 110 and 160 kg, was measured. The diets were utilized at 90–92% as DE and at up to 90% as metabolizable energy (ME) by pregnant sows, and at 91.5–95.5% as DE and at 90–92.9% as ME by lactating sows. The energy requirement for maintenance and pregnancy (y) is given by the equation:

$$y = 376 + 0.022e^{0.081t}$$

where t = days of pregnancy. The net efficiency of ME utilization for lactation is 71.5 + 7% and the maintenance requirement is 514 kJ ME/kg$^{0.75}$.

Introduction

Studies were made on the efficiency of food energy utilization by pregnant and lactating sows fed with diets drawn up as a part of cooperative work with C.R.Z.V., Jouy-en-Josas (France).

Materials and methods

The trials were made on two groups of six sows (Large White × Landrace), each sow weighing at farrowing between 110 and 160 kg. After insemination, until the 85th day of pregnancy, they received the same diet: 40% maize, 30% barley, 15% sunflower meal, 10% wheat bran, 2% alfalfa meal, 0.5% salt, 1% chalk, 0.5% bone meal and 1% premix, to which 0.05% lysine and 0.025% methionine were added. After this period, one group was fed with a diet composed of 83% maize, 15% soya bean meal and 3% premix, with 14% crude protein, given 1.9 kg/sow per d, while the second group received 500 g maize starch as well. After farrowing, sows were fed with the first-group diet supplied *ad lib*. There were 79 balance trials, each lasting five or three days, during which food intake, faeces and urine excreted and respiratory exchanges were determined in two 24-h periods

Burlacu, Iliescu and Stavri, 1978). The total energy retained as fetus (y) was computed using the equation:

$$y = 12e^{0.032t}$$

where t = days of pregnancy, proposed by De Villier *et al.* (quoted by Henry and Etienne, 1978). The energy so determined was subtracted from the total balance energy value to estimate the maternal-tissue balance. The ME intake was arranged to produce zero maternal-tissue balance, considering the efficiency of ME above and below the maintenance level of 70% and 82%, respectively (Verstegen, van Es and Nijkamp, 1971). The energy balance for lactating sows was made separately for sows and for the litter, the litter size being balanced after farrowing so that sows' milk production was solicited uniformly by the piglets. For this purpose, special pens with two compartments were used, one for the sow and one for her litter, which were allowed to suckle once an hour in the first week, every two hours in the second week, every three hours in the third week and every four hours in the fourth week of lactation. The respiratory exchanges for lactating sows were determined as follows. On the first day, the energy metabolism of the sows and litter together was determined for 24 h. The sows were then removed from the chamber and on the second day respiratory exchanges of the whole litter were measured in smaller respiration chambers (300 ℓ capacity) for short periods of two, three or four hours. On the third day the sows and litter together were again introduced into the respiration chambers, and the metabolism was determined for 24 h. Therefore, a trial lasted three days. The milk production was determined in two ways: by weighing the litter before and after suckling and by litter weight gain, allowing a heat of combustion of between 3.71 and 4.55 MJ/kg (depending on age), taking into account an efficiency of milk utilization of 95% as ME, that of the prestarter of 85% as ME and the total heat production of the litter. The trials were made at 22–23°C.

Results

In the first part of pregnancy the diet was utilized by sows at 77 ± 1% as DE and at 75.3 ± 0.9% as ME from gross energy (GE), the gross efficiency being 26.8 ± 1.2% with a weight gain composed of 29.5% protein and 70.5% fat. In the last part of pregnancy, Group 1 received a diet composed only of maize and soya bean meal and had an average intake of 640 kJ ME/kg$^{0.75}$, the digestibility of the diet being 90.2 ± 0.9% and the metabolizability 88.4 ± 0.7%. The heat of combustion was greater than that of sows in the first part of pregnancy: 72.2 ± 2% in comparison with 48.5 ± 3% of GE. Therefore, the balance energy was 16.2%, the weight gain being composed of 35.7% protein and 64.3% fat. Group 2, who received an addition of starch, had an increased digestibility (92.3 ± 0.8%) and also a higher value of ME (90.3 ± 7% of GE). The energy balance was greater (24.3% of GE) and the weight gain showed a greater quantity of fat (76.5%) in comparison with Group 1, the protein quantities in absolute values being slightly higher for Group 2 (50 kJ/kg$^{0.75}$ in comparison with 42 kJ/kg$^{0.75}$ for Group 1). The energy requirement for maintenance and pregnancy in terms of ME (y), dependent on the period of pregnancy, is given by the equation:

$$y = 376 + 0.0222e^{0.081t} \qquad s_b = \pm 0.04$$

where t = days of pregnancy. The average number of piglets in Group 1 litters was 9.66 (6–14) and the average weight of each piglet was 1.393 kg; for Group 2 the values were 8.33 (7–12) and 1.471. Group 2, fed with an addition of starch in the last part of pregnancy, had no increase in milk production. Energy balance values determined during weeks of lactation showed at first a slight decrease of sows' live weight from 146 ± 3 kg (first week) to 142 ± 4 kg (third week), then in the fourth week an increase (144 ± 3 kg), an increased food intake from 1052 MJ ME/kg$^{0.75}$ (first week) to 1550 MJ ME/kg$^{0.75}$ (third week), utilized as 91.7–95.5% DE and 90–92.9% ME. The heat production ranged between 67.2 ± 2% (first week) and 47.2 ± 3% (third week), and milk production between 30.6 ± 3% (first week) and 23.2 ± 2% (fourth week) of GE, with a negative balance in the first two weeks (–5% and –1.7%, respectively), and a positive balance in the third and fourth week of lactation (15.5% and 19.1%, respectively). Thus, for a litter weight gain from 1.360 kg (average weight at farrowing) to 5.47 kg (average weight at four weeks), sows had an average milk production from 4.15 kg (third week of lactation) to 3.48 kg (fourth week of lactation), considering the heat of combustion of 1 kg milk = 4.44 MJ (Henry and Etienne, 1978). The net efficiency of ME for lactating sows, computed by simple regression, is 514 kJ ME/kg$^{0.75}$ or 369 kJ ME/kg$^{0.75}$.

Discussion

The results concerning energy requirement for maintenance and pregnancy correspond to those obtained by Brody (quoted by Henry and Etienne, 1978), who points out an energy gain of 30% in the last days of pregnancy, and to those obtained by Verstegen, van Es and Nijkamp (1971) who found an average gain of 22% for the last three weeks of pregnancy. In lactating sows, for a food intake a little lower than that noted by Etienne, Duee and Pastuszewska (1975), there was less milk production, but in compensation a smaller reduction in live weight after farrowing, on average 4 kg instead of 27 kg. As with Salmon-Legagneur (quoted by Henry and Etienne, 1978), there was a higher cereals digestibility for lactating sows of 5–6% more than for growing and fattening pigs and 2.6% more than for pregnant sows. The net efficiency of ME utilization is between the values noted by Smith (1959) (40–53%), Lodge (1957) (88%), Gutte and Lenkeit (1960) (80.4%) and Pielok (quoted by Henry and Etienne, 1978) (81.2%).

References

BURLACU, G., ILIESCU, M. and STAVRI, J. (1978). *J. agric. Sci., Camb.*, **90**, 165–172

ETIENNE, M., DUEE, P.H. and PASTUSZEWSKA, B. (1975). *J. Rech. Porcine France, Paris,* **7**, 115–120

GÜTTE, J.O., and LENKEIT, W. (1960). *Z. Tierphysiol. Tierernähr. Futtermittelk.*, **15**, 165–191

HENRY, Y. and ETIENNE, M. (1978). *J. Rech. Porcine France, Paris,* 119–166

LODGE, G.A. (1957). *J. agric. Sci., Camb.*, **49**, 200–210

SMITH, D.H. (1959). *N.Z. J. agric. Res.*, **2**, 1057–1083

VERSTEGEN, M.W.A., van ES, A.J.H. and NIJKAMP, N.J. (1971). *Anim. Prod.*, **13**, 677–684

75

PRODUCTION AND PHYSIOLOGICAL CRITERIA IN LAYING RHODE ISLAND RED HENS IN RELATION TO PRENATAL AND POSTNATAL ENVIRONMENTAL TEMPERATURES

H. MICHELS, E. DECUYPERE and R. GEERS
Laboratorium voor Ecofysiologie der Huisdieren, Faculteit der Landbouwwetenschappen, Katholieke Universiteit Leuven, België

Summary

Laying hens hatched from eggs incubated at different temperatures were used at ages 6–10 months in one-week metabolic experiments. Data on heat production and metabolizable energy showed the relationship of energy balance to early incubation temperatures as well as to the temperatures during trials. In addition to this, data on food consumption, egg production, haematocrit values and hormonal criteria suggested the existence of a postnatal differential hormonal status as a consequence of early incubation temperatures. A lower food efficiency was found in animals incubated at lower temperatures.

Introduction

It has been shown that different incubation temperatures can affect postnatal growth in chickens (Romanoff, 1936; Muambi, 1974; Decuypere, 1979). Michels, Geers and Muambi (1974) found a higher body weight at 12 weeks of postnatal age in cocks incubated at 36.8 °C and reared at 22 °C, in comparison with a control group incubated at 37.8 °C. Their results on food-conversion efficiency showed that sex was significant but also suggested that incubation temperatures may influence post-hatching energy utilization.

In the present study an attempt is made to examine the effects of incubation temperatures on energy balance, egg production and other physiological criteria (TSH, prolactin, haematocrit) in Rhode Island Red pullets between six and 10 months of age kept at different environmental temperatures

Materials and methods

During 1973–1974 and 1974–1975 six groups of 240 eggs were subjected to different incubation temperatures (*Table 75.1*).

After hatching, 12 chickens per group were reared in constant environmental conditions. Prehatching, incubation and rearing conditions were described in detail by Geers, Michels and Decuypere (1978b) and by Decuypere (1979).

Table 75.1 INCUBATION TEMPERATURES FOR SIX GROUPS OF 240 EGGS

Group	*Days 0–10*	*Days 11–17*	*Days 18–hatch*
1	35.8	37.8	33.5
2	35.8	37.8	37.5
3	36.8	37.8	33.5
4	36.8	37.8	37.5
5	37.8	37.8	33.5
6	37.8	37.8	37.5

From 6 to 10 months of postnatal age, two animals from each group were taken 'at random' for one-week metabolic experiments, the week being divided into three subperiods, I, II and III. Subperiods I and II were, respectively, days 0–2 (25 °C) and days 2–4 (at one of the temperatures: 25 °C, 20 °C, 15 °C, 10 °C and 5 °C), during which times the chickens were fed. Subperiod III was days 4–7, when the animals were fasted at one of the temperatures mentioned. Metabolic experiments were carried out as described by Geers, Michels and Decuypere (1978b), in two open-circuit respiration chambers, which had been formerly used as incubators. Oxygen uptake and carbon dioxide output were measured by the diaferometer method, according to Romijn and Lokhorst (1966), but were calculated as described by Geers, Michels and Decuypere (1978a). Ventilation rate was calculated for standard conditions (van Es, 1961) and heat production was calculated according to Romijn and Lokhorst (1961) but adapted for kJ.

Before and after each subperiod, the animals, their food consumption, eggs and faeces were weighed, stored, sampled and analysed, as described by Michels, Geers and Muambi (1974) for food and eggs, although an adiabatic calorimeter was used for energy analysis. Faeces were dried at 65 °C in a vacuum oven for 24 h and weighed again in order to determine dry matter content.

Food consumption (kg dry matter during 6–10 months) and egg production (cumulative weight in grams and % hen-day production for the first four months after onset of lay) were measured in all animals. Haematocrit values (in the two experimental years) and TSH and prolactin (only in 1973–1974) were measured every month. Plasma prolactin and TSH levels were determined using an heterologous radioimmunoassay. In order to produce a standard curve for prolactin and TSH, several dilutions (up to 1:64) of chicken plasma samples were made. The competitive inhibitions of rat prolactin -^{131}I and TSH -^{131}I to their antisera by the chicken plasma dilutions were plotted and the relative differences of all other chicken plasma samples to these curves were used for comparison.

Results and discussion

The efficiency of the utilization of food energy during subperiod II, the experimental animals taken together over the two years, with respect to the effect of early incubation temperatures, is given in *Figure 75.1*. The normal distribution of data on metabolizable energy, heat production and energy balance and the

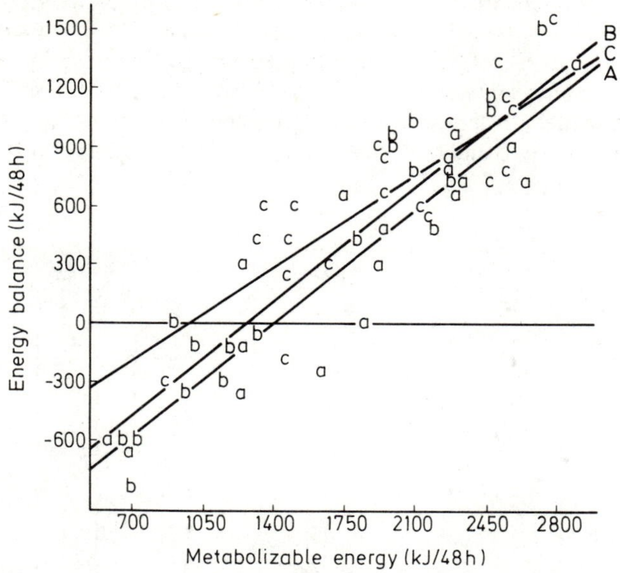

Figure 75.1 the effects of early incubation temperatures on the relationship between energy balance (EB) and metabolizable energy (ME) 1973–1974 and 1974–1975.
A: 35.8°C; EB = 0.8023 ME –1136; s_b = 0.082; r = 0.9207; P⩽0.001
B: 36.8°C; EB = 0.9708 ME –1268; s_b = 0.068; r = 0.9605; P⩽0.001
C: 37.8°C; EB = 0.7074 ME – 711; s_b = 0.108; r = 0.8391; P⩽0.001

homogeneity of variance between groups were proved by statistical analysis (Dixon, 1975). The energy balance was calculated as metabolizable energy minus heat production (Farrell, 1974). Comparison of linear regression equations (Dixon, 1975) gave a lower energy balance for animals incubated at 35.8 °C, as proved by statistically significant lower adjusted group means ($P<0.05$). Regression coefficients were not statistically different. The lower energy balance in the 35.8 °C group can probably be explained by a higher maintenance requirement, which was proved by a statistically higher heat production ($P\leqslant0.01$) in this group. Fasting heat production (subperiod III) showed the same trend. With respect to the postnatal environmental temperatures, the lowest energy balance occurred at 5 °C ($P\leqslant0.02$). This can probably be explained again by a higher

Table 75.2 ADJUSTED GROUP MEANS OF HEAT PRODUCTION (HP IN kJ/48 h) AND ENERGY BALANCE (EB IN kJ/48 h) FOR EARLY PRENATAL AND POSTNATAL TEMPERATURES 1973–1974 AND 1974–1975

Temp. (°C)	*35.8°C*		*36.8°C*		*37.8°C*		*Mean*	
	HP	*EB*	*HP*	*EB*	*HP*	*EB*	*HP*	*EB*
5	1756	31	1561	291	1406	385	1574	236
10	1491	317	1259	570	1299	498	1349	461
15	1470	365	1456	432	1291	522	1405	440
20	1474	397	1311	519	1151	690	1312	535
25	1387	433	1247	626	1058	790	1230	616
Mean	1515	308	1366	487	1241	577		

maintenance requirement of these animals, as shown by a significantly higher heat production ($P \leqslant 0.05$). Group means, with respect to early prenatal and to postnatal temperatures, of heat production and energy balance both adjusted for metabolizable energy intake are given in *Table 75.2*. Data have been taken together for late prenatal temperatures and for the two years, since no statistically significant differences could be found with respect to these experimental factors.

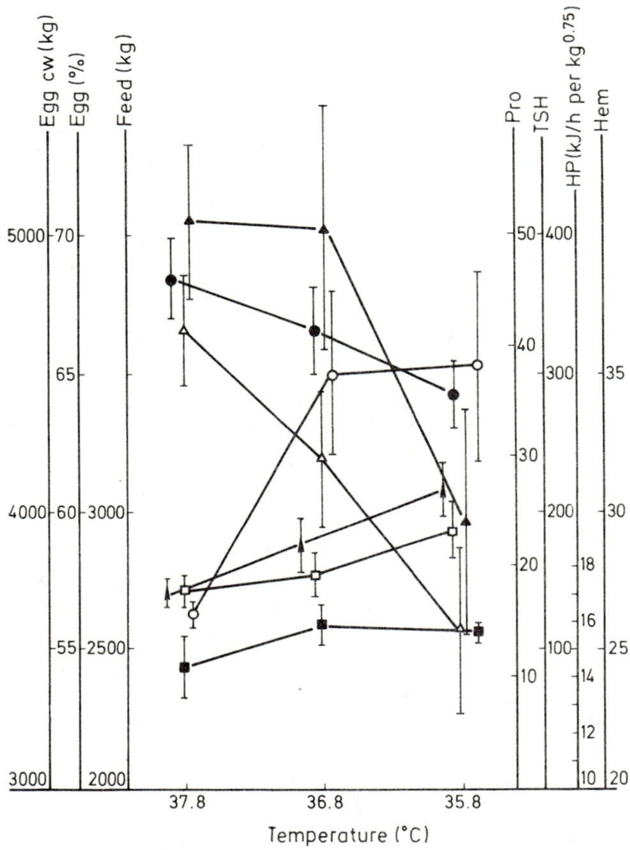

Figure 75.2 Production and physiological criteria in relation to incubation temperatures, 0–10 days, 1973–1974. Egg cw: △ cumulative egg weight; Egg: ▲ hen-day production; TSH: ○ thyroid-stimulating hormone (relative difference to the standard curves); Pro: ● prolactin (relative difference to the standard curves); Feed: □ average monthly feed consumption; HP: ■ heat production; Hem: ♦ mean haematocrit value per month

In both experimental years, haematocrit values were statistically different ($P \leqslant 0.05$ and $P \leqslant 0.01$) as proved by analysis of variance (Dixon, 1975). This corresponds to differences concerning TSH ($P \leqslant 0.025$). Heat production (sub-period 1) and food consumption also showed significant differences in 1974–1975 ($P \leqslant 0.05$ for heat; $P \leqslant 0.01$ for food). All these data, when measured over the two years, showed the same trend. Inverse trends were observed for prolactin and egg production records. For cumulative egg weight in 1973–1974 the difference was statistically significant ($P \leqslant 0.01$), indicating again a lower food efficiency in animals incubated at lower early prenatal temperatures. Trends and differences for 1973–1974 are shown in *Figure 75.2*. Research at more widely differing incubation temperatures using greater numbers of chickens deserves further study. The results are also stimulating for continuing research on relationships between energy balance and endocrinological criteria, depending on pre- and postnatal environmental variables, taking into account possible interindividual variabilities.

Acknowledgements

The authors are indebted to the Laboratory of Comparative Endocrinology, Catholic University of Leuven, for analysis of hormonal criteria, supported by the NIH rat pituitary program. E. Decuypere would also like to acknowledge support given by the Belgian Nationaal Fonds voor Wetenschappelijk Onderzoek (NFWO).

References

DECUYPERE, E. (1979). *Agricultura*, PhD thesis

DIXON, W.J. (1975). *Biomedical Computer Programs.* Berkeley; University of California Press

FARRELL, D.J. (1974). 'General principles and assumptions of calorimetry.' In *Energy Requirements of Poultry.* Eds. T.R. Morris and B.M. Freeman. Edinburgh; Br. Poult. Sci. Ltd.

GEERS, R., MICHELS, H. and DECUYPERE, E. (1978a). *Annls Biol. anim. Biochim. Biophys.*, 18, 1309–1315

GEERS, R., MICHELS, H. and DECUYPERE, E. (1978b). *Annls Biol. anim. Biochim. Biophys.*, **18**, 1363–1369

MICHELS, H., GEERS, R. and MUAMBI, S. (1974). *Br. Poult. Sci.*, **15**, 517–523

MUAMBI, S. (1974). *Agricultura*, **22**, 3–60

ROMANOFF, A.L. (1936). *Poult. Sci.*, **15**, 311–315

ROMIJN, C. and LOKHORST, W. (1961). *Energy Metabolism*, pp. 49–59. Eds. E. Brouwer and A.J.H. van Es. EAAP Publication No. 10. The Netherlands: Wageningen

ROMIJN, C. and LOKHORST, W. (1966). *Acta physiol. pharmac. neerl.*, **14**, 1–11

van ES, A.J.H. (1961). PhD thesis, Pudoc, Wageningen

76

ENERGY METABOLISM AND FEED EVALUATION AT HIGH ENVIRONMENTAL TEMPERATURES WITH RESPECT TO LAYING HENS

C.W. SCHEELE and N.A. MUSHARAF*
Spelderholt Institute for Poultry Research, Beekbergen, The Netherlands

Summary

Results of balance experiments with laying hens at high environmental temperatures show a beneficial effect of a high fat and limiting amino acid content in diets. Differences in energy density of the diets had no effect on measured parameters. Possibilities for an effective characterization of dietary fat and amino acids for lipogenesis and protein synthesis in these circumstances are discussed.

Introduction

It has been demonstrated that reductions in performance of laying hens in warm climatic conditions are not necessarily due to temperatures *per se*, but may be due to reductions in the intake of essential nutrients other than the main energy-yielding ingredients because of the decreased appetite (Payne, 1966; Reid and Weber, 1973). Little attention has been paid to the effect of differences between the main energy-yielding nutrients on poultry metabolism in warm environments.

Materials and methods

Balance studies according to the comparative slaughter method were carried out in a climate chamber with laying hens (Shaver 288), 24 weeks old. The experiments lasted four weeks with a preliminary period of six weeks. The ambient temperature was cyclic, 38 °C by day and 27 °C at night, at 50% constant relative humidity. Light was 17 hours a day.

A basal diet as it is applied in some tropical countries, low in metabolizable energy (ME) and limiting amino acid (lysine) content, was compared with diets higher in ME and lysine content and with isoenergetic diets higher in fat content. It was designed as a complete factorial experiment 2^3, with the factors ME, ME:

* Present address: Ministry of Agriculture, Animal Resources Department, P.O. Box 293, Khartoum, Sudan

lysine ratio and ME:fat ratio each at two levels. There were 30 hens per ration, five hens per cage and six replicates. The dietary ME levels and the digestible lysine and fat contents determined in the experiment were as shown in *Table 76.1.*

Table 76.1 DIETARY ME LEVELS AND DIGESTIBLE LYSINE AND FAT CONTENTS IN DIETS A–H

	A	*B*	*C*	*D*	*E*	*F*	*G*	*H*
ME (MJ/kg DM)	11.8	12.1	11.9	12.0	13.5	13.3	13.5	13.6
Dig. lysine (g/kg DM)	4.7	4.5	5.3	5.3	4.6	4.9	6.25	6.25
Dig. fat (g/kg DM)	37.0	77.4	37.6	75.4	42.9	85.6	42.6	87.7

DM = dry matter

Results and discussion

No influence of ME level or fat content of the diets was found on the daily ME consumption (*Table 76.2*). This is in agreement with results reported for poultry under moderate climatic conditions (Hill and Dansky, 1954; Hill, Anderson and Dansky, 1956). In these experiments a significant effect ($P<0.01$) of the ME: lysine ratio on the ME intake was observed. In practice a given energy value of a poultry ration is normally accompanied by the limiting amino acid content. Consequently it will be still possible to predict the feed intake for the hens as the ratio ME:lysine content is known. This is an important item as poultry are normally fed *ad lib.* It is also demonstrated that a higher lysine content had a beneficial influence on nearly all recorded parameters. Dietary fat level had a direct and clear effect on fat deposition in the carcass ($P<0.001$) but had no effect on egg production or deposition of fat in eggs. No influence of the ME level of the diets on the recorded results could be determined.

For two groups of diets different in ME content, regression equations were calculated. ME intake – 0.9 energy balance (EB) in carcass was plotted against energy deposition in eggs. The maintenance requirements of the birds were estimated as EB=0 for the high and low energy diets.

The results shown in *Table 76.2* indicate that there was no clear difference between the two estimated maintenance requirements. Next the efficiency of the ME used for egg production (k_o) was calculated:

$$k_0 = \frac{\text{Energy deposition in eggs}}{\text{ME intake} - 0.9\ \text{EB in carcass} - \text{Maintenance requirements}} \times 100\%$$

All quantities of energy are expressed in $kJ/kg^{0.75}$ per day.

Net energy (NE) values for production for the diets were calculated by means of these k_o values.

In spite of the constant maintenance requirements estimated for different groups of diets, the NE values of groups of diets which produced practically the same results showed a large variation. The main criticism of these calculations of NE values is a statistical one. Caused by the high degree of cross-correlation between the measured energetic variables, a large variation in NE values for the same kind of diets perhaps will be inevitable in experiments in which maintenance

Table 76.2 PRODUCTION CHARACTERISTICS AND PARAMETERS OBTAINED FROM BALANCE EXPERIMENTS WITH LAYING HENS. VALUES ARE GIVEN PER HEN PER DAY

	A	*B*	*C*	*D*	*E*	*F*	*G*	*H*
ME intake ($kJ/kg^{0.75}$)	524.7^{a}*	560.2^{a}	608.6^{b}	654.4^{b}	583.4^{a}	592.1^{a}	628.2^{b}	617.2^{b}
Egg production (g)	27.5^{a}	29.3^{a}	33.0^{b}	36.0^{b}	30.0^{a}	30.1^{a}	32.5^{b}	33.5^{b}
Energy balance (EB) ($kJ/kg^{0.75}$)	$47.7^{a,x}$	$86.2^{a,y}$	$102.5^{b,x}$	$139.3^{b,y}$	$76.2^{a,x}$	$94.5^{a,y}$	$106.7^{b,x}$	$128.5^{b,y}$
Protein deposition in eggs (g)	3.0^{a}	3.1^{a}	3.5^{b}	3.9^{b}	3.3^{a}	3.2^{a}	3.5^{b}	3.7^{b}
Fat deposition in carcass (g)	$-2.1^{a,x}$	$-1.1^{a,y}$	$-1.0^{b,x}$	$-0.6^{b,y}$	$-1.4^{a,x}$	$-1.0^{a,y}$	$-1.1^{b,x}$	$-0.6^{b,y}$
Fat deposition in eggs (g)	2.2^{a}	2.3^{a}	2.5^{b}	2.8^{b}	2.4^{a}	2.3^{a}	2.5^{b}	2.7^{b}
Maintenance requirements ($kJ/kg^{0.75}$)	←	435 (A–D)		→	←	444 (E–H)		→
k_o (%)	57^{x}	71^{y}	61^{x}	65^{y}	54^{x}	61^{y}	57^{x}	73^{y}
Net energy (NE) values (MJ/kg DM)	6.2	7.7	6.6	7.2	6.6	7.4	6.9	8.9
NE intake ($kJ/kg^{0.75}$)	299	398	371	425	315	361	358	451
Lysine:NE ratio (g/MJ)	0.76	0.58	0.80	0.74	0.70	0.66	0.91	0.70
Utilization of digestible lysine for lysine deposition in eggs (%)	74	76	78	76	85	78	77	83

* Values in row not followed by the same letter are significantly different ($P<0.05$)
The letters a and b stand for lysine level
The letters x and y stand for fat level

requirements are high relative to energy quantities needed for production. The clear beneficial effect of dietary fat could not be characterized effectually by means of an NE value of fat, most probably because of the overwhelming effect of the maintenance requirements.

It also seems that it is not possible to predict the feed consumption independently of feed composition by means of the NE values, also because of a large variation in NE intake. The NE:lysine ratios of the diets are an important characterization of feed, and in practice this is accompanied by a strange phenomenon. From these ratios it can be concluded that less lysine per MJ NE is accompanied by better results if diets are compared with different fat levels. Comparing diets with the same fat level it appears that a higher lysine content per MJ NE produces the best results.

In practice this kind of confusion is not desirable. It means that, because of the differences in NE intake, the fat level of the diets has to be known independently of the energy value of the diets in order to estimate the true value of diets for egg production.

CHARACTERIZATION OF DIETARY FAT AND PROTEIN FOR EGG PRODUCTION MORE INDEPENDENTLY FROM MAINTENANCE REQUIREMENTS

It has been shown that the measured efficiency of lysine for lysine deposition in eggs is quite constant. Rations with lysine as the first limiting amino acid are common in normal poultry feeding.

The minimum amount of dietary fat in rations for layers can be calculated if preferential use for deposition of fat in eggs is desired. This depends on the costs of dietary fat relative to carbohydrates. The apparent efficiency of lysine in the diets for deposition of that amino acid in egg protein is constant at 70%: 1 g digestible lysine produces 0.7 g lysine deposited in eggs. The lysine content of egg protein is also constant at 7%: 0.7 g deposited lysine $= 0.7 \times \frac{100}{7} = 10$ g deposited protein. The ratio deposited protein: deposited fat = 1.4. Therefore, $\frac{10}{1.4} = 7.14$ g fat is deposited per gram of digestible lysine in the diet. The efficiency of digestible fat for lipogenesis is assumed to be 80%: 7.14 g egg fat can be synthesized from $\frac{7.14}{0.8} = 8.9$ g digestible dietary fat.

Consequently, per gram of digestible lysine the rations should contain at least 9 g digestible fat. If dietary fat is too expensive compared with carbohydrates the recommended fat level in poultry diets can be excluded. The ME values of diets can still be used to estimate the energy requirements of poultry for deposition of energy and for total heat production together (Scheele, Janssen and Verstegen, 1976). The ME values also provide the possibility in practice of predicting reliably the daily feed consumption of poultry in relation to the limiting amino acid content of the diets.

References

HILL, F.W. and DANSKY, L.M. (1954). *Poult. Sci.*, **33**, 112–119

HILL, F.W., ANDERSON, D.M. and DANSKY, L.M. (1956). *Poult. Sci.*, **35**, 54–59

PAYNE, C.G. (1966). *Physiology of the Domestic Fowl*, pp. 235–241. Eds. C. Horton-Smith and E.C. Amoroso. Edinburgh, London: Oliver and Boyd

REID, B.L. and WEBER, C.W. (1975). *Poult. Sci.*, **52**, 1335–1343

SCHEELE, C.W., JANSSEN, W.M.M.A. and VERSTEGEN, M.W.A. (1976). *Energy Metabolism in Farm Animals*, pp. 277–281. Ed. M. Vermorel. EAAP Publication No. 19. Clermont-Ferrand: G. de Bussac

77

DISCUSSION PAPER: USE OF ENERGY IN REPRODUCTION AND LACTATION

W.N. GARRETT
Department of Animal Science, University of California, Davis, California 95616, USA

It may be instructive to compare, in a general way for the major farm species, the energy deposition during fetal development with the energy output during lactation. *Table 77.1* gives the approximate composition of the young of several species of interest. *Table 77.2* lists approximate values for the composition of

Table 77.1 APPROXIMATE COMPOSITION OF THE YOUNG OF SEVERAL SPECIES AT BIRTH

Species	*Water* (%)	*Protein* (%)	*Fat* (%)	*Ash* (%)	*Energy* (kJ/g)
Cattle[1]	75.0	17.0	4.0	4.0	5.6
Sheep[2]	80.0	14.0	2.5	3.5	4.3
Swine[3]	83.0	12.0	1.5	3.5	3.4
Chicken[4]	78.0	15.5	4.6	1.9	5.5
Man[5]	74.0	11.5	12.0	2.5	7.4

[1] Haigh, Moulton and Trowbridge, 1920; Farrell, Garrett and Hinman, 1976; Prior and Laster, 1979
[2] Rattray *et al.*, 1974; McDonald *et al.*, 1979
[3] Schmidt *et al.*, 1973; Elliott and Lodge, 1977
[4] Mitchell, Card and Hamilton, 1931a,b; Brody, 1945; De Schrijver, 1974; Romanoff and Romanoff, 1967
[5] Widdowson, 1968; Apte and Iyengar, 1972

Table 77.2 APPROXIMATE COMPOSITION[1] OF THE MILK OF SEVERAL SPECIES AND OF CHICKEN EGGS

Species	*Water* (%)	*Protein* (%)	*Fat* (%)	*Carbohydrate*[2] (%)	*Ash* (%)	*Energy* (kJ/g)
Cattle	88.0	3.2	3.4	4.6	0.7	2.9
Sheep	81.3	5.0	8.0	5.0	0.7	5.1
Swine	82.5	5.1	6.0	5.5	0.9	4.4
Man	88.0	1.1	4.2	7.0	0.2	3.0
Chicken egg	66.5	11.7	10.5	0.8	10.5	6.7

[1] Brett, Corbett and Inskip, 1972; Gyorgy, 1971; Hull, 1979; Macy, Kelley and Sloan, 1953; O'Grady *et al.*, 1973; Peart, Edwards and Donaldson, 1975; Romanoff and Romanoff, 1967; Sibbald, 1979
[2] Lactose except in eggs

Table 77.3 APPROXIMATE DEPOSITION OF ENERGY IN THE FETUS COMPARED TO THAT SECRETED IN THE MILK AND IN COMPARISON TO FASTING HEAT PRODUCTION (FHP) OF THE FEMALE

Species	*Fetal energy deposition*				*Milk or egg production*	
	Total (MJ)	*Maximum* (MJ/d)	*Average*[6] FE_i	*Maximum*[6] $FE_{i(max)}$	*Daily*	*Daily (MJ)* *FHP*
Cattle[1]	225	3.0	0.022 (0.122)	0.082 (0.44)	73	2.0
Sheep[2]						
Singles	22	0.7	0.024 (0.124)	0.11 (0.55)	9.5	1.5
Twins	35	1.1	0.038 (0.198)	0.17 (0.87)	13.0	2.0
Triplets	41	1.3	0.044 (0.233)	0.20 (1.02)	14.5	2.2
Swine[3] (10)	42	1.1	0.029 (0.125)	0.09 (0.37)	26.4	2.1
Human[4]	23	0.25	0.013 (0.060)	0.04 (0.17)	1.8	0.27
Chicken[5]	0.20	0.023	–	–	0.39	0.65

[1] 550 kg cow, 25 kg milk/d, 40 kg calf, FHP 320 $kJ/W^{0.75}$

[2] 70 kg ewe, 1.5, 2.1, 2.3 kg milk/d and 5.0, 4.0, 3.1 kg for single, twin and triplet births, FHP 265 $kJ/W^{0.75}$

[3] 135 kg sow, 6 kg milk/d, 10 piglets of 1.25 kg, FHP 315 $kJ/W^{0.75}$

[4] 60 kg female, 600 g milk/d, 3.1 kg infant, FHP 295 $kJ/W^{0.75}$

[5] Chicken 1.9 kg, chick weight 35 g, egg weight 58 g, FHP 355 $kJ/W^{0.75}$

[6] See text for explanation of FE_i and $FE_{i(max)}$. Figures in parenthesis are ME required for fetal growth ÷ ME_m for non-pregnant female

milk from these same species (the egg included for comparative purposes). *Table 77.3* was constructed from *Tables 77.1* and *77.2*, using various assumptions regarding animal weights and energy requirements as shown in the footnotes.

The two centre columns of *Table 77.3* were calculated in order to compare the average (FE_i) and maximal ($FE_{i(max)}$) daily deposition of energy in the fetus(es) with the fasting heat production of the mother. These values were calculated from the relationships:

$$FE_i = \frac{\text{Energy in fetus at term}}{\text{Fasting heat production of the female} \times \text{days of gestation}}$$

or

$$FE_{i(max)} = \frac{\text{Maximal daily energy deposition in fetus}}{\text{Fasting heat production of the female}}$$

If FE_i and $FE_{i(max)}$ are considered to represent relative terms, giving a measure of the nutrition stress of pregnancy compared to the maintenance requirement of the dam, then some species comparisons are possible. It appears that cattle, sheep bearing single lambs and swin giving birth to 10 young are quite closely ranked by either FE_i or $FE_{i(max)}$. The ewe bearing twins or triplets is under considerably higher nutritional stress. The human mother, because of a long gestation and relatively small fetus, is under much less stress according to these indices.

The last two columns of *Table 77.3* show similarly calculated indices for the comparative stress of lactation. The numbers obtained depend to a large extent of the level of milk production assumed, but two points are particularly relevant to this discussion. Lactation in all species is at least 10 times as stressful as pregnancy even if compared at the peak of fetal growth and 50–100 times more stressful if average deposition over the period of gestation forms the basis for comparison. The hen (if egg production is considered comparable to lactation) and the lactating human are under considerably less nutritional stress.

The use of net energy deposited in the fetus or milk and fasting heat production of the dam may not be the most suitable basis for these comparisons. Perhaps it would be preferable to use metabolizable energy (ME) requirements, which would also account for the energy needed to maintain the fetus during the gestation period. When comparisons are made on the basis of ME requirements, the ratios calculated for lactation change very little since the efficiency of energy use for milk production is similar to that for maintenance. The FE_i and $FE_{i(max)}$ calculated from ME required for pregnancy compared to that required for maintenance of the dam are shown in parentheses in *Table 77.3* (centre two columns). These numbers indicate about a 12% increase in ME required over the whole of gestation for cattle, swine and single-bearing ewes and about a 6% increase for the human. At the time of maximum fetal growth the ME requirement of a ewe bearing triplets doubles but appears unlikely to exceed 1.5 times maintenance in cattle or swine. This maximum requirement during gestation is for a very short time compared to the extended period for the increased requirements during lactation.

There are many questions yet to be answered regarding the utilization of energy during pregnancy and lactation. One of these, the efficiency with which energy is stored in the dam's tissues during gestation, is considered by Tissier, *et al.* (*see* page 329) for sheep and by Noblet and Close (*see* page 335) for the sow. The findings of Noblet and Close suggest no change in the maintenance requirement for the pregnant sow, but possibly a higher (not statistically significant) efficiency of energy utilization for protein (but not fat) synthesis in the pregnant animal. A question raised by the finding of Cristian *et al.* (*see* page 363) that the maintenance requirement during lactation in sheep is greater than for dry sheep relates to the definition of maintenance. It can be argued that any increase in requirements as a result of production is a production requirement, even if more energy is needed to maintain a larger liver and a larger gut which may be associated with a production level of feeding. If this latter assumption is used it seems probable that the efficiency of protein and fat synthesis by the pregnant or lactating animal may not be greater than in non-pregnant or non-lactating animals since less feed would be allocated to maintenance and more to the production functions, thus lowering the calculated efficiency of production. The high efficiency of energy utilization for reproduction in the sow (compared to the ruminant), as suggested by Noblet and Close, may be due, at least in part, to the weight of the sow not being corrected for the products of conception. The question of the maintenance requirement was also raised by Burlacu *et al.* (*see* page 367). Their equations indicate the maintenance requirement of the pregnant sow for the first third of gestation is about 0.38 $MJ/W^{0.75}$ but during lactation the estimate is 0.52 $MJ/W^{0.75}$. A clear definition of 'maintenance' or a standardized procedure to calculate maintenance requirements could be helpful in comparing the results of different investigations.

The efficiency of ME use for maternal gain in pregnant sheep, estimated as 90% by Tissier *et al.* (*see* page 329), appears to be within the range expected for diets of intermediate metabolizability. However, a general question not specifically related to this chapter can be raised as to whether the efficiency of energy utilization for fetal growth varies according to the source of the ME, i.e. the constituents of the diet, as does the efficiency of ME use for growth and lactation.

Does the information reported by Iwasaki, Haryu and Kameoka (*see* page 307) that heat production/$W^{0.75}$ in Japanese Black cattle was not affected by the advance of pregnancy imply that the efficiency of utilization of energy for maintenance and gain of the products of conception is the same as for maintenance and gain of the dam? The work of Corbett *et al.* (*see* page 319) with grazing ewes shows a significant increase in ME required to maintain zero energy balance in pregnant compared to non-pregnant Merinos during the last stage of gestation (130 days). Differences in maintenance requirements which might be related to weather or perhaps to seasonal variation in metabolic rate were also observed by Corbett *et al.*

The data of Rattray, Trigg and Urlich (*see* page 325) indicate that the energy stores of the ewe at mid-pregnancy exert an influence on the extent to which the dietary energy will be used to replenish maternal tissue. This finding is not only interesting practically but also from an energy balance viewpoint. What is the mechanism which retards fetal growth or permits maternal reserves to be replenished at the expense of fetal growth? The finding implies that the priority a fetus has for nutrients is subject to some control, perhaps monitored via the dam's energy stores.

Schneider *et al.* (*see* page 341) raise some important questions concerning the relationship between metabolizability (q), the protein content of the diet and the efficiency of ME utilization for lactation (k_ℓ). Their results indicate grams of protein per unit of ME can account for a significant part of the variation in k_ℓ but q (range of 0.55–0.67) had no effect on the utilization of ME for lactation or for fattening. These results are contrary to the finding of Tyrrell and Moe (*see* page 311) that protein level had no effect on the partial efficiency of ME use for milk plus body-tissue synthesis. Moe and Tyrrell did, however, find that protein level had an influence on q which reached a maximum at 17% crude protein. The coefficients associated with the protein:energy ratio in the regression equations of Schneider *et al.* have negative signs, indicating a lowering of k_ℓ with increasing protein:ME ratios. The situation with regard to protein is further complicated by interactions observed between energy and protein (supplied by barley and groundnut cake) by Thomas *et al.* (*see* page 357). The conflicting results of these experiments together with work of Ørskov and MacDonald (*see* Part 2, page 147) and Ørskov *et al.* (1979), dealing with the efficiency of utilization of the volatile fatty acids (VFAs), and that of Sutton, Oldham and Hart (*see* page 303), dealing with the influence of diet (and the VFAs) on milk composition, sites of digestion and changes in growth hormone and insulin, indicate that much is yet to be learned about the efficiency complex. Bull *et al.* (*see* page 351) have also shown a trend toward more tissue deposition in dairy cows fed varying levels (4–11%) of ammonium lactate with no observed difference in the efficiency of energy utilization. A similar observation was made by van der Honing (*see* page 315) where cows fed diets containing tallow or soya bean oil retained more energy than those fed a control diet. There was no significant difference in the utilization of ME for milk plus tissue energy and maintenance between diets, but cows fed soya bean oil had a significant depression in milk-fat synthesis and gained the most body energy.

The use of monozygous twins to investigate the carry-over effects of underfeeding of pasture-fed dairy cows (Trigg *et al.*, *see* page 345) convincingly demonstrates that the prior plane of nutrition can influence energy metabolism. These investigators demonstrated that previously well fed cows had about 7 MJ/d additional ME available for milk production compared to cows previously on a low plane of nutrition. The 'source' of most of this additional energy was tissue reserves and from a mechanistic viewpoint this result may have a basis similar to that mentioned with regard to Rattray, Trigg and Urlich's chapter (*see* page 325).

A very interesting finding is that of Michels, Decuypere and Geers (*see* page 371) of a lower food efficiency in laying hens produced from eggs incubated at lower temperatures. The results of a regression analysis of energy balance (EB) against ME intake did not show significant differences in the efficiency of ME use for EB (as measured by the regression coefficients) but a statistically higher heat production was calculated for the hens from eggs incubated at the lower temperature. A graph showing the average heat production/$W^{0.75}$ per h in a different experiment, however, does not show an increased heat production on a scaled body weight basis. Feed intake was slightly higher for the low temperature group and egg production was lower for these hens. This suggests that the differences in heat production could be partly the result of a different partitioning of ME to egg production and body weight gain rather than just a difference in maintenance requirement as suggested by the authors. Scheele and Musharaf (*see* page 377) did not find a difference in the maintenance requirement of laying hens kept at a high environmental temperature and fed diets varying in fat and lysine content, although the energy use for egg production (k_o) averaged across diets by fat level was 0.57 for diets containing the lower level of fat (3.2–4.3%) and 0.67 for those with the higher level (7.5–8.8%). The digested lysine was used at a constant level of efficiency. However, an interaction existed in that less lysine per MJ of NE appeared appropriate across diets with different fat levels, but higher lysine per MJ of NE produced the best results if the comparison was from diets with equal fat levels. A finding of this nature is difficult to handle practically but occurs in other phases of animal production.

Many important questions have been raised in this section, perhaps more than have been answered. This is not unusual in any active, important area of investigation, but since controversy can lead to refinement of knowledge it must be considered that significant progress is being made in this area of research.

References

APTE, S.V. and IYENGAR, L. (1972). 'Composition of the human foetus.' *Br. J. Nutr.*, 27, 305–312

BRETT, D.J., CORBETT, J.L. and INSKIP, M.W. (1972). 'Estimation of the energy value of ewe milk.' *Proc. Austr. Soc. Anim. Prod.*, 9, 286–291

BRODY, S. (1945). *Bioenergetics and Growth.* New York: Reinhold

BURLACU, G., MOISA, D., IONILA, D., TASCENCO, V. and BALTAC, M. (1974). In *Energy Metabolism of Farm Animals*, pp. 265–268. Eds. K.H. Menke, H.-J. Lantsch and J.R. Reichl. EAAP Publication No. 14. Stuttgart: Universität Hohenheim Dokumentationsstelle

De SCHRIJVER, R. (1974). In *Energy Metabolism of Farm Animals*, pp. 281–284. Eds. K.H. Menke, H.-J. Lantsch and J.R. Reichl. EAAP Publication No. 14. Stuttgart: Universität Hohenheim Dokumentationsstelle

DONALD, H.P. and RUSSELL, W.S. (1970). 'The relationship between live weight of ewe at mating and weight of newborn lamb.' *Anim. Prod.,* **12**, 273–280

ELLIOT, J.I. and LODGE, G.A. (1977). 'Body composition and glycogen reserves in the neonatal pig during the first 96 hours postpartum.' *Canad. J. Anim. Sci.,* **57**, 141–150

FERRELL, C.L., GARRETT, W.N. and HINMAN, N. (1976). 'Growth, development and composition of the udder and gravid uterus of beef heifers during pregnancy.' *J. Anim. Sci.,* **42**, 1477–1489

GYORGY, P. (1971). 'Biochemical aspects of human milk.' *Am. J. clin. Nutr.,* **24**, 970–975

HAIGH, L.D., MOULTON, C.R. and TROWBRIDGE, P.F. (1920). 'Composition of the bovine at birth.' *Res. Bull. Univ. Mo. agric. Exp. Stn.,* No. 38

HULL, B. (1979). 'Uniformity of human milk.' *Am. J. clin. Nutr.,* **32**, 304–312

McDONALD, I., ROBINSON, J.J., FRASER, C. and SMART, R.I. (1979). 'Studies on reproduction in prolific ewes. 5. The accretion of nutrients in the foetuses and adnexa.' *J. agric. Sci.,* **92**, 591–603

MACPHERSON, R.M., ELSLEY, F.W.H. and SMART, R.I. (1969). 'The influence of dietary protein intake during lactation on the reproductive performance of sows.' *Anim. Prod.,* **11**, 443–451

MACY, I.G., KELLEY, H.J. and SLOAN, R.E. (1953). *The Composition of Milks.* National Research Council Publication 254. Washington, DC: National Academy of Sciences

MITCHELL, H.H., CARD, L.E. and HAMILTON, T.S. (1931). 'Technical study of the growth of white leghorn chickens.' *Bull. Ill. agric. Exp. Stn.,* No. 367

MITCHELL, H.H., CARROLL, W.E., HAMILTON, T.S. and HUNT, G.E. (1931). 'Food requirements of pregnancy in swine.' *Bull. Ill. agric. Exp. Stn.,* No. 375

O'GRADY, J.F., ELSLEY, F.W.H., MACPHERSON, R.M. and McDONALD, I. (1973). 'The response of lactating sows and their litters to different dietary energy allowances. I. Milk yield and composition, reproductive performance of sows and growth of litters.' *Anim. Prod.,* **17**, 65–74

ØRSKOV, E.R., GRUBB, D.A., SMITH, J.S., WEBSTER, A.J.F. and CORRIGALL, W. (1979). 'Efficiency of utilization of volatile fatty acids for maintenance and energy retention by sheep.' *Br. J. Nutr.,* **41**, 541–551

PEART, J.N., EDWARDS, R.A. and DONALDSON, E. (1975). 'The yield and composition of milk of Finnish Landrace × Blackface ewes.' *J. agric. Sci.,* **85**, 315–324

PRIOR, R.L. and LASTER, D.B. (1979). 'Development of the bovine fetus.' *J. anim. Sci.,* **48**, 1546–1553

RATTRAY, P.V., GARRETT, W.N., EAST, N.E. and HINMAN, N. (1974). 'Growth, development and composition of the ovine conceptus and mammary gland during pregnancy.' *J. anim. Sci.,* **38**, 613–626

ROMANOFF, A.L. and ROMANOFF, A.J. (1967). *Biochemistry of the Avian Embryo.* New York: Interscience Publishers

SCHMIDT, M.T., VEUM, T.L., CLARK, J.L. and KRAUSE, G.F. (1973). 'Chemical composition of crossbred swine from birth to 136 kg with two planes of nutrition from 53 to 136 kg.' *J. anim. Sci.,* **37**, 683–687

SIBBALD, I.R. (1979). 'The gross energy of Avian eggs.' *Poult. Sci.,* **58**, 404–409

TASAKI, I. and SASA, Y. (1970). In *Energy Metabolism of Farm Animals*, pp. 197–200. Eds. A. Schürch and C. Wenk. EAAP Publication No. 13. Zurich: Juris Druck

THOMSON, A.M., HYTTEN, F.E. and BILLEWICZ, W.Z. (1970). 'The energy cost of human lactation.' *Br. J. Nutr.*, **24**, 565–572

WIDDOWSON, E.M. (1968). 'Growth and composition of the fetus and newborn.' In *Biology of Gestation*, Vol. 2 *The Fetus and Neonate*, p. 23. Ed. N.S. Assali. New York: Academic Press

78

GENERAL DISCUSSION

Two main topics were discussed.

Efficiency of conversion of feed for reproduction

The main question considered was why this value is so low. It was reported that in sows values of about 30% were found, and in sheep they were even lower at about 12%. Possible reasons for these low values were advanced. One was the method of estimation: fetal growth accounts for only a small part of the total energy metabolism, so that there may be considerable error, which could give rise to misleading apparent differences between species. Other factors referred to were that the maintenance requirement of the fetus may not be assessed correctly by adding it to that of the mother, that protein turnover is very high, that the metabolic rate of the placenta was quoted to be very high, and that hormones associated with reproduction may be associated with high metabolic levels.

Effects of diets upon lactation

If a diet is constant in protein content then the protein supply at each feeding level will not be similar; more work is required on nitrogen recycling in relation to passage rate, feed intake and the stability of fermentation. More research is required with surgical techniques to study digestion and with animal calorimetry to follow energy balances. Extreme diets are often employed in laboratory experiments, but some participants felt that these should be followed up with more normal diets.

Physiologically there are several reasons for avoiding too large a negative energy balance in early lactation: because of the inefficiency of first transforming feed to body reserves and then later to milk, ketosis may result or there may be adverse effects on fertility. There may be a relationship between seasonal effects on the hormones associated with reproduction and seasonal effects on metabolism.

The use of regression coefficients to determine energy costs and partial efficiencies during pregnancy was discussed, but in general it was felt that these coefficients were unrealistic. Because of the relatively small retention during pregnancy it is difficult to obtain precise estimates of partial efficiencies for pregnancy.

PART 6

METABOLIC RESPONSES IN THE GROWING ANIMAL

79

EFFECT OF FEEDING LEVEL AND FEED COMPOSITION ON ENERGY UTILIZATION, PHYSICAL ACTIVITY AND GROWTH PERFORMANCE OF PIGLETS

H.M. HALTER, C. WENK and A. SCHÜRCH
Institute for Animal Production, ETH Zürich, Switzerland

Summary

In five respiration experiments with 36 castrates (5–40 kg W) housed in groups of three, the influence of feeding level on protein and fat deposition, energy utilization and physical activity was studied.

Of the retained energy (RE) of *ad lib.*-fed piglets, 50–35% was deposited as protein and 50–65% as fat (5–40 kg W). Feed restriction caused a reduction of RE and of the protein: fat ratio. Restrictively fed animals deposited about 75% of the RE as protein at the beginning and about 50% at the end of the experiments. Maintenance needs amounted to 718 kJ/$W^{0.569}$ or 510 kJ/$W^{0.75}$, of which an average 22% was due to physical activity. Partial efficiency for growth depended on the exponent of metabolic weight.

Introduction and experimental design

Today, piglets are frequently weaned when only three to four weeks old and then housed in flat-deck batteries. This system makes it possible to increase the reproductive performance of the sow on the one hand, and, on the other, to exploit the high growth capacity of the piglet before the initiation of the fattening period.

In a series of five respiration experiments the energy metabolism of 36 Large White castrates in the live-weight interval of 5–40 kg was investigated. Groups of three piglets were held in batteries in order to simulate the housing environment found in practice, thus minimizing the excitement in the animals arising from experimental conditions. Grouping the piglets made it impossible to collect faeces and urine separately. The results are therefore expressed as averages of three animals. Feed intake was measured three times weekly and faecal and urine excretion, gaseous exchange and physical activity on four days every week, except in Experiment 2 in which 6-day-collection periods were used. Gaseous exchange and physical activity of the piglets were measured using the equipment described by Wenk, Prabucki and Schürch (1971) and Wenk and van Es (1976) respectively. The design of the five experiments is shown in *Table 79.1.*

Table 79.1 EXPERIMENTAL DESIGN

	Group 1	Group 2	Group 3
Experiment 1: feed 1			
feeding level	*ad lib.*	restricted (60% production feed)	–
Experiment 2: feed 1			
feeding level	*ad lib.*	restricted (40% production feed)	–
Experiment 3 and 4:			
feed	2	3	2
feeding level	(90% of *ad lib.*)	(90% of *ad lib.*)	(70% of group 1)
Experiment 5:			
feed	2	3	
feeding level	*ad lib.*	*ad lib.*	
Feed 1:	14.2 MJ ME/kg	201 g XP/kg	standard feed
Feed 2:	13.8 MJ ME/kg	258 g XP/kg	high protein feed
Feed 3:	14.7 MJ ME/kg	202 g XP/kg	feed 2 + 25% starch

Results

Soon after the beginning of the experiments the *ad lib.*-fed piglets showed high daily live-weight gains and high rates of protein and fat deposition (13–155 g protein and –7 to 347 g fat/d). Of the RE 35–50% was deposited as protein and 50–65% as fat (*Figure 79.1*). Some of the piglets were initially up to four days in negative energy balance as a result of the effects of weaning and adaptation to the experimental environment. The daily gain of both *ad lib.* and restrictively-fed piglets was not markedly affected by the protein content of the rations. These all supplied equivalent amounts of energy, but the ratio of RE in the form of protein to energy in the form of fat was slightly increased at higher levels of protein intake. Restrictively-fed piglets had a lower rate of growth, protein deposition and fat deposition when compared with *ad lib.*-fed animals. By limiting energy supply in these young animals the protein deposition was reduced but the fat deposition still represented over 30% of the total energy deposition.

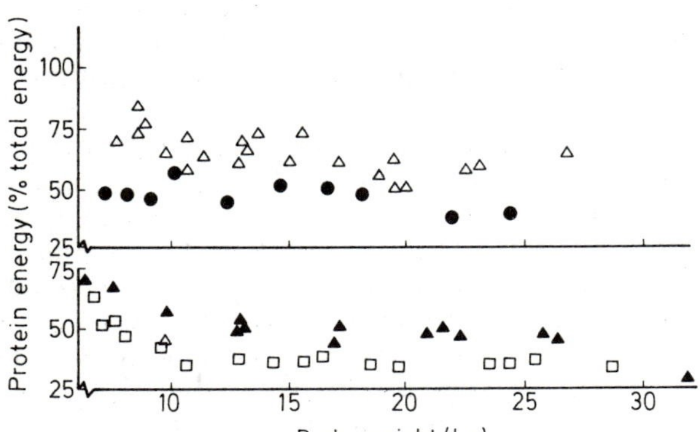

Figure 79.1 Protein deposition as a percentage of total energy deposition. △ = *restricted feeding;* ● = *ad lib. feeding;* ▲ = *feed 2 (see Table 79.1);* □ *feed 3 (see Table 79.1).*

It was possible to estimate daily body-weight gain (ΔW; g) from protein (ΔP; g) and fat (ΔF; g) deposition using the following equation:

$$\Delta W = 3.43\ \Delta P + 0.85\ \Delta F + 0.136 \qquad r = 0.942;\ RSD = 0.072$$

The energy requirement for activity of *ad lib.*-fed piglets amounted to 10–17% of the total heat production or to 18–25% (mean 22%) of the maintenance requirement. In the restrictively-fed piglets the energy requirement for activity showed more variation (10–28% of total heat production) and was on average about 30% higher than in *ad lib.*-fed animals. The change of the protein content of the ration had no effect on activity.

Maintenance requirement and partial efficiency of metabolizable energy (ME) utilization for body-tissue gains, and protein and fat deposition, were estimated by linear regression analysis after pooling the results from Experiments 2–5 (n = 53). Energy retention and ME intake (MEI) were scaled by dividing by the metabolic body weight using the exponent 0.569. The relationship between the two scaled parameters RE* and MEI* is shown in *Figure 79.2.* Also shown in

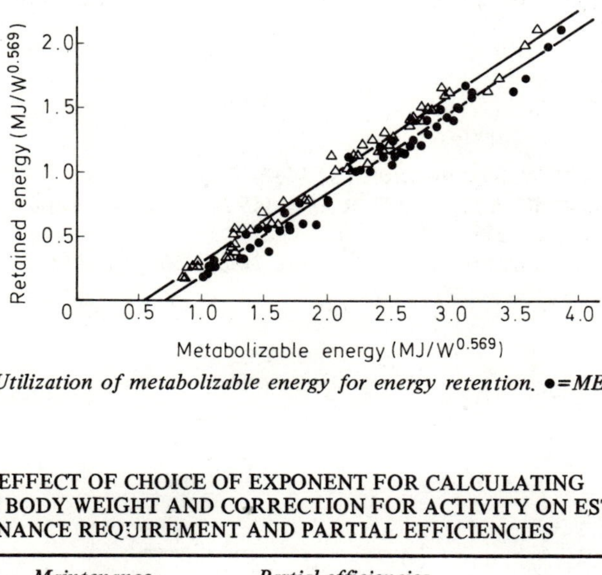

Figure 79.2 Utilization of metabolizable energy for energy retention. ●=MEI; △ = MEI*–AE*.*

Table 79.2 EFFECT OF CHOICE OF EXPONENT FOR CALCULATING METABOLIC BODY WEIGHT AND CORRECTION FOR ACTIVITY ON ESTIMATES OF MAINTENANCE REQUIREMENT AND PARTIAL EFFICIENCIES

Exponent	*Maintenance requirement (with or without* AE* *correction;* kJ/W*)		*Partial efficiencies*			*Correlation coefficient*
			k_g	k_p	k_f	r
0.569	+ AE*	718	0.642			0.989
	– AE*	560	0.655			0.991
	+ AE*	717		0.598	0.677	0.989
	– AE*	547		0.603	0.690	0.991
0.75	+ AE*	510	0.698			0.991
	– AE*	397	0.698			0.991
	+ AE*	510		0.661	0.728	0.991
	– AE*	389		0.644	0.735	0.991

Figure 79.2 is the relationship between RE* and MEI* after correcting for the scaled energy requirement for activity (AE*). The linearity between RE* and MEI* or (MEI* − AE*), over a wide range of feeding levels, is demonstrated in *Figure 79.2.* Attention should also be drawn to the equality of the slopes.

The effect of the scaling exponent 0.569 or 0.75, and the correction of MEI for activity, on estimates of maintenance requirement (ME_m) and the partial efficiency of ME utilization for body-tissue gain (k_g) and for protein (k_p) and fat deposition (k_f) is shown in *Table 79.2.*

Maintenance requirement of the grouped piglets was either 718 $kJ/W^{0.569}$ or 510 $kJ/W^{0.75}$. The activity correction reduced ME_m by about 22%. Using the exponent 0.569 proposed by Breirem (1939), instead of 0.75, reduced the estimates of the k values. The activity correction in the regression using the exponent 0.569 increased the estimates of k_p and k_f whereas using the exponent 0.75 caused a decrease of k_p and an increase of k_f.

Acknowledgement

The authors acknowledge the advice of Dr. E.F. Thomson during the preparation of this chapter.

References

BREIREM, K. (1939). *Tierernährung,* **11**, 487–528

WENK, C., PRABUCKI, A.L. and SCHÜRCH, A. (1971). *Mitt. Lebensmitteluntersuchung Hyg.,* **61**, 378–387

WENK, C. and van ES, A.J.H. (1976). *Schweiz. landw. Mh.,* **54**, 232–236

THE ENERGY REQUIREMENTS FOR GROWTH IN THE EARLY-WEANED PIG

W.H. CLOSE and M.W. STANIER
ARC Institute of Animal Physiology, Babraham, Cambridge CB2 4AT, England

Summary

The energy cost of maintenance (ME_m) and the energetic efficiency of protein (k_p) and fat (k_f) synthesis have been determined in groups of early-weaned piglets at environmental temperatures of 18°C, 23°C and 28°C. A decrease occurred in ME_m from 721 kJ/$kg^{0.75}$ per day at 18°C to 623 at 23°C and 544 at 28°C. Below thermoneutrality, that is at 18°C and 23°C, k_p and k_f were both approximately 0.80; within thermoneutrality (28°C) k_p was estimated as 0.74 and k_f as 0.72.

Introduction

Rapid growth and development of the early-weaned pig can be obtained only under environmental conditions which allow maximum utilization of dietary energy for growth purposes. At low environmental temperatures, the animal's maintenance energy requirement is increased and there is a reduction in the supply of energy available for growth. Although this reduction of energy supply has been discussed for older pigs (Close, 1978), no such information is available for the early-weaned piglet and the experiments reported here were designed to provide this information for piglets maintained at several planes of nutrition when exposed to a range of environmental temperatures.

Materials and methods

The experiments were carried out at three environmental temperatures with three feeding levels at each temperature. There were, therefore, nine temperature-feeding level combinations and, as each treatment was duplicated, this involved a total of 18 groups of piglets. The environmental temperatures were 18 °C, 23 °C and 28 °C (± 0.5). At each temperature one of three feeding levels was offered: at 18 °C and 23 °C these were 30, 45 and 60 g/kg body weight per day. The three feeding levels employed at 28 °C were 20, 30 and 45 g/kg body weight per day. The food supplied 16.0 MJ (± 0.1) of metabolizable energy (ME) and 243 (± 4) g of crude protein (CP) per kg dry matter. Water was available *ad lib.*

Piglets weaned at 14 days of age were transferred in groups of six to mobile pens in a temperature-controlled room for a seven-day-habituation period. At 21 days of age two piglets from each group were slaughtered and the remaining four animals were removed together to a calorimeter (Close and Mount, 1975) where their heat loss and energy and nitrogen balances were measured for a 14-day period. To facilitate the balance procedures each experiment was divided into two separate seven-day-collection periods. At 35 days of age, the animals were slaughtered. The partition of the ME intake of the piglets into heat loss (H), energy retention (ER) and protein (P) and fat (F) deposition was determined between the 21st and 35th day of age at each temperature-feeding level combination from the calorimeter and balance trials and compared with that determined from the comparative-slaughter technique.

Results and discussion

With the exception of one of the two groups of piglets at the highest level of feeding at 18 °C, there was good agreement in the estimates of ER, P and F deposition from the comparative-slaughter studies and that determined from the calorimeter and balance procedures. In the majority of cases the variation between estimates from the two procedures was less than 10%, with the greatest difference occurring at 18 °C.

Because of the good agreement between estimates determined from the two procedures, the results of H, ER, P and F deposition for each treatment have been combined and the mean values are presented in *Table 80.1*. These results have been expressed per unit of metabolic body size ($kg^{0.75}$).

The low ME intakes of the animals at 18 °C resulted from partial restriction of food when scouring occurred. At 23 °C, however, the lower than anticipated feed intake at the highest level offered was due to the 60 g/kg feeding level being

Table 80.1 MEAN BODY WEIGHT, FOOD INTAKE, BODY-WEIGHT GAIN, METABOLIZABLE ENERGY INTAKE, HEAT LOSS, ENERGY RETENTION AND PROTEIN AND FAT DEPOSITION OF THE EARLY-WEANED PIGLETS AT EACH TEMPERATURE-FEEDING LEVEL COMBINATION

Environmental temperature (°C)	*Body weight* (kg)	*Food intake* (g/d)	*Body-weight gain* (g/d)	*ME*	*H*	*ER*	*P*	*F*
				(kJ/$kg^{0.75}$ per day)				
18	4.68	136	77	635	671	−36	126	−162
	6.62	258	190	904	798	106	181	−75
	5.64	228	173	888	809	79	163	−84
23	5.00	145	88	603	620	−17	131	−148
	6.64	292	252	1038	692	346	222	124
	7.68	399	345	1272	735	537	244	293
28	5.44	107	75	426	517	−91	83	−174
	6.33	179	145	656	566	90	157	−67
	6.24	261	213	963	668	295	190	105

ME = metabolizable energy; ER = energy retention; P = protein deposition; F = fat deposition; H = heat loss

above the animal's *ad lib.* intake. The *ad lib.* intake of food at this temperature was equivalent to 52 g/kg body weight per day.

The relationship between ER and ME intake at each temperature is shown in *Figure 80.1.* From the linear regression equations relating ER (y) to ME (x) the following relations were determined:

$$18\,^\circ C;\ y = 0.56\ (\pm 0.07)\ x - 404\ (\pm 58) \qquad (r = 0.95) \quad (1)$$
$$23\,^\circ C;\ y = 0.82\ (\pm 0.03)\ x - 511\ (\pm 33) \qquad (r = 0.99) \quad (2)$$
$$28\,^\circ C;\ y = 0.71\ (\pm 0.05)\ x - 386\ (\pm 33) \qquad (r = 0.98) \quad (3)$$

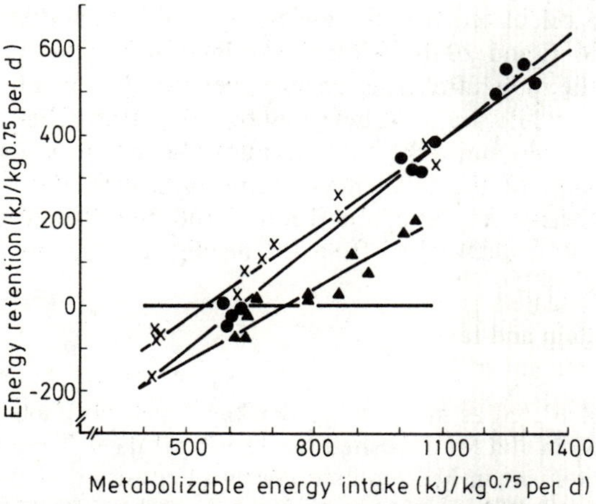

Figure 80.1 The relation between energy retention (kJ/kg$^{0.75}$ *per day) and metabolizable energy intake* (kJ/kg$^{0.75}$ *per day) at environmental temperatures* (°C) *of 18* (▲ – ▲), *23* (● – ●) *and 28* (x – x).

The partial efficiencies (k) of 0.82 and 0.71 at 23 °C and 28 °C, respectively, are typical of values obtained just below and within the zone of thermoneutrality (Close, 1978). The value of 0.56 for k at 18 °C is lower than anticipated, but equally low values have previously been reported for pigs kept at low environmental temperatures (Fuller and Boyne, 1972). In the present experiments, this unexpectedly low value may have arisen for the following reasons:

1. The mildly pathological state of the piglets.
2. The addition of an antibiotic (Framomycin) to the feed to reduce the incidence of gastroenteritis.
3. Possible compensatory growth effects arising in the second week of the experiment following the abnormally low feed intakes in the first week when food was partially restricted due to the high incidence of gastroenteritis.

As in experiments reported with older pigs (Close, Mount and Brown, 1978), P deposition appeared to be independent of environmental temperature and dependent only on the level of food intake. Thus, at each temperature P increased by 0.19 kJ/kJ increment in ME intake. Unlike P deposition, F deposition was dependent on both the environmental temperature and the level of feeding. The increment in F deposition, per unit increment in ME intake, increased from

0.37 at 18 °C to 0.63 and 0.52 at 23 °C and 28 °C, respectively. It would, therefore, appear that the low efficiency of fat retention is responsible for the low value of k determined at 18 °C (equation 1).

From the equations relating ER to ME intake (equations 1, 2 and 3) it is possible to calculate the animal's maintenance energy requirement (ME_m) as that intake at which ER = 0. The maintenance energy requirement decreased with increase in temperature from 721 kJ/kg$^{0.75}$ per day at 18 °C to 623 at 23 °C and 544 at 28 °C. The temperature of 28 °C is within the animals' thermoneutral zone at their maintenance energy intake. From the determination of the extra thermoregulatory heat production required at 23 °C, described by Verstegen *et al.* (1973), it can be calculated that the piglets' critical temperature (T_c) decreased from 27 °C to 24 °C and 20 °C as ME intake increased from once to twice and three times the thermoneutral maintenance energy requirement.

Estimates of k, while useful in determining the partial efficiency with which energy is retained, do not take into account the composition of the tissue deposited. Estimates of the efficiency of utilization of energy for protein (k_p) and fat (k_f) synthesis take the composition of the gain into account. Values of k_p and k_f have been determined from the results at 23 °C and 28 °C from the equation:

$$ME_p = aP + bF \qquad (4)$$

where ME_p is calculated as ME – ME_m (kJ/kg$^{0.75}$ per day) and P and F are the quantities of protein and fat deposited (kJ/kg$^{0.75}$ per day). The values of a and b represent the increment in ME required per unit deposition of P and F and their reciprocals represent the respective values of k_p and k_f. At 23 °C, estimates of k_p of 0.84 and k_f of 0.82 have been calculated by the above method; at 28 °C the respective values were 0.74 and 0.72. At each level of feeding at 18 °C the animals were in negative fat balance. If fat is catabolized with an efficiency of 0.80, and substituting this in equation (5), it can be calculated that the energetic efficiency of protein synthesis is 0.78. Taking the mean of the estimates of k_p and k_f as equivalent to k, then a mean k value of 0.80 can be calculated at 18 °C. This efficiency is characteristic of values below an animal's critical temperature. At 18 °C and 23 °C the values of k_p and k_f have been determined at intakes below T_c. At 28 °C, however, the estimates of k_p and k_f were determined above T_c and these suggest that the efficiencies of protein and fat synthesis are higher than those determined under similar conditions for older animals (Close, 1978).

References

CLOSE, W.H. (1978). *Br. J. Nutr.*, **40**, 433–438
CLOSE, W.H. and MOUNT, L.E. (1975). *Br. J. Nutr.*, **34**, 279–290
CLOSE, W.H., MOUNT, L.E. and BROWN, D. (1978). *Br. J. Nutr.*, **40**, 423–431
FULLER, M.F. and BOYNE, A.W. (1972). *Br. J. Nutr.*, **28**, 373–384
VERSTEGEN, M.W.A., CLOSE, W.H., START, I.B. and MOUNT, L.E. (1973). *Br. J. Nutr.*, **30**, 21–35

81

ENERGY RETENTION, PROTEIN AND FAT GAIN IN EARLY-WEANED PIGLETS IN RELATION TO ENVIRONMENTAL TEMPERATURE

J. LE DIVIDICH, J. NOBLET and A. AUMAITRE
I.N.R.A. Station de Recherches sur l'Elevage des Porcs, 78350 Jouy-en-Josas, France
M. VERMOREL
Laboratoire d'Etudes du Métabolisme Energétique, 63110 Beaumont, France

Considerable attention has been given to the influence of environmental temperature on energy metabolism in the newborn and the growing pig (Holmes and Close, 1977). However, there is a lack of information on the energy metabolism of piglets immediately after weaning in spite of the obvious effects of the thermal environment on performance (Le Dividich, 1979). Thus the present experiment was designed to investigate the effects of environmental temperature on energy and nitrogen metabolism in the early-weaned piglet during a period of 12 consecutive days following weaning.

Materials and methods

Seventy-two piglets were used in six balance trials, and each one involved two groups of six piglets. The mean age at weaning was 24.8 days (21–28 days) and the mean live weight was 6.32 ± 1.24 kg. Three environmental temperatures were chosen, 20, 24 and 28 °C, and it was assumed that 28 °C was close to the critical temperature (Cairnie and Pullar, 1957). The experimental design was an incomplete balanced block design with four replications for each temperature. Each trial was divided into two main periods of six days (P_1 and P_2) and each main period was divided into two sub-periods of three days p_1 and p_2, p_3 and p_4, respectively.

Piglets were kept in groups in totally wired cages. They were group-fed pellets supplying 0.245 g crude protein, 17.71 kJ gross energy and 14.34 kJ of metabolizable energy (ME) per gram of dry matter (DM). The mean DM content of the food was 92.7%. The food was offered according to an increasing scale of feeding, on an average of 120 g/d per piglet at day 3, 160 at day 6, 230 at day 9 and 310 at day 12.

Gas exchanges were continuously recorded during the 12 experimental days. Heat production (HP) was estimated from carbon dioxide production, oxygen consumption and urinary-nitrogen losses (Brouwer, 1965). Energy retained (ER) was obtained as the difference between ME intake and HP. Protein deposition (P) was calculated from the nitrogen balance. The energy value of P was determined, taking the energy equivalent of protein as 23.8 kJ/g. The difference between ER and P provided an estimate of fat deposition (F).

Results

There was no significant effect of environmental temperature on the weight gain in piglets which averaged 120 g/d over the 12-day period.

The results (*Table 81.1*) show that HP increased linearly ($P<0.05$) by 12.5 and 11.9 $kJ/kg^{0.75}$ per d for each 1 °C as environmental temperature was lowered from 28 °C to 20 °C during P_1 and P_2, respectively. Taking p_2 as a reference period (from day 4 the piglets were quiet, consumed more food and their energy balance was either positive or nil), it appeared that the percentage increase in HP ($kJ/kg^{0.75}$ per d) between p_2 and p_4 was similar for piglets raised at 24 °C and 28 °C and averaged 23%; it was 11% for those kept at 20 °C.

Table 81.1 EFFECT OF ENVIRONMENTAL TEMPERATURE ON HP, ER AND PARTITION OF ER BETWEEN P AND F ($kJ/kg^{0.75}$ PER d)

Experimental temperature (°C)	*Period*	*ME intake*	*HP*	*ER*		
				Total	*as P*	*as F*
20	1	395	543	−149	64	−212
	2	887	606	282	193	89
24	1	405	491	−86	70	−155
	2	843	573	269	182	87
28	1	419	441	−24	68	−92
	2	854	520	332	185	147

HP = heat production; ME = metabolizable energy; ER = energy retained; P = protein deposition; F = fat deposition

During P_1, the HP exceeded ME intake and consequently energy balance was negative. The calculated mobilized body energy increased linearly ($P<0.05$) with decreasing environmental temperature and amounted to 24, 86 and 149 $kJ/kg^{0.75}$ per day at 28, 24 and 20 °C, respectively. In addition, the energy balance became positive from days 3, 4 and 5 in piglets raised at 28, 24 and 20 °C, respectively.

From the data of p_2, p_3 and p_4 sub-periods, energy retention was related to ME at each temperature. The relevant equations are shown below:

$$20\,°C;\ Y = 0.91\ (\pm 0.07)\ X - 520\ (\pm 49.1) \qquad (r = 0.97) \quad (1)$$
$$24\,°C;\ Y = 0.76\ (\pm 0.07)\ X - 378\ (\pm 45.6) \qquad (r = 0.96) \quad (2)$$
$$28\,°C;\ Y = 0.79\ (\pm 0.09)\ X - 325\ (\pm 52.5) \qquad (r = 0.94) \quad (3)$$

where Y is energy retained ($kJ/kg^{0.75}$ per d), X is ME intake ($kJ/kg^{0.75}$ per d) and r is the correlation coefficient.

Nitrogen retention was not influenced by environmental temperature. It amounted to 0.45 and 1.26 $g/kg^{0.75}$ per d on average during P_1 and P_2, respectively. Consequently, P was similar between 20°C and 28°C, but it increased linearly ($P<0.05$) with food intake according to the following equation:

$$Y = 0.27\ (\pm 0.02)\ X - 44.3 \qquad (n = 24;\ r = 0.94) \quad (4)$$

where Y is protein gain (kJ/kg$^{0.75}$ per d), X is ME intake (kJ/kg$^{0.75}$ per d) and r is the correlation coefficient.

When protein retention (Y; g/kg$^{0.75}$) was related to protein intake (X; g/kg$^{0.75}$) the following regression was obtained:

$$Y = 0.77\ (\pm\ 0.15)\ X - 2.42 \qquad (n = 24;\ r = 0.91) \qquad (5)$$

Extrapolation to protein retention = zero provides an estimate of protein requirement for maintenance of 3.14 g/kg$^{0.75}$ per d corresponding to an energy cost for protein maintenance of 183 kJ/kg$^{0.75}$ per d of ME.

During P_1 the piglets gained protein although their energy balance was negative. Consequently, energy requirements for maintenance, thermoregulation and protein deposition involved a partial mobilization of lipid reserves which amounted to 92, 155 and 212 kJ/kg$^{0.75}$ per d, at 28, 24 and 20 °C, respectively. During P_2 the piglets gained both F and P; F increased with air temperature whereas P was similar between treatments.

Discussion

The linear increase in HP when environmental temperature falls from 28 °C to 20 °C during P_1 and P_2 implies that probably the critical temperature lies at or above 28 °C. Thus the present results clearly show that piglets raised in groups in totally wired cages require a relatively high environmental temperature (28 °C), particularly during a period of about two weeks following weaning between three and four weeks of age.

It is interesting to note that between p_2 and p_4 the increase in HP was approximately 50% higher in piglets kept at 28 °C and 24 °C than in those kept at 20 °C, although the increase in ME intake was slightly lower in the former groups of piglets. This would indicate that the heat increment of feeding was, partly at least, used for thermoregulation in the cold (20 °C), which is in agreement with the results of Verstegen *et al.* (1973). These observations could be related to the variations in the efficiency of ME utilization for growth : 0.91 at 20 °C, 0.76 and 0.79 at 24 °C and 28 °C, respectively. From equation (3) the thermoneutral estimation of energy requirement for maintenance is calculated to be 411 kJ/kg$^{0.75}$ per d. This value agrees with that of 425 kJ/kg$^{0.75}$ per d found by Jordan (1973) for piglets and those of Verstegen *et al.* (1973) and Close and Mount (1976) who obtained values of 418 and 440 kJ/kg$^{0.75}$ per d for growing pigs, respectively. The corresponding values at 24 °C and 20 °C are 497 and 571 kJ/kg$^{0.75}$ per d.

For p_2, p_3 and p_4 sub-periods the ME intake in piglets was 744 kJ/kg$^{0.75}$ per d. The calculated values of ER (from equations 1, 2 and 3) for that level of ME intake are 157, 188 and 263 kJ/kg$^{0.75}$ at 20, 24 and 28 °C, respectively. This means that an increase of the environmental temperature of 1 °C involves an increase of ER or a decrease of HP of 13.2 kJ/kg$^{0.75}$ per d which for an 8 kg piglet corresponds to the energy supplied by 4.7 g food/°C per d.

Between 20 °C and 28 °C, P was not influenced by air temperature in our piglets, and this was in good agreement with the results of Fuller (1965). Consequently, nitrogen retention appears to be affected only under extreme environmental conditions, either in very cold (Close and Mount, 1976) or in very hot conditions (Holmes, 1973). Whereas nitrogen retention was not affected by air temperature, the fat deposition closely depended on air temperature. In addition, fat reserves appeared to be easily mobilized during the weaning period, and this was in good accordance with the previous results of Baur and Filler (1959), Whittemore, Aumaitre and Williams (1978).

Present results suggest that, during the weaning period, protein is deposited with priority and independently of environmental temperature, whereas fat deposition depends on environmental temperature.

References

BAUR, L.S. and FILLER, L.J. Jr. (1959). *J. Nutr.,* **69**, 128

BROUWER, E. (1965). In *Energy Metabolism of Farm Animals*, p. 441. Ed. K.L. Blaxter. EAAP Publication No. 11. New York and London: Academic Press

CAIRNIE, A.B. and PULLAR, J.D. (1957). *J. Physiol.,* **139**, 15P

CLOSE, W.H. and MOUNT, L.E. (1976). In *Energy Metabolism of Farm Animals*, p. 176. Ed. M. Vermorel. EAAP Publication No. 19. Clermont-Ferrand: G. de Bussac

FULLER, M.F. (1965). *Br. J. Nutr.,* **19**, 531

HOLMES, C.W. (1973). *Anim. Prod.,* **16**, 117

HOLMES, C.W. and CLOSE, W.H. (1977). In *Nutrition and the Climatic Environment*, p. 51. Eds. W. Haresign, H. Swan and D. Lewis. London: Butterworth

JORDAN, J.W. (1973). *Studies on the Energy Metabolism of the Early Weaned Pig.* PhD Thesis. The Queen's University, Belfast

LE DIVIDICH, J. (1979). In *French Swine Days Research*, p. 133. Paris: I.N.R.A.–I.T.P.

VERSTEGEN, M.W.A., CLOSE, W.H., START, I.B. and MOUNT, L.E. (1973). *Br. J. Nutr.,* **30**, 21

WHITTEMORE, C.T., AUMAITRE, A. and WILLIAMS, I.H. (1978). *J. agric. Sci., Camb.,* **91**, 681

RESTRICTION OF FEED INTAKE AND COMPENSATORY GROWTH RESPONSES IN PIGS

D. GÄDEKEN, H. BÖHME and H.J. OSLAGE
Institute of Animal Nutrition of the Federal Research Centre of Agriculture, Braunschweig-Völkenrode, Federal Republic of Germany

Summary

In a respiration trial with two groups each of four pigs the effect of early nutrition on subsequent performance and nitrogen and energy metabolism was studied. The restriction of feed intake of about 45% in the period from 5 to 25 kg W improved daily weight gain and feed efficiency in the fattening period combined with a higher daily protein and water deposition. As a result of the faster growth the pigs gained less fat in the whole experimental period.

Introduction

Several workers have shown that pigs respond to an adequate diet following periods of restricted nutrition by gaining weight (W) faster and more efficiently (Elsley, 1963; Nielsen, 1964; Kirchgeβner and Roth, 1976) and by depositing more lean and less fat than their contemporaries which had been fed adequate diets during the whole period (Wyllie *et al.*, 1969; Zimmerman and Khajarern, 1973). This phenomenon of so-called compensatory growth may be influenced by several factors, such as severity of restriction or period of growth during which it is imposed. The physiological background for the compensatory response is still under discussion.

The aim of our study was to investigate the influence of a high and a low level of nutrition during the rearing period on daily live-weight gain, nitrogen metabolism, protein and fat deposition and efficiency of metabolizable energy (ME) available for growth during the subsequent growing-finishing period.

Materials and methods

Eight early-weaned, castrated male pigs of the German Landrace, divided into two groups, were included in this experiment. Throughout the rearing period from 5 to 25 kg W the piglets of group I were fed on a high-standard, nearly *ad lib.* diet, whereas the piglets in group II received approximately 55% of that amount. This procedure resulted in a restriction of energy as well as of protein intake. In the following growing-finishing period up to 100 kg W the animals of both groups were fed to the same level which was based on live weight.

During the rearing period from 5 to 25 kg W the piglets were fed a pelleted diet with a crude protein content based on dry matter decreasing from 25.6% at the beginning to 21.5% at the end of the period. In the growing-finishing period from 25 to 100 kg W the pigs were fed a barley/soybean-meal diet with a crude protein content on a dry matter basis of 19.8% in the growing period up to 55 kg W and 17.2% in the finishing period from 55 to 100 kg W.

Throughout the experiment the pigs were kept individually in metabolism cages and held at an environmental temperature decreasing from 29 °C to 26 °C in the rearing period and further decreasing to 22 °C in the fattening period. Each balance period consisted of a collection period of 6–9 days with the two 24-hour respiration measurements in an open-circuit unit in the middle of each period. The energy balance was calculated according to the carbon-nitrogen method described by Brouwer (1965).

Results and discussion

The results are given in *Table 82.1*

Table 82.1 RESULTS OF PERFORMANCE, NITROGEN AND ENERGY BALANCE

Live-weight range (kg)	5 – 25		25 – 55		55 – 100	
Group	I	II	I	II	I	II
Average live weight (kg)	13.6		39.4		77.2	
Daily feed intake (g)	658	355	1774	1788	2778	2794
Days in the period	41.6	92.8	40.9	35.9	51.7	46.6
Daily gain (g)	465	208	772	856	882	967
Feed : gain (kg:kg)	1.41	1.73	2.28	2.11	3.19	2.87
Daily nitrogen intake (g)	21.3	11.5	50.0	50.3	68.0	68.3
Nitrogen digestibility (%)	85.5	88.9	80.4	81.7	80.5	82.5
Daily nitrogen gain (g)	11.7	6.0	20.9	22.2	22.9	25.3
Nitrogen utilization (R_N/D_N) 100 (%)	64.1	59.4	52.0	53.7	41.6	44.7
Daily ME intake (kJ)	8914	4843	22321	22494	34712	35285
Daily energy gain (kJ)	4368	1366	11263	11263	17695	17772
Daily fat gain (g)	66.6	12.0	204.5	198.2	357.0	351.8
Efficiency of ME for growth (%)	76.9	86.3	71.1	70.0	69.7	68.7

During the rearing period the piglets of group I fed according to standards gained weight more rapidly ($P<0.01$) and had a better feed conversion rate than the restricted-fed pigs. In the fattening period from 25 to 100 kg, however, the pigs of group II previously restricted in feed intake gained faster and required less feed per unit gain. This is in agreement with the results of other workers, although the differences did not reach statistical significance. However, the tendency towards a better performance of the pigs previously fed restricted amounts of feed is very strong. The differences just failed to reach significance at the 5% level because of the small number of pigs and the relatively large variation within each group.

PROTEIN AND FAT DEPOSITION

As shown in *Table 82.1* restricted feed intake resulted in a higher nitrogen digestibility that was possibly due to the reduced endogenous losses. In the rearing period this difference was significant ($P<0.01$). In the fattening period there was also an indication of a better nitrogen digestibility. From 5 to 25 kg W the piglets of group II retained protein at a rate of 38 g/d. That was only 52% of the daily retention in group I which deposited 73 g/d on average. This lower protein deposition in the restricted-fed animals was compensated for partly during the fattening period from 25 to 100 kg W. Group II gained protein on an average of 149 g/d compared with 137 g in group I.

In the rearing period, nitrogen utilization was lower in the restricted-fed piglets. Considering the higher nitrogen digestibility this seems to be caused by the limited energy supply of this group. Respective calculations in the fattening period suggest a better nitrogen utilization in the previously restricted-fed pigs. A possible explanation is that the protein turnover rates in the fattening period might be influenced by the restricted protein and energy intake in the rearing period. Vaughan, Filer and Churella (1962) found that pigs fed a high protein level in the starter period subsequently had higher protein turnover rates.

Over the live-weight range of 5–25 kg the fat deposition of the restricted-fed piglets amounted to a daily average of only 12 g compared with 67 g in group I. In the period from 25 to 100 kg W the pigs previously restricted in feed intake gained fat at an average of 282 g/d compared with 288 g in group I, but the difference was not significant.

During the rearing period the gain of energy in fat in relation to the total-energy gain was fairly constant, about 60%, for the piglets in group I, while it increased from 0 to 45% in the restricted-fed piglets. In the fattening period this proportion increased from 68 to 82%. The difference between the groups was small, but the previously restricted-fed pigs tended to have lower proportions of fat energy.

Calculations of the whole fat and protein accretion show that the amounts of fat deposited during the whole experiment was affected by the treatment. In the rearing period the restricted-fed pigs deposited 50% less fat and nearly the same amount of protein as the pigs fed according to standards. At the end of the fattening period both groups had deposited similar amounts of protein but mainly as a result of the shorter fattening period the previously restricted-fed pigs had deposited approximately 5 kg or 16% ($P<0.05$) less fat than the pigs fed adequately during the whole experimental period. These findings are in agreement with those of Wyllie *et al.* (1969) and Zimmerman and Khajarern (1973).

ENERGETIC EFFICIENCY

In the rearing period the total-energy gain in relation to the intake of ME was only 28% for the restricted-fed piglets while it was 49% for the intensively-fed piglets. In the fattening period the pigs retained approximately 51% of the ME with no difference between both groups.

The efficiency of ME available for growth was calculated using a value of 460 kJ/kg $W^{0.75}$ per d for the maintenance requirement (ME_m) in the rearing period because the piglets were kept at comparatively high environmental temperatures. For the fattening period, the figures reported by Breirem (1939; ME_m = 821 kJ per $W^{0.56}$) have been used. As shown in *Table 82.1* the partial energy efficiency of ME for growth in the piglets from 5 to 25 kg W was obviously affected by the feeding level. The restricted-fed piglets utilized ME for growth more efficiently ($P<0.05$). This result is somewhat surprising as the piglets of group II had higher amounts of protein in their total gain. Furthermore, higher maintenance requirements could be expected for the restricted-fed piglets because these animals had been a little more restless and nervous.

In the fattening period the difference in the efficiency of utilization of ME available for growth was not significant, but the pigs that were previously fed a restricted diet tended to a somewhat lower level which was probably due to the different costs of protein and fat deposition.

The results of this study are in good agreement with those of other workers. There is strong evidence that pigs are able to compensate for slow growth during early stages of development through enhanced growth during later stages. The effect of compensation is due to a higher nitrogen retention during the later stages of growth.

References

BREIREM, K. (1939). *Tierernährung,* **11**, 487–528

BROUWER, E. (1965). In *Energy Metabolism of Farm Animals*, pp. 441–443. Ed. K.L. Blaxter. EAAP Publication No. 11. New York and London: Academic Press

ELSLEY, F.W.H. (1963). *J. agric. Sci., Camb.*, **61**, 243–251

KIRCHGEßNER, M. and ROTH, F.X. (1976). *Züchtungskunde,* **48**, 45–55

NIELSEN, H.E. (1964). *Anim. Prod.*, **6**, 301–308

VAUGHAN, O.W., FILER, L.J. and CHURELLA, H. (1962). *Pediatrics,* **29**, 90–96

WYLLIE, D., SPEER, V.C., EWAN, R.C. and HAYS, V.W. (1969). *J. Anim. Sci.*, **29**, 433–438

ZIMMERMAN, D.R. and KHAJARERN, S. (1973). *J. Anim. Sci.*, **36**, 189–194

83

EFFECTS OF DIETARY COMPOSITION ON ENERGY METABOLISM AND RATE OF UTILIZATION OF THYROXINE

D.L. INGRAM and M.J. DAUNCEY*
ARC Institute of Animal Physiology, Brabraham, Cambridge CB2 4AT, England

Summary

Pigs were fed various diets for one week. The rate of resting oxygen consumption between 12 and 20 h after the last meal was then measured. In other pigs the disappearance rate of labelled thyroxine (K) was determined. Both metabolic rate (MR) and K increased in pigs on a high intake of standard feed by comparison with a low intake. This was not related simply to the energy content of the food. A diet high in protein was associated with an increase in MR and K, while one of equivalent energy but high in glucose did not increase either MR or K. A high-energy diet with additional fat was associated with a high K value, but the rise in MR was not large. When the bulk of food was increased but not the metabolizable energy (ME) content, the value of K did not change and the increase in MR was not significant.

Introduction

The MR of animals in a thermally neutral environment depends on energy intake (Graham *et al.*, 1959; Close, Mount and Start, 1971; Dauncey, *see* p. 269). The higher the energy intake the greater the energy expenditure over a 24 h period. Furthermore, in the absence of any change in environmental temperature the rate of utilization of thyroxine (T_4) increases when energy intake is increased (Yousef and Johnson, 1968; Ingram and Kaciuba-Uscilko, 1977). Even on exposure to a low temperature the initial increment in the utilization rate of T_4 and release of T_4 from the thyroid gland is maintained only if energy intake is increased (Evans and Ingram, 1977).

In this chapter consideration has been given to the role played by the composition of the diet in influencing the MR at rest between 12 and 20 h after the last meal, and the rate of utilization of T_4. Comparisons have been made between young pigs on different energy intakes, on the same energy intake but with different nutrient composition and on the same energy intake but with different bulk.

*Formerly of the MRC Dunn Calorimetry Group, Babraham

Materials and methods

ANIMALS

Young pigs aged 6–12 weeks and weighing between 10 and 20 kg were used. Different animals were used for the determination of MR and the rate of utilization of T_4. The animals were housed separately in a thermally neutral environment at 25 °C.

ENERGY INTAKE AND NUTRIENT COMPOSITION

The quantity of food the animals received was calculated as a percentage of their body weight and presented at 0830 h each day. Water was available *ad lib.* Each diet was fed for one week before determinations of MR or utilization of T_4 were made. Each pig received the following diets in different weeks:

(1) 20 g commercial pig feed/kg body weight
(2) 60 g feed/kg body weight in studies on MR, or 40 g feed/kg body weight in studies on T_4.
(3) 20 g feed/kg body weight plus a supplement of fish meal, glucose, desiccated coconut or bran. The energy value of the fish meal, glucose or coconut was equivalent to 40 g feed/kg body weight in the studies on MR, and 20 g feed/kg body weight in the studies on T_4. Similarly, the quantity of bran was the equivalent in volume of either 40 or 20 g feed/kg body weight. Details of the diet are given in *Table 83.1* and elsewhere (Dauncey and Ingram, 1979).

Table 83.1 COMPOSITION OF PIG FEED AND SUPPLEMENTS (FROM DAUNCEY AND INGRAM, 1979)

	Gross energy (kJ/100 g)	*Composition* (g/100 g)			
		Protein	*Fat*	*Carbohydrate*	*Dietary fibre*
Pig feed	1580	18	2	62	5
Coconut	2850	6	62	6	24
Fish meal	1790	66	6	0	0
Glucose	1560	0	0	100	0

MEASUREMENT OF MR

The rate of oxygen consumption was determined at 25 °C using an open-circuit chamber and a paramagnetic oxygen analyser. Each pig was placed in the chamber between 1200 and 1600 h and the rate of oxygen consumption overnight between 2200 and 0800 h was measured.

RATE OF UTILIZATION OF T_4

Thyroxine labelled with ^{125}I was injected into the jugular vein and the disappearance of radioactivity from the blood was followed by sampling via an indwelling catheter beginning 12 h after injection. The fractional disappearance rate of ^{125}I (K) was calculated using a single exponential decay model as previously described (Evans and Ingram, 1977; Ingram and Kaciuba-Uscilko, 1977).

STATISTICAL ANALYSIS

The results for MR were analysed using the method described by Rowell and Walters (1976). The values of K were compared using paired *t* tests.

Results

EFFECT OF ENERGY INTAKE AND DIETARY COMPOSITION ON THE RATE OF OXYGEN CONSUMPTION BETWEEN 12 AND 20 h AFTER THE LAST MEAL

Pigs fed 60 g feed/kg body weight had a higher rate of resting metabolism over the whole period 12–20 h after the last meal than when they received only 20 g feed/kg ($P<0.001$). When the pigs were presented with the same energy value as 60 g feed/kg, in the form of 20 g feed/kg plus a supplement, MR depended on the supplement. Fish meal, which is high in protein, increased MR by comparison with 20 g feed/kg alone ($P<0.01$), but glucose did not. When additional energy was provided by coconut, which is high in fat, the MR was intermediate between that associated with 60 and 20 g feed/kg and was not statistically different from either. The addition of bran to the diet, which increased the bulk but not the ME value, was associated with a mean MR greater than the value for 20 g feed/kg alone, but the result was equivocal because of the wide variation and consequently large standard error (*Table 83.2*).

Table 83.2 RATES OF OXYGEN CONSUMPTION BETWEEN 12 AND 20 h AFTER THE LAST MEAL ON VARIOUS DIETS (MEANS ± SEM FOR 8 OR 9 PIGS; VALUES FOR EXPERIMENTS ON FISH MEAL, GLUCOSE AND COCONUT FROM DAUNCEY AND INGRAM, 1979)

Group of pigs and supplement	*Oxygen consumption* (ml/kg$^{0.67}$ per min)		
	20 g feed/kg + supplement	*20 g feed/kg*	*60 g feed/kg*
1. Fish meal	17.8 ±0.44	14.0 ±0.49	16.3 ±0.71
2. Glucose	15.2 ±0.80	16.5 ±0.74	18.6 ±0.63
3. Coconut	15.7 ±0.62	14.0 ±0.31	16.7 ±0.84
4. Bran	17.7 ±3.6	15.1 ±0.77	17.5 ±1.09

EFFECT OF ENERGY INTAKE AND NUTRIENT COMPOSITION ON UTILIZATION OF T_4

The fractional disappearance rate (K) was greater when the pigs were fed 40 g feed/kg than 20 g feed/kg ($P<0.001$). The addition of a supplement of fish meal or coconut to 20 g feed/kg increased the value of K ($P<0.02$), but the addition of glucose did not. An increase in the bulk of food did not increase the value of K above that for 20 g feed/kg alone ($P>0.05$). Values of K for each diet are presented in *Table 83.3.*

Table 83.3 RATES OF UTILIZATION OF THYROXINE (FRACTIONAL DISAPPEARANCE RATE K) FOR PIGS ON VARIOUS DIETS (MEANS ± SEM)

Group of pigs and supplement	*Value of K*		
	20 g feed/kg + supplement	*20 g feed/kg*	*40 g feed/kg*
5. Fish meal	0.0349 ±0.0012	0.0312 ±0.0014	0.0362 ±0.0013
6. Glucose	0.0271 ±0.0018	0.0282 ±0.0012	0.0361 ±0.0042
7. Coconut	0.0369 ±0.0072	0.0262 ±0.0004	0.0338 ±0.0072
8. Bran	0.0305 ±0.0014	0.0321 ±0.0009	0.0338 ±0.0009

Discussion

The present results show that the MR at rest between 12 and 20 h after the last meal and the rate of disappearance of T_4 are influenced not only by the energy content of the diet but also by the form in which the food is presented. There is also a close relation between the effects on MR and the value of K of particular constituents of the diet. The two diets in which this was not clear cut were those involving supplements of coconut and bran where MRs were intermediate between those for 20 and 60 g feed/kg and not statistically different from either. It therefore appears that certain nutrients are associated with an increase in MR which persists for a long period after a meal and that even in animals fed only once a day the MR can be permanently raised.

One explanation for an increased turnover of T_4 in animals eating large quantities of food has been advanced by Van Middlesworth (1960). It was suggested that the extra faecal bulk associated with eating more food resulted in an increased loss of T_4 in the faeces and that this in turn led to an increased rate of secretion from the thyroid gland. In the present study and that of Ingram and Kaciuba-Uscilko (1977), the addition of bran or chopped straw to the diet, which considerably increased the bulk of faeces, did not change the value of K. It therefore seems unlikely that the failure of glucose to influence the rate of disappearance of T_4 was related to the faecal bulk being the same as in the diet of 20 g feed/kg.

The effect of energy intake on MR and the value of K may simply be related to the time taken to digest the meal. In man the ingestion of glucose increases MR (Pittet *et al.*, 1976), but the effect is short lasting and may not persist for long enough to influence the utilization of T_4. The origin of the increased MR of animals on a high plane of nutrition is not completely understood. It has been shown both in growing pigs (Dauncey and Ingram, 1979) and in adult obese humans (Jung, Shetty and James, 1979) that the β-blocker propranolol reduces the increased metabolism. The possibility is, therefore, that catecholamines are involved in diet-induced thermogenesis and that the part played by thyroid hormones may involve a potentiating effect.

Acknowledgement

We thank Mr D.E. Walters and Mr D. Brown, ARC Statistics Unit, Cambridge, for their help and Mr K.F. Legge for his skilled technical assistance.

References

CLOSE, W.H., MOUNT, L.E. and START, L.B. (1971). *Anim. Prod.*, **13**, 285–294

DAUNCEY, M.J. and INGRAM, D.L. (1979). *Br. J. Nutr.*, **41**, 361–370

EVANS, S.E. and INGRAM, D.L. (1977). *J. Physiol., Lond.*, **264**, 511–521

GRAHAM, N.McC., WAINMAN, F.W., BLAXTER, K.L. and ARMSTRONG, D.G. (1959). *J. agric. Sci., Camb.*, **52**, 13–24

INGRAM, D.L. and KACIUBA-USCILKO, H. (1977). *J. Physiol., Lond.*, **270**, 431–438

JUNG, R.T., SHETTY, P.S. and JAMES, W.P.T. (1979). *Proc. Nutr., Soc.*, **57A**

PITTET, P., CHAPPUIS, P., ACHESON, D., de TECHTERMAN, F. and JÉQUIER, E. (1976). *Br. J. Nutr.*, **35**, 281–292

ROWELL, J.G. and WALTERS, D.E. (1976). *J. agric. Sci., Camb.*, 87, 423–432

VAN MIDDLESWORTH, L. (1960). *Recent Prog. Horm. Res.*, **46**, 405–438

YOUSEF, M.F. and JOHNSON, H.D. (1968). *Endocrinology*, **82**, 353–358

84

ROLE OF GLUCAGON IN THE REGULATION OF METABOLISM IN GROWING CHICKENS

J.S. KELLER
Institute of Genetics and Animal Breeding, Jastrzebiec, Poland

Summary

Three series of experiments were performed on 126 broiler chickens of both sexes, weighing between 100 and 2000 grams, at two levels of feeding. The chickens were injected subcutaneously with glucagon 50 or 150 μg/kg body weight. No calorigenic effect of glucagon was found in chickens up to two weeks old. Glucagon exerted an ever stronger influence on the metabolic rate in growing chickens with the peak effect (an increase of about 30%) occurring at seven weeks old when the body weight was approximately 1000 g. The calorigenic effect of glucagon was greater in birds maintained on a lower feeding level. The effect can be attributed mainly to a lipolytic effect, as evidenced by the rapid fall in respiratory quotient after glucagon injection and by the increase of glycogen content in the liver and pectoral muscle of chickens on a lower feeding level. The highest increase in the metabolic rate was observed about 30 min after the glucagon injection.

Introduction

The calorigenic effect of exogenous glucagon has been thoroughly described in mammals (Davidson, Salter and Best, 1960; Hoch, 1971), but in birds it is still poorly known (Freeman, 1971; Palokangas, Vihko and Nuuja, 1973). It was the aim of the present work to study the calorigenic action of glucagon and its influence on the glycogen content in the liver and muscle of growing chickens maintained on different feeding levels.

Materials and methods

Three series of experiments were performed on 126 broiler chickens of both sexes which weighed between 100 and 2000 g.

All chickens were kept in electrically heated metal cages and received a commercial mixture for broilers. The method of forced feeding was applied (Keller and Wagner, 1973) to ensure a precise programming of food intake and growth rate. The chickens were injected subcutaneously with 50 μg/kg of glucagon (Eli Lilly Co., Indianapolis) in experiments 1 and 2, or 150 μg/kg in experiment 3. The glucagon was dissolved in a 0.02 M glycine buffer at pH 9.5 which was also used in experiments 2 and 3 as a control solution. The metabolic rate was measured under standard conditions using an MG-4 Kipp and Zonen Diaferometer with a BD-5 Micrograph for recording. The glycogen content in the liver and pectoral muscles was determined by the method of Travalyn and Harrison (1952) as modified by Dowgiałło (1975). The changes in the metabolic rate or in the glycogen content were calculated as a percentage of the initial values, that is, of the measurements conducted before the chickens were injected with the tested hormone.

EXPERIMENT 1: CALORIGENIC EFFECT OF EXOGENOUS GLUCAGON IN GROWING CHICKENS AT DIFFERENT BODY WEIGHTS

In the first experiment a moderate feeding level was applied, that is a level at which the chickens reached a weight of 1 kg at the age of seven weeks. Starting from the second week of life the metabolic rate was determined in different chickens at two-week intervals. The starvation period was 18–20 hours.

The results obtained show that the increase in metabolic rate after a single injection of glucagon was very small in a 10-day-old chicken weighing about 100 g, and it amounted to about 4% of the initial value. In chickens weighing 400–800 g the metabolic rate increased to 13–19%, and it remained at a level of about 29–30% in those aged seven weeks and weighing about 1000 g. The highest increase was observed about 30 min after the glucagon injection.

EXPERIMENT 2: INFLUENCE OF THE GROWTH RATE OF CHICKENS ON THE CALORIGENIC ACTION OF GLUCAGON

In the second experiment two feeding levels ('A' higher and 'B' lower) were applied, and in consequence the mean body weight of the birds at the age of seven weeks was 1300 g in group A and 750 g in group B. The metabolic rate was determined in the chickens when they weighed at least 800 g. Three starvation periods were used: two, 12 and 24 h in each group respectively.

In chickens starved for 2 h the increase of metabolic rate after glucagon injection was only 16.6% in group A, but in group B it was as much as 26.9% (this difference was significant at $P<0.05$). In chickens starved for 12–24 h, the glucagon treatment caused an increase in metabolic rate by about 24–30% in both groups A and B. The increase of the metabolic rate in control birds receiving a glycine buffer ranged from –2.4% to +8%.

In chickens starved for 2 h the respiratory quotient (RQ) dropped after glucagon injection by 0.173 in group A and 0.170 in group B. Analogous changes in control birds were +0.033 and –0.047, respectively. In chickens starved for 12–24 h no significant changes in RQ were observed after the glucagon treatment.

EXPERIMENT 3: INFLUENCE OF GLUCAGON ON THE GLYCOGEN CONTENT IN LIVER AND PECTORAL MUSCLE OF CHICKENS MAINTAINED ON DIFFERENT FEEDING LEVELS

In the third experiment two feeding levels ('A' higher and 'B' lower) were applied as in experiment 2. In consequence the mean body weight of about 800 g was obtained in group A at the age of five weeks and in group B at the age of seven weeks. Two starvation periods were used of 1–1.5 h and 18–20 h in each experimental group respectively, creating a total of four subgroups. Half of the chickens from each subgroup were injected with glucagon and the other half with the control solution. The glycogen content in the liver and pectoral muscle, obtained *in vivo* by the biopsy technique, was determined 30, 60 and 90 min after the glucagon treatment.

In chickens starved for 1–1.5 h, the glycogen content in the liver was twice as high in group B when compared with group A. An analogous difference concerning the pectoral muscle reached about 20%. Glucagon treatment decreased the glycogen content in the liver of chickens from group A and increased it in group B. The glycogen content in the pectoral muscle decreased only in chickens from group A as a result of glucagon treatment.

In chickens from group B, starved for 18–20 h, the glycogen content in the liver was 63% higher when compared with group A. An analogous difference concerning the pectoral muscle reached about 20%. Glucagon treatment increased the glycogen content in the liver of chickens from group A but did not influence it in group B. The glycogen content in the pectoral muscle did not change in either group under the influence of glucagon treatment.

Conclusions

No calorigenic effect of glucagon was found in the youngest chickens up to two weeks old. However, it was observed that with the growth and development of chickens glucagon exerted an ever stronger influence on the metabolic rate, with its peak effect (an increase of about 30%) occurring in chickens seven weeks old and weighing approximately 1000 g. The calorigenic effect of glucagon depended also on the feeding level of the chickens, and appeared more clearly in birds maintained on a lower feeding level. The highest increase in the metabolic rate was observed about 30 min after the glucagon injection. It seems that the calorigenic effect of glucagon in chickens may be attributed mainly to its lipolytic effect, as evidenced by the rapid fall of the value after glucagon injection, as well as by the increase of the glycogen content in the liver and pectoral muscle of chickens maintained on a lower feeding level.

References

DAVIDSON, J.W.F., SALTER, J.M. and BEST, C.H. (1960). *Am. J. Clin. Nutr.*, 8, 540

DOWGIAŁŁO, A. (1975). *Chemical Composition of an Animal's Body and of its Food*, p. 160. IBP Handbook 44. Oxford: Blackwell

FREEMAN, B.M. (1971). *Physiology and Biochemistry of the Domestic Fowl*, p. 1115. Eds. D.J. Bell and B.M. Freeman. London: Academic Press

HOCH, F.L. (1971). *Energy Transformation in Mammals. Regulatory Mechanisms.* Philadelphia: Saunders

KELLER, J. and WAGNER, J. (1973). *Ann. Zootech.*, 22, 93

PALOKANGAS, R., VIHKO, V. and NUUJA, I. (1973). *Comp. Biochem. Physiol.*, **45A**, 489

TRAVALYN, W.Z. and HARRISON, J.S. (1952). *Biochem. J.*, **50**, 298

85

THE MEASUREMENT OF GLUCOSE KINETICS IN SHEEP – WHAT RELEVANCE DO SUCH MEASUREMENTS HAVE TO THE GLUCOSE REQUIREMENTS OF AN ANIMAL?*

J.C. MacRAE
Rowett Research Institute, Bucksburn, Aberdeen AB2 9SB, Scotland
A.R. EGAN
Waite Agricultural Research Institute, University of Adelaide, South Australia 5064

Introduction

The glucose economy of ruminants has elicited considerable interest, especially in situations where they require substantial amounts of the metabolite (e.g. pregnancy and lactation). Lindsay (1979a) and Elliot (1979) have recently reviewed many of the studies of the gluconeogenic potential of various metabolites. This chapter reports data on the glucose kinetics of pregnant ewes and attempts to consider how the ruminant utilizes the glucose it produces.

Materials and methods

Measurements of glucose and carbon dioxide production rates were made in six pregnant, twin- or triplet-bearing, Greyface ewes at approximately 100 d, 120 d and 140 d of gestation. Initially all six ewes received a basal diet of medium quality hay (1100 g/d). As their demand for nutrients increased during the last six weeks of gestation three were given a supplement of a proprietary concentrate pellet (Ewemax; 450 g/d at 120 d and 600 g/d at 140 d) to maintain plasma ketones at low levels (plasma non-esterified fatty acids, NEFA, <600 μmol/ℓ; D-3 hydroxybutyrate, D-3HB, <800 μmol/ℓ). The other three ewes were kept on the basal diet throughout; their plasma ketones rose to a subclinical level towards the end of gestation (plasma NEFA, 1000 μmol/ℓ; D-3HB 1500 μmol/ℓ). During each experimental period each ewe received 16-h continuous intravenous infusions of U-^{14}C-glucose and ^{14}C-bicarbonate. The 'production rate' of each metabolite was measured from the relationship between plasma specific activity at 12–16 h and infusion rate (*see* Annison *et al.*, 1967).

* The practical work reported in this chapter was carried out at the HFRO Bush Estate, Edinburgh, when J.C.M. was on the staff and A.R.E. was on study leave aided in part by an ARC Underwood Award.

Results and discussion

THE PREGNANCY EXPERIMENT

Glucose production rates obtained at each stage of pregnancy are given in ***Figure 85.1.*** Those obtained in the supplemented ewes increased from 171.5 ± 4.77 g/d of glucose at 100 d gestation to 259 ± 8.91 g/d at 140 d. Corresponding glucose values in the non-supplemented animals were only from 154.5 ± 2.28 to 182.7 ± 2.66 g/d.

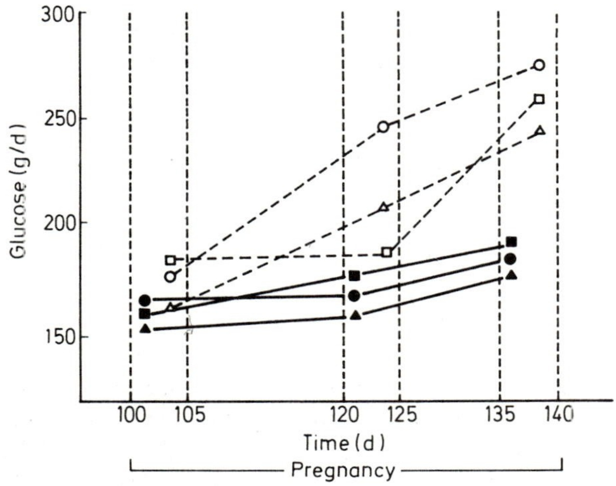

Figure 85.1 Glucose production rates of Greyface ewes given high (○,□,△) *or low* (●,■,▲) *levels of feed during pregnancy. Lamb birth weights (Nos.):* ○ = *11.8 (2);* □ = *11.9 (3);* △ = *10.6 (3);* ■ = *3.2 + 1 dead;* ● = *10.4 (2);* ▲ = *9.3 (2).*

It has always been assumed that the glucose requirement of ewes increases substantially during the latter stages of pregnancy (Kronfeld, 1958; Armstrong, 1965; Reid, 1968). However ***Figure 85.1*** seems to support a postulate made by Lindsay (1970) that whole-animal glucose entry rates increase little during pregnancy other than as a direct result of an increased dietary intake; a glucose increase of 30 g/d appeared to be sufficient to meet the increased demand for glucose to supply fetal needs.

THE RUMINANTS NEEDS FOR GLUCOSE *PER SE*

In view of this finding it seems appropriate to consider carefully what is meant by the 'glucose requirement of a ruminant'. Available information on glucose requirement and gluconeogenesis in ruminants (e.g. see Ballard, Hanson and Kronfeld, 1969 and Ramberg, 1977) implies that the animal has a considerable obligatory need for glucose. However, in a recent review Lindsay (1979b) suggested that any substantial need for glucose *per se* is arguably confined to the pregnant and lactating animal, and that other functions for which glucose appears to be the main metabolite in non-ruminants (e.g. provision of glycerol-P and the reducing equivalent, $NADPH_2$, for fatty acid synthesis and in muscle metabolism) are provided for mainly by alternative metabolites such as acetate, lactate or amino acids in the ruminant.

This sparing of glucose in ruminants is one important consideration which is sometimes overlooked when the glucose economy of the ruminant is assessed. For example, it can be seen from *Table 85.1* that whenever the transfer of ^{14}C-labelled

Table 85.1 THE RELATIONSHIP BETWEEN GLUCONEOGENESIS FROM AMINO ACIDS AND AMINO ACID OXIDATION

Amino acid	*Physiological state of animal*	*% of amino acid flux to glucose*	*% of amino acid flux to carbon dioxide*
Threonine*	Pregnant sheep	4	12
Threonine[1,2]	Maintenance fed sheep	3	10–12
Methionine[2]	Maintenance fed sheep	<1	5–8
Isoleucine[2]	Maintenance fed sheep	<1	14–18
Glutamate[3]	Lactating goats	22	44
Valine[3]	Lactating goats	9	15
Arginine[3]	Lactating goats	10	17
Glutamate[4]	Maintenance fed sheep	11	58
Serine[4]	Maintenance fed sheep	11	24

* Present experiment
[1] Egan and MacRae (1978)
[2] Egan and MacRae (1979)
[3] Egan, Moller and Black (1970)
[4] Heitman, Hoover and Sniffen (1973)

carbon from a potentially gluconeogenic precursor to glucose and carbon dioxide is measured simultaneously, the amount appearing in carbon dioxide is always greater than that appearing in glucose; admittedly during any conversion of precursor to glucose carbon dioxide will be produced in the intermediary steps, but the ratio of carbon dioxide-carbon thus produced to the glucose-carbon yielded will be in most cases less than unity.

The two situations where there is a significant requirement for glucose *per se* in ruminants are in pregnancy and lactation. Setchell *et al.* (1972) and Christenson and Prior (1978) estimated from arteriovenous-difference studies that the uptake of glucose by the pregnant uterus of sheep can be from 40 to 120 g/d. Similarly, Annison, Bickerstaffe and Linzell (1974) and Bickerstaffe, Annison and Linzell (1974) estimated that 800–2200 g/d of glucose were taken up by the udder of lactating cattle.

INTERPRETATION OF THE TRACER KINETIC DATA

If we accept Lindsay's (1979b) conclusion that the ruminant's demand for glucose *per se* is not large, apart from that in the pregnant and lactating female, then the relevance that the quite substantial production rates, measured in tracer experiments on non-pregnant non-lactating animals (50–100 g/d in mature wether sheep; White *et al.*, 1969; Egan and MacRae, 1978; Wilson, MacRae and Buttery, 1979; 700–800 g/d in fasted non-lactating cattle; Kronfeld, Ramberg and Shames, 1971), may have to the glucose or indeed the carbon economy of the animal needs to be considered.

These irreversible losses of glucose are usually equated with the animal's net production and/or utilization rates (e.g. *Figure 85.1*). However, only approximately one-third of this glucose carbon can be accounted for in the oxidative production of carbon dioxide (Bergman, 1963; Annison *et al.*, 1967; Ford and Reilly, 1970), and this value falls to less than 20% in lactating animals, where another 50% of the glucose is exported out of the animal as milk lactose (Brown and Lindsay, 1966; Bickerstaffe *et al.*, 1974).

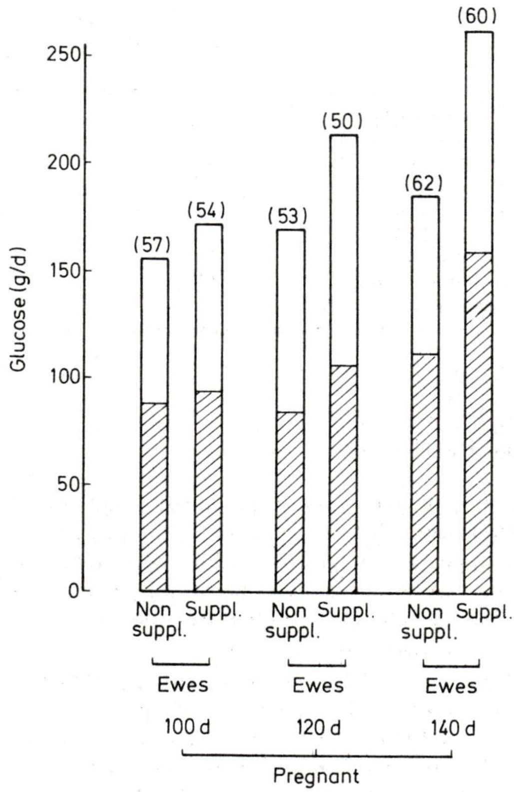

Figure 85.2 Amounts of glucose produced and oxidized in carbon dioxide ▨ in the twin- and triplet-bearing pregnant ewes. Values in parenthesis above each histogram are the percentage of glucose which appears in carbon dioxide.

Data from the pregnant ewes substantiate these findings. *Figure 85.2* gives an indication of the proportions of the 'production rates' which were oxidized to carbon dioxide. In animals where glucose is 'exported' either as lactose (these ewes in their subsequent lactation; Egan and MacRae, unpublished) or in urine, after phloridzin treatment of wethers (Egan and MacRae, 1978), the values are lower (42–48%). Similarly, Setchell *et al.*, (1972) reported that only 40% of the glucose taken up by the gravid uterus was oxidized.

It would appear, therefore, that 40–70% of the glucose produced is not completely oxidized. If so, what happens to this glucose? In the present experiment the sheep were fed continuously and so presumably would need little short-term storage of glucose as glycogen. On the other hand metabolites of glucose are intermediates for anabolic processes such as protein and fatty-acid synthesis and thus glucose carbon could be diverted to these products. If the ^{14}C-glucose donates a proportion of its carbon to non-essential amino acids some of these molecules will become incorporated into protein and over the time scale of the 16-h infusion very few will be returned to the glucose pool. This loss of ^{14}C into protein will appear as part of the irreversible loss of glucose carbon even though the ^{14}C may reappear in glucose much later when the protein is subsequently degraded and the amino acids are used as gluconeogenic precursors.

This incorporation of ^{14}C into protein has been demonstrated previously in dairy cows (Black, Kleiber and Baxter, 1955) and rats (Shipley, Gibbons and Chudzik, 1974). In a subsequent experiment performed on the present ewes, during their lactation, 5–9% of infused ^{14}C was recovered from the first 24-h sample of milk casein and albumin (virtually none was present in milk fat).

Thus it should be considered whether the ruminant animal maintains its glucose and its non-essential amino acid economies in an integrated interconversion system where the net input of each particular metabolite to the system, from dietary precursors (e.g. propionate in the case of glucose), accounts for only a small fraction of the total flux of each metabolite. Indeed, there is already some evidence to suggest that this appears to be the case for the amino-acid economy of both ruminants and non-ruminants (MacRae and Reeds, 1980).

If this is so, then the plateau-specific activities measured in infusion experiments will presumably measure 'net flux' through the glucose pool, and the difference between this flux and the rate of oxidation (i.e. appearance of ^{14}C-carbon dioxide) might give an indication of the net amount of glucose carbon diverted to anabolic processes. It seems unlikely, however, that such measurements will provide information on the animal's 'net production' or 'net utilization' of the metabolite and hence furnish any indication of the 'glucose requirement' of the animal.

It is perhaps more important, therefore, in future studies to attempt to investigate the dynamics of this interconversion system rather than collect yet more data on irreversible-loss rates in animals given different diets and subjected to different physiological or environmental stresses.

References

ANNISON, E.F., BICKERSTAFFE, R. and LINZELL, J.L. (1974). *J. agric. Sci., Camb.*, **82**, 87

ANNISON, E.F., BROWN, R.E., LENG, R.A., LINDSAY, D.B. and WEST, C.E. (1967). *Biochem. J.*, **104**, 135–147

ARMSTRONG, D.G. (1965). In *Physiology of Digestion in Ruminants*, p. 272. Eds. R.W. Dougherty, R.S. Allen, W. Burroughs, N.L. Jacobson and A.D. McGilliard. London: Butterworths

BALLARD, F.J., HANSON, R.W. and KRONFELD, D.S. (1969). *Fedn Proc. Fedn Am. Socs exp. Biol.*, **28**, 218

BERGMAN, E.N. (1963). *Am. J. Physiol.*, **204**, 147

BICKERSTAFFE, R., ANNISON, E.F. and LINZELL, J.L. (1974). *J. agric. Sci., Camb.*, **82**, 71

BLACK, A.L., KLEIBER, M. and BAXTER, C.F. (1955). *Biochem. biophys. Acta.*, **17**, 346

BROWN, R.E. and LINDSAY, D.B. (1966). *Fedn Proc. Fedn Am. Socs exp. Biol.*, **25**, 343

CHRISTENSON, R.K. and PRIOR, R.L. (1978). *J. Anim. Sci.*, **46**, 189

EGAN, A.R. and MacRAE, J.C. (1978). *Proc. Nutr. Soc.*, **37**, 15A

EGAN, A.R. and MacRAE, J.C. (1979). *Annls Rech. Vet.*, No. 2. **10**, 376

EGAN, A.R., MOLLER, F. and BLACK, A.L. (1970). *J. Nutr.*, **100**, 419

ELLIOT, J.M. (1979). *Proceedings of the Fifth International Symposium on Ruminant Physiology.* Eds. Y. Ruckebusch and P. Thivend. In press

FORD, E.J.H. and REILLY, P.E.B. (1970). *Res. Vet. Sci.*, **11**, 575

HEITMAN, R.N., HOOVER, W.H. and SNIFFEN, C.J. (1973). *J. Nutr.*, **103**, 1587

KRONFELD, D.A. (1958). *Cornell Vet.*, **48**, 394

KRONFELD, D.S., RAMBERG, C.F. and SHAMES, D.M. (1971). *Am. J. Physiol.*, **220**, 886

LINDSAY, D.B. (1970). In *Digestion and Physiology of Ruminants*, p. 438. Ed. A.T. Phillipson. Newcastle upon Tyne: Oriel

LINDSAY, D.B. (1979a). In *Protein Metabolism in the Ruminant.* Ed. P.J. Buttery. London: Agricultural Research Council

LINDSAY, D.B. (1979b). *Proc. Nutr. Soc.*, **38**, 295

MacRAE, J.C. and REEDS, P.J. (1980). In *Protein Deposition in Animals.* Eds. P.J. Buttery and D.B. Lindsay. London: Butterworths

RAMBERG, C.F. Jr. (1977). *Fedn Proc. Fedn Am. Socs exp. Biol.*, **36**, 225

REID, R.L. (1968). *Adv. Vet. Sci.*, **12**, 163

SETCHELL, B.P., BASSETT, J.M., HINKS, N.T. and GRAHAM, N.McC. (1972). *Q. J. exp. Physiol.*, **57**, 257

SHIPLEY, R.A., GIBBONS, A.P. and CHUDZIK, E.B. (1974). *Can. J. Physiol. Pharmacol.*, **52**, 797

WHITE, R.G., STEEL, J.W., LENG, R.A. and LUICK, J.R. (1969). *Biochem. J.*, **114**, 203

WILSON, S., MacRAE, J.C. and BUTTERY, P.J. (1979). *Res. Vet. Sci.*, **26**, 256

86

ENERGY METABOLISM OF SHEEP DURING NUTRITIONAL LIMITATION AND REALIMENTATION

E.F. THOMSON*, M. GINGINS, J.W. BLUM, H. BICKEL and A. SCHÜRCH
Institute of Animal Production, Swiss Federal Institute of Technology, Zürich, Switzerland

Summary

In two experiments the energy metabolism of adult sheep during undernutrition and realimentation and of lambs during compensatory growth was measured. Fasting heat production of sheep decreased by 11% following undernutrition, but quickly returned to initial levels during realimentation. Dietary energy was used with a high efficiency for tissue deposition during realimentation of sheep. Compensating lambs deposited 17% more protein and 11% less fat than controls. Serum levels of thyroxine and triiodothyronine were positively correlated and reverse triiodothyronine negatively correlated with energy balance.

Introduction

Animal species have been exposed during evolution to periodic fluctuations in their nutrient supply. This has necessitated the development of flexible metabolic mechanisms which enables them to reduce their energy demands during periods of scarcity and to replenish rapidly their energy stores during periods of nutrient abundance. Thus cattle can reduce their energy demands when feed supply is limited (Ledger and Sayers, 1977) and express compensatory or catch-up growth when the feed-energy supply is restored (Wilson and Osbourn, 1960; Allden, 1970). Since doubt still remains concerning the metabolic adaptations involved in ruminants during nutritional limitation and realimentation, metabolism experiments were conducted using sheep and lambs subjected to markedly differing levels of feeding. Details of the experiments reported here are presented elsewhere (Gingins, 1978; Thomson, 1979).

Materials and methods

Two experiments were conducted in which similar high-cereal pelleted diets were fed, which had a dry matter (DM) content of 18 MJ gross energy per kg, 2.4–2.7% nitrogen and 6–8% crude fibre. The rations contained 12–13 MJ metabolizable energy (ME)/kg DM. In experiments 1 and 2, 77 and 46 successful determinations of gas exchange were made using open-circuit respiration chambers.

* Present address: International Centre for Agricultural Research in the Dry Areas, P.O. Box 5466, Aleppo, Syria

EXPERIMENT 1

Two groups, each of four two year-old wethers that initially weighed 47 kg, were subjected to the following levels of feeding: 28 days maintenance (600 g diet/d), 126 days undernutrition (alternately 150 or 300 g diet/d) and 51 days realimentation (either 800 or 1200 g diet/d). Following each regimen the animals were fasted for five days. The hormones thyroxine (T_4), 3,5,3′-triiodothyronine (T_3) and 3,3′,5′-triiodothyronine (rT_3) were measured by radioimmunoassay in the serum of blood samples taken by jugular puncture preprandially between 0800 and 0900 h on the last day of each balance period.

EXPERIMENT 2

Eight wether lambs, which initially weighed 22.5 kg, were divided into two groups. The lambs of the control group C were fed near to *ad lib.* (about 80 g DM / metabolic body size; MBS) until they reached 44 kg after 120 days of the experiment. The lambs of the second group RR were subjected to a 144-day-restriction period (70 g daily gain) and were thereafter realimented for 56 days at the same level of feeding as C lambs over the live-weight interval 32–44 kg.

Results and discussion

The live weight of the sheep in experiment 1 decreased from an initial fasted value of 44 kg to 33 kg after undernutrition and recovered from 33 kg to a final fasted weight of either 40 kg or 47 kg when the sheep were fed a diet of 800 or 1200 g/d respectively. In experiment 2 the daily gains of compensating RR lambs (219 g) were significantly higher ($P<0.05$) than those of C lambs (167 g).

The results in *Table 86.1* show that the fasting heat production decreased as a consequence of undernutrition and increased again to the initial levels following realimentation. The estimated maintenance requirement, not shown in *Table 86.1*, had increased again to the initial levels after only 14 days realimentation. Nitrogen balance during fasting showed a trend similar to that of fasting heat production.

Table 86.1 MEASURED FASTING HEAT PRODUCTION, NITROGEN BALANCE AND ESTIMATED MAINTENANCE REQUIREMENT IN EXPERIMENT 1

Phase of experiment	*Before undernutrition*	*After undernutrition*	*After realimentation*
Fasting heat production (kJ/MBS)	279A[1]	249B	294A
Nitrogen balance[2] (g N/MBS)	−0.39a	−0.27b	−0.47a
		During undernutrition	*During realimentation*
Maintenance requirement (kJ/MBS)	–	275	374

[1] Means followed by similar letters of the same case do not differ significantly ($P>0.05$)
[2] Measured during fasting
[3] Estimated from linear regression; RE = a + bMEI, where RE = retained energy (kJ/MBS) and MEI = intake of metabolizable energy (kJ/MBS)

Energy deprivation was associated with decreased T_3 and T_4 and increased rT_3 concentrations, and with realimentation increased T_3 and T_4 and decreased rT_3 levels occurred. Thus, T_3 and rT_3 were significantly correlated with energy balance (T_3, $r = 0.73$; rT_3, $r = 0.50$).

The daily intake of ME, the deposition of protein and fat, the ratio of protein to fat deposition and the partial efficiency of ME utilization of sheep and lambs during realimentation are shown in *Table 86.2.*

Table 86.2 DAILY INTAKE OF ME, DEPOSITION OF PROTEIN AND FAT, RATIO OF PROTEIN : FAT DEPOSITION AND PARTIAL EFFICIENCY (k_g) of ME UTILIZATION FOR ENERGY DEPOSITION OF SHEEP AND LAMBS DURING REALIMENTATION

Group *Level of feeding/group*	*Sheep* *800 g*	*Sheep* *1200 g*	*Lambs* *C*	*Lambs* *RR*
Intake of ME (kJ/MBS)	594A[1]	825B	992a	956a
Protein deposition (kJ/MBS)	62A	104B	86a	101a
Fat deposition (kJ/MBS)	107A	226B	275a	244a
Protein:fat ratio	1:1.7	1:2.2	1:3.2	1:2.4
Partial efficiency[2] k_g	– 0.73 –		0.56	0.59

[1] Means followed by similar letters of the same case do not differ significantly ($P > 0.05$)
[2] Estimated from linear regression

An increase in the level of realimentation in experiment 1 was associated with a greater increase of fat then protein deposition. In experiment 2 the RR lambs deposited 17% more protein and 11% less fat than C lambs from slightly less ME intake. It is of interest that even adult sheep were depositing a similar amount of protein as lambs during realimentation. The protein:fat ratio in the energy deposited by C lambs was wider than that found in the compensating RR animals, the ratio of the latter group being similar to that of the sheep at the high level of realimentation.

The estimates of k_m by regression (0.75, not shown in *Table 86.2*) during undernutrition and k_g (0.73) during realimentation were similar, a finding also reported by Searle and Graham (1975). These results show that, during the realimentation of sheep, dietary ME is utilized for protein and fat deposition at a higher efficiency than predicted from metabolizability by the Agricultural Research Council (1965) conventions.

Since the energy cost of protein deposition is higher than that of fat deposition, the higher rate of protein deposition of the compensating animals should have reduced k_g. This was not the case. On the contrary, k_g in experiment 1 was remarkably high, and in experiment 2 RR lambs were marginally more efficient than C lambs. This observation suggests that the efficiency of protein deposition (k_p) may have increased. Multiple regression analysis of the intake of ME on

Table 86.3 ESTIMATES OF EFFICIENCY OF PROTEIN (k_p) AND FAT (k_f) DEPOSITION OF SHEEP AND LAMBS

	k_p	k_f
Experiment 1, sheep	0.68	0.80
Experiment 2, C lambs	0.33	0.76
Experiment 2, RR lambs	0.49	0.67

protein and fat deposition was made in order to test this hypothesis. The results, expressed as partial efficiencies, are presented in *Table 86.3.*

The limitations of this form of regression analysis have been discussed by Koong, Garrett and Canolty (1976). However, it is still considered possible to interpret the results in *Table 86.3* as an indication of an improvement in the efficiency of protein deposition. The value of k_p in experiment 1 is well above the values reported in the literature. The values of k_p in experiment 2, even though they may have been influenced by autocorrelation between the dependent variables, correspond with the values published by Ørskov and MacDonald (1976).

It is concluded that during undernutrition, sheep reduce their energy requirements not only by inevitably losing body weight but also by decreasing their maintenance requirement per unit metabolic body size. Efficiency of dietary energy utilization for tissue deposition may increase during realimentation, but the extent of the increase will depend on the degree of the previous nutritional limitation. Shifts away from fat deposition towards protein deposition during realimentation will reduce the energy value of the gains of compensating animals.

References

AGRICULTURAL RESEARCH COUNCIL (1965). *The Nutrient Requirements of Farm Livestock, No. 2: Ruminants.* Technical Reviews and Summaries. London: HMSO

ALLDEN, W.G. (1970). *Nutr. Abstr. Rev.*, **40**, 1167–1184

GINGINS, M. (1978). *Thesis No. 6228, ETH Zürich*

KOONG, L.J., GARRETT, W.N. and CANOLTY, N.L. (1976). In *Energy Metabolism of Farm Animals*, pp. 149–152. Ed. M. Vermorel. EAAP Publication No. 19. Clermont-Ferrand: G. de Bussac

LEDGER, H.P. and SAYERS, A.R. (1977). *J. agric. Sci., Camb.*, **88**, 11–26

ØRSKOV, E.R. and MacDONALD, I. (1976). In *Energy Metabolism of Farm Animals*, pp. 133–136. Ed. M. Vermorel. EAAP Publication No. 19. Clermont-Ferrand: G. de Bussac

SEARLE, T.W. and GRAHAM, N.McC. (1975). *Aust. J. Agric. Res.*, **26**, 355–361

THOMSON, E.F. (1979). *Thesis No. 6382, ETH Zürich*

WILSON, P.N. and OSBOURN, D.F. (1960). *Biol. Rev.*, **35**, 324–363

87

CHANGES IN FASTING METABOLISM OF CATTLE AS A CONSEQUENCE OF SELECTION FOR GROWTH RATE

J.E. FRISCH and J.E. VERCOE*
CSIRO, Division of Animal Production, Tropical Cattle Research Centre, Rockhampton, Queensland, Australia

Summary

Fasting metabolism, and live weight maintained on a constant feed intake, were measured in bulls from a control line and from a line selected for growth rate in the presence of environmental stresses. Bulls from the selected line had lower maintenance requirements and fasting metabolism per kilogram live weight than the control line bulls. Possible explanations for these reductions are advanced.

Introduction

It has previously been reported (Frisch and Vercoe, 1977) that the *Bos indicus* breeds so far examined have lower metabolic rates than comparable European *Bos taurus* breeds. The *Bos indicus* breeds have evolved in the presence of high levels of environmental stress due to parasites, diseases, heat and generally low planes of nutrition, and this may be the reason for their lower metabolic rates. It is therefore possible that selection for growth rate in unadapted breeds in the presence of environmental stresses may also develop animals with relatively low metabolic rates. While this may improve their level of adaptation to tropical conditions, it can also be expected to have the undesirable consequence of lowering inherent growth rate that is, growth rate in the absence of environmental stresses (Frisch and Vercoe, 1977, 1978).

Changes that occur as a result of selection for growth rate are being investigated at the Tropical Cattle Research Centre. This study reports on the changes that have occurred in metabolic rate.

Materials and methods

Twelve bulls were selected at about 200 days of age for high weight for age

* Present Address: Joint FAO/IAEA, Division of Atomic Energy, P.O. Box 590, Vienna, A-1011, Austria

(corrected for dam age) from a closed line of crossbred Hereford × Shorthorn (HS). This line (selected line) has been under selection principally for increased growth rate under environmentally stressful conditions. Selection procedures culled approximately 50% of males for low corrected 200-day weight and a further 15% of those remaining were culled for low tick resistance. Final selection of bulls used for mating was for high weight per day of age at about two years. A further 10 HS bulls were selected at random from a random breeding control line (control line) from which the base population of the selected line was originally derived.

The 12 bulls were selected from the top of the selected line rather than from near the mean to accentuate any likely differences between the lines associated with the weight difference and to make the detection of those differences easier. The mean difference between the lines in the 200-day weight of the total male population was 10 kg, and this was 10 kg less than the difference between the two groups of animals used in this study. Thus, 50% of the weight difference between the lines at the start of the experiment was due to the long-term effects of selection, and the remaining 50% was the result of immediate selection. Some of the 12 bulls were to be the sires of the next generation. Consequently, although not identical in weight to the population mean, the bulls represented the direction in which that population mean was being moved by selection for growth rate under environmentally stressful conditions.

Both lines have been grazed together since selection commenced in 1966 with minimal control of ectoparasites and endoparasites, no alleviation of heat loads and no treatment for diseases other than those readily controlled by vaccination. They have normally lost weight for several months of each year, but even during these periods supplementary feeding was not practised.

After a preliminary 100-day period of *ad lib.* feeding, the bulls were fasted for 48 hours, weighed and then fed 5 kg air-dry lucerne chaff per day individually in shaded outdoor pens. Ectoparasites and endoparasites were rigidly controlled during the entire period of feeding. All animals were weighed at the same time each week and gains calculated from these fed weights.

The mean period of restricted feeding was 144 days. For the first 60 days after fast, lucerne quality was low and both groups lost weight. Lucerne quality was then changed and for the next 28 days both lines gained weight. Gains decreased over the last 49 days and over the last 14 days they were 50 and 40 g/d for the selected and control lines respectively. Live weights had then equilibrated sufficiently to be considered close to maintenance. Fasted live weights and metabolic rates were then measured after a 96-hour fast using the methods of Vercoe (1970). Total gains were the difference between initial and final fasted live weights. Fasting metabolism was calculated per kilogram fasted live weight for reasons expressed previously (Frisch and Vercoe, 1977). Since only fasted live weights were recorded at the start and finish of the experiment, the number of days over which fed gains were calculated was 14 fewer than for fasted gains.

The data were analysed by the method of least squares (Harvey, 1960) with the variance partitioned into that due to lines, lines × groups and error. The interaction term failed to reach significance ($P>0.10$). The variance due to groups arose since animals were measured in four groups over a four-week period at the start and end of the experiment. Only line means are presented.

Results and discussion

The least-squares mean values for fasted live weights at the start and end of restricted feeding, gains over the entire 144 days, gains for the first 81 days and last 49 days over which fed gains were recorded and fasting-heat production for each line are shown in *Table 87.1.*

Table 87.1 LEAST-SQUARES MEANS FOR FASTED LIVE WEIGHT, GAINS AND FASTING-HEAT PRODUCTION

	Fasted Live weight (kg)		*Gains*			*Fasting heat production*	
Line	*Start*	*Finish*	*144 days*	*First* 81 days*	*Last* 49 days*	*MJ/d*	*kJ/kg per d*
Selected	201[a]	217[a]	16[b]	12.8[b]	6.4	18.1	83.3[b]
Control	175	198	23	19.2	6.5	17.7	89.1

[a] Selected > control ($P<0.05$)
[b] Selected < control ($P<0.05$)
* Calculated from fed weights

The bulls from the selected line were significantly heavier ($P<0.05$) at the start of the experiment but gained less over the entire 144 days ($P<0.05$). However, the difference between lines in total gains arose through significantly higher ($P<0.05$) gains in the control line during the first 81 days. Over the last 49 days, as both lines approached their equilibrium weights, gains were similar. Consequently, the differences in fasting metabolism cannot be ascribed to differences in previous rates of gain. This similarity of gains and the significantly higher live weight maintained by the selected line at the end of the period ($P<0.05$) indicate a lower maintenance requirement in this line.

Differences between lines in daily fasting metabolism were small and not significant ($P>0.20$). Consequently the selected line had significantly lower ($P<0.05$) fasting heat production per unit of live weight. This parameter is in general associated with low inherent growth rate (Frisch and Vercoe, 1977) which has previously been reported for this line (Frisch and Vercoe, 1978).

A number of reasons can be advanced to explain why maintenance requirements or fasting metabolic rate should decline when selection has been directed towards, and has achieved (Frisch and Vercoe, 1978), increases in growth rate in the presence of moderate to high levels of environmental stress. First, under conditions of low feed availability alone, growth rate is highest in animals or breeds with low maintenance requirements (Frisch and Vercoe, 1977). Secondly, in hot climates growth rate is related to the level of heat tolerance (Turner, 1962): breeds of high heat tolerance have low heat production (Frisch and Vercoe, 1978). Thirdly, where animals are near maintenance or losing weight their resistance to many parasites and diseases is reduced (Turner and Short, 1972). On low planes of nutrition the animals with the highest maintenance requirements could be expected to be the first to lose weight and parasite and disease resistance. Consequently, growth would be depressed even further by the increased parasite or disease load. Selection for growth rate in the presence of any or all of these environmental stresses would, therefore, almost invariably favour animals with low maintenance requirements.

The lowering of maintenance requirements per animal could be achieved under these stressful conditions by a reduction in mature size of the breed. However, where selection is for increased growth rate (i.e. on the basis of weight/age or 'size') it seems more probable that gross maintenance requirements would be reduced either by reducing maintenance requirements per unit of live weight as reported here or by increasing the efficiency of utilization of ME for maintenance (Vercoe, 1970). It is also probable that a combination of these phenomena may occur. Conversely, selection for growth rate at low levels of environmental stress could be expected to increase mature size (Cartwright and Fitzhugh, 1974).

In many situations the effect of any or all of these correlated changes may be undesirable. An understanding of the reasons for the changes and of their consequences should be useful in the development of selection programmes aimed at achieving the desired increase in growth rate without any undesirable side effects.

References

CARTWRIGHT, T.C. and FITZHUGH, H.A. (1974). *First World Congress on Genetics applied to Livestock Production,* **1**, 643–653

FRISCH, J.E. and VERCOE, J.E. (1977). *Anim. Prod.,* **25**, 343–358

FRISCH, J.E. and VERCOE, J.E. (1978). *Fourth World Congress on Animal Production.* In press

HARVEY, W.R. (1960). *United States Department of Agriculture Agricultural Research Service Publication* No. 20, p. 8

TURNER, H.G. (1962). *Aust. J. Agric. Res.,* **13**, 180–192

TURNER, H.G. and SHORT, A.J. (1972). *Aust. J. Agric. Res.,* **23**, 177–193

VERCOE, J.E. (1970). In *Energy Metabolism of Farm Animals,* pp. 85–88. Eds. A. Schürch and C. Wenk. EAAP Publication No. 13. Zurich: Juris Druck

88

COLD EXPOSURE AND FAT METABOLISM IN YOUNG CATTLE

A.W. BELL
School of Agriculture, Le Trobe University, Bundoora, Victoria 3083, Australia

Summary

Acute cold exposure of young steers caused resting metabolic heat production to double, and calculations based on changes in respiratory quotient suggested that lipid oxidation contributed 61% and 73% of the extra heat in fed and fasted steers, respectively. Uptake of plasma free fatty acids (FFA) by the shivering hind limb, estimated from measurements of arteriovenous differences in ^{14}C-FFA and leg blood flow, increased 5–8 fold, as did oxygen uptake, but simultaneous release of ^{14}C-carbon dioxide accounted for only 17–32% of ^{14}C-FFA uptake during cold exposure. It is concluded that oxidation of fatty acids mobilized from body stores, although important, is not the sole source of the metabolic response to cold, and that plasma FFA are not necessarily a preferred fuel for shivering muscle.

Introduction

The notion that catabolism of body fat is the exclusive or preferred source of energy for the metabolic response to cold in laboratory animals arose from the calorimetric studies of Kayser (1937) and, later, Pagé and Chénier (1953). It was extended to the domestic ruminants by the classic work at the Hannah Institute on the effects of ambient temperature on energy exchanges in sheep (Graham *et al.*, 1959) and cattle (Blaxter and Wainman, 1961). With the advent of more direct methods for measuring oxidative metabolism of energy substrates in the whole animal, oxidation of non-lipid substrates, such as glucose, was shown to contribute significantly to cold thermogenesis in non-ruminants (Depocas and Masironi, 1960; Minaire *et al.*, 1973). However, despite similar observations in sheep (McKay, Young and Milligan, 1974), the idea that plasma free fatty acids (FFA) mobilized from adipose tissue are the primary immediate fuel for the thermogenic tissues during cold exposure, even in well-fed ruminants, has persisted.

In this paper the importance of fat catabolism during short-term cold exposure of young steers is examined, in terms of both indirect estimations based on measurements of respiratory-gas exchange in the whole animal, and direct measurement of FFA oxidation in the major thermogenic tissue, i.e. shivering muscle.

Materials and methods

Eleven Ayrshire steers aged 8–12 months and weighing 122–189 kg were used. They were housed in an insulated byre and individually fed 1.5–2 kg of a mixture of chopped straw, rolled barley, cattle pellets and molasses twice daily. Growth rates during the 4–6 weeks prior to experiment were 0.3–0.5 kg/d. All animals were surgically prepared with a carotid loop and permanent indwelling catheters in the external iliac artery and vein of one hind leg (Bell, Thompson and Findlay, 1974), while five steers had an additional catheter placed in the abdominal aorta.

In each experiment an animal was exposed for 5–6 h to either a thermoneutral (13–17 °C, still air) or a moderately cold (–1 to +3 °C, wind speed 2–3 m/s, clipped hindquarters) environment, and was fed 3–6 h (termed 'fed') or 20–23 h (termed 'fasted') before the experiment started. In all, 28 experiments were performed, consisting of seven replications of each of the four possible combinations of environmental and feeding treatments. All experiments included measurement of whole-animal heat production and respiratory quotient (RQ) by open-circuit calorimetry, plasma FFA concentration by gas-liquid chromatography, and oxygen uptake and net exchanges of plasma FFA in the hind limb by simultaneous measurement of leg blood flow, using a dye-dilution method, and arteriovenous concentration differences. In addition, in 16 experiments a mixture of [1-^{14}C]-labelled palmitic, stearic and oleic acids, complexed in bovine plasma and in relative proportions similar to those normally found in bovine plasma, was infused into the abdominal aorta at the rate of about 1.2 μc/min for 3.5–4 h. During the last hour of this infusion several paired samples of arterial (A) and external iliac venous (V) blood was taken for analysis of the specific radioactivity of FFA and carbon dioxide and concentration of oxygen. Leg blood flow was measured twice during this sampling period.

Data for rate of entry in the whole animal and net uptake, actual uptake and fractional oxidation in the hind limb of plasma FFA, and the contribution of FFA oxidation to total oxidative metabolism in the limb, were calculated from the above measured values, using the equations of Dagenais, Tancredi and Zierler (1976).

Results

HEAT PRODUCTION AND LIPID METABOLISM IN THE WHOLE ANIMAL (*Table 88.1*)

Cold exposure doubled heat production and significantly lowered RQ in both fed and fasted steers. Assuming no cold-induced change in protein oxidation during these relatively short periods of measurement, RQ values were used to estimate the contribution of fat catabolism to heat production. Substantial increases in these derived values were accompanied by 2–3 fold increases in both arterial concentration and rate of entry of plasma FFA.

OXYGEN UPTAKE AND FFA METABOLISM IN THE HIND LIMB (*Table 88.2*)

Persistent shivering of moderate intensity in the leg muscles was associated with a 6–8 fold increase during cold exposure in oxygen uptake by the hind limb. Although the net release of FFA from leg tissues during thermoneutrality

Table 88.1 EFFECTS OF COLD AND FEEDING ON WHOLE-BODY HEAT PRODUCTION (M) AND LIPID METABOLISM IN STEERS (VALUES ARE MEANS ± SEM)

	n	*Thermoneutral*		*Cold*	
		Fasted	*Fed*	*Fasted*	*Fed*
M (kJ/kg$^{0.75}$ per h)	7	20.3 ± 0.6	23.3 ± 1.2	39.0 ± 2.2	38.8 ± 3.0
RQ	7	0.90 ± 0.02	0.99 ± 0.01	0.84 ± 0.02	0.91 ± 0.02
Estimated contribution of fat catabolism to M (%)	7	33 ± 6	6 ± 2	52 ± 5	28 ± 6
Arterial plasma FFA (μmol/ℓ)	7	375 ± 31	132 ± 15	919 ± 65	452 ± 68
FFA entry rate (μmol/kg$^{0.75}$ per h)	4	357 ± 27	178 ± 33	722 ± 66	583 ± 82

n = number of animals

Table 88.2 EFFECTS OF COLD AND FEEDING ON OXYGEN UPTAKE AND FFA METABOLISM IN THE HIND LIMB IN STEERS (VALUES ARE MEANS ± SEM)

	n	*Thermoneutral*		*Cold*	
		Fasted	*Fed*	*Fasted*	*Fed*
Oxygen uptake (ml/kg$^{0.75}$ per h)	7	32.7 ± 5.1	24.7 ± 6.0	200.3 ± 32.8	198.0 ± 26.1
Net FFA uptake (μmol/kg$^{0.75}$ per h)	7	−24 ± 10	−13 ± 5	28 ± 26	44 ± 14
Actual FFA uptake (μmol/kg$^{0.75}$ per h)	4	47 ± 6	22 ± 7	248 ± 41	178 ± 43
Fractional oxidation of FFA uptake (%)	4	13.5 ± 1.6	2.8 ± 0.6	17.3 ± 2.6	31.9 ± 5.8
FFA oxidation/oxygen uptake (%)	4	8.9 ± 1.5	2.0 ± 0.6	15.4 ± 3.2	13.0 ± 2.4

n = number of animals

shifted to a net uptake in the cold, mean changes were relatively small and very variable. In contrast, actual FFA uptake, calculated from arteriovenous differences in ^{14}C-FFA across the limb, increased approximately in proportion to oxygen uptake. However, simultaneous measurements of ^{14}C-carbon dioxide release, which had reached fairly steady levels after 2.5 h of ^{14}C-FFA infusion, showed that although FFA oxidation increased during cold exposure most of the FFA taken up by leg tissues were not immediately oxidized to carbon dioxide and water.

Discussion

Calculations based on changes in RQ indicate that increased lipid oxidation contributed 73% of the metabolic response to cold in the fasted steers and 61% in the fed steers. Associated changes in arterial concentration and entry rate of plasma FFA were consistent with a marked increase in the mobilization and utilization of long-chain fatty acids derived from body stores. Also, the uptake of plasma FFA by shivering muscle, the major site of cold thermogenesis in young cattle (Bell, Thompson and Findlay, 1974), was closely related to the level of oxidative metabolism in this tissue. However, most of these FFA were not quickly oxidized to completion. Thus, during short-term cold exposure of moderate severity, increased fat catabolism, although important, was by no means the sole source of the extra heat produced by the whole animal; nor did plasma FFA appear to be a 'preferred' fuel for shivering muscle. These observations are in accord with direct evidence for increased utilization of non-lipid substrates during cold exposure of ruminants (McKay, Young and Milligan, 1974; Bell and Thompson, 1976).

The metabolic fate of the large fraction of FFA taken up by shivering muscle but not quickly converted to carbon dioxide is not known. Possibly, much was partially oxidized to short-chain derivatives, such as acetate, to be later oxidized to completion *in situ* or metabolized elsewhere in the body. Ruminant muscle has recently been shown to be a significant producer of endogenous acetate (Pethick and Lindsay, 1978), and there is limited evidence that endogenous acetate production in the whole animal is increased during short-term cold exposure (Bell, Gardner and Thompson, 1974).

References

BELL, A.W., GARDNER, J.W. and THOMPSON, G.E. (1974). *Br. J. Nutr.*, **32**, 471–477

BELL, A.W. and THOMPSON, G.E. (1976). In *Energy Metabolism of Farm Animals*, pp. 41–44. Ed. M. Vermorel. EAAP Publication No. 19. Clermont-Ferrand: G. de Bussac

BELL, A.W., THOMPSON, G.E. and FINDLAY, J.D. (1974). *Pflügers Archiv*, **346**, 341–350

BLAXTER, K.L. and WAINMAN, F.W. (1961). *J. Agric. Sci., Camb.*, **56**, 81–90

DAGENAIS, G.R., TANCREDI, R.G. and ZIERLER, K.L. (1976). *J. clin. Invest.*, **58**, 421–431

DEPOCAS, F. and MASIRONI, R. (1960). *Am. J. Physiol.*, **199**, 1051–1055

GRAHAM, N.McC., WAINMAN, F.W., BLAXTER, K.L. and ARMSTRONG, D.G. (1959). *J. Agric. Sci., Camb.*, **52**, 13–24

KAYSER, C. (1937). *C. r. Séanc. Soc. Biol.*, **126**, 701–704

McKAY, D.G., YOUNG, B.A. and MILLIGAN, L.P. (1974). In *Energy Metabolism of Farm Animals*, pp. 39–42. Eds. K.H. Menke, H.-J. Lantzsch and J.R. Reichl. EAAP Publication No. 14. Hohenheim: Universität Hohenheim Dokumentationsstelle

MINAIRE, Y., VINCENT-FALQUET, J.-C., PERNOD, A. and CHATONNET, J. (1973). *J. appl. Physiol.*, **35**, 51–57

PAGÉ, E. and CHÉNIER, L.-P. (1953). *Rev. Can. Biol.*, **12**, 530–541

PETHICK, D.W. and LINDSAY, D.B. (1978). *Proc. Nutr. Soc.*, **37**, 101A

89

ENERGY METABOLISM, SUBSTRATE RESERVES AND GLUCOSE UTILIZATION IN THE PIGLET

H.S. BAYLEY, G.L. CROZIER and J.L. ATKINSON
Department of Nutrition, University of Guelph, Guelph, Ontario, Canada N1G 2W1

Summary

The total rates of heat production of fasting piglets were measured by indirect calorimetry. The glycogen contents of suckled piglets were determined and compared to fasted piglets of the same age. Total and net entry rates of glucose were measured in fasting piglets. The amounts of glucose needed to support the observed rates of energy expenditure were compared to the amounts of glycogen available for mobilization and to the metabolic utilization of glucose. The catabolism of the amino acids alanine and leucine indicated that alanine may make a substantial contribution to the energy metabolism of the piglet by direct oxidation rather than via gluconeogenesis.

Introduction

Mortality of the neonate remains the major biological problem of pig production. Goodwin (1957) showed that there is a close relationship between vital functions and the levels of glucose circulating in the blood; thus, the supply of glucose in the first days of life may be critical in determining the viability of a piglet. The experiments were carried out to elucidate the relationship between the demands of energy metabolism for fuel and the amounts of fuel stores available for mobilization.

Materials and methods

An open-circuit indirect calorimeter was used to measure rates of heat production in piglets fasted continuously for several days. Groups of four piglets were removed from the sow at birth or after suckling for between 2 and 13 days.

Piglets weaned at birth had a lower rate of energy expenditure than any of the other groups during the first day of the fast (*Table 89.1*), even though they were the same weight as the piglets which were weaned at two days of age. Piglets which had suckled for at least two days produced between 15 and 22 kJ/kg per h with a mean of 18 kJ/kg per h in the first day of the fast. This rate of heat production could be sustained by the combustion of glucose at a rate of 1.15 g/kg per h if the glucose was the sole metabolic substrate. Heat production of the piglets weaned at birth was greatly reduced on the second and subsequent days of fasting and none of these piglets survived for the fifth day of the experiment. Heat production of the older piglets was sustained at higher levels for a longer period. The rapid development of the ability to maintain their metabolic rate during a fast confirms the earlier report of Bayley and McDonald (1970).

To investigate the loss of body tissue to provide fuel to sustain energy metabolism piglets were taken from the sow either as newborns or after they had suckled for several days. They were killed either immediately or after a 48 hour fast at 32°C. Glycogen was determined in samples of several tissues, but the only

Table 89.1 HEAT PRODUCTION OF PIGLETS WEANED AT BIRTH OR BETWEEN TWO AND 13 DAYS OF AGE AND FASTED FOR FIVE CONSECUTIVE DAYS AT 30°C. EACH VALUE DETERMINED WITH A GROUP OF FOUR PIGLETS AND EXPRESSED AS kJ/kg per h

Age weaned (d)	*Weight of pig at weaning* (kg)	*Day of fasting*				
		1st	*2nd*	*3rd*	*4th*	*5th*
0	1.6	13	8	8	6	–
2	1.5	18	12	–	10	7
5	2.2	22	14	9	8	–
7	2.3	18	16	11	11	10
9	2.5	15	14	–	12	11
11	3.2	17	16	–	13	9
13	3.6	19	15	13	11	11

tissues making a substantial contribution to the total body store were liver and muscle. Glycogen concentration in the sample of longissimus dorsi was related to total muscle glycogen by assuming that the sample was typical of all the muscle, and that muscle constituted 29% of body weight.

The most noteworthy effect of the fast (*Table 89.2*) was the almost complete utilization of the liver glycogen reserves, even though the muscle contained a greater total amount of glycogen. The newborn animals contained 36 g glycogen, those killed after suckling for one day contained 26 g and the older piglets contained less than 20 g glycogen. There was substantially less muscle glycogen in the fasted piglets than in the fed piglets of the same age, except for the 21-day-old group. In no case was there an absence of glycogen from the muscle which indicated that it is not as labile during a fast in a thermoneutral environment as was the liver glycogen. The similarity between the amounts of muscle glycogen in the fed and fasted 21-day-old pigs indicates that this energy reserve is unimportant to pigs of this age which have accumulated substantial amounts of fat. The newborn piglets contained more muscle glycogen than the one-day-old piglets which had been allowed to suckle. This may be an indication of the importance of glycogen in the neonate as a fuel for muscular activity: shivering thermogenesis, exploration and competition with siblings.

Quantitative comparison of liver and muscle glycogen between fed and fasted piglets of the same age is complicated by the differences in live weights of the animals, but the fasted two-day-old piglets contained 21.9 g less glycogen than the newborns; this loss of an energy reserve should be compared to the 13 and 8 kJ/kg per h of heat production during the first and second day of fasting for the piglets fasted from birth. This rate of heat production would be sustained by the combustion of 0.67 g glucose/kg per h or 45 g glucose for the 1.3 kg piglet fasting for 48 hours, indicating that in the unsuckled newborn piglet the glycogen reserves are adequate to supply only half of the energy loss during a two day fast

Table 89.2 TOTAL GLYCOGEN IN LIVER AND MUSCLE OF PIGLETS FROM 0–21 DAYS OF AGE, EITHER IMMEDIATELY AFTER REMOVAL FROM THE SOW (FED) OR AFTER A 48 h FAST (FASTED). VALUES ARE MEANS OF FIVE PIGLETS PER GROUP

Age (d)	*Status*	*Body weight* (kg)	*Glycogen* (g)	
			Liver	*Muscle*
0	––	1.3	5.0	36
1	Fed	1.2	2.5	26
2	Fasted	1.2	0.1	19
3	Fed	1.8	4.0	13
3	Fasted	1.2	0.3	4
6	Fed	2.1	4.7	17
6	Fasted	1.7	0.3	13
10	Fed	2.5	3.8	12
10	Fasted	1.8	0.3	5
21	Fed	3.1	5.9	16
21	Fasted	4.2	0.5	17

in a thermoneutral environment. Fasted piglets of three, six or 10 days of age contained approximately 10 g less glycogen than the fed animals. This difference was much less than the amount that would be needed to sustain the levels of energy metabolism observed in the first study. For instance the 1.8 kg, three-day-old piglet fasting for 48 hours would metabolize the equivalent of 100 g of glucose if this were the sole substrate.

Glucose metabolism in young pigs was studied directly using [6-^{3}H] and [U-^{14}C] glucose in a primed constant infusion procedure to estimate total and net glucose entry rates. Piglets were taken from the sow after suckling for three hours, two days or 21 days, and then fasted for 24 hours. Specific activity of blood glucose with respect to both ^{14}C and ^{3}H stabilized between 90 and 120 minutes after the start of the infusion and glucose entry rates were calculated for both labels to indicate net and total entry rates respectively.

Table 89.3 shows that total glucose entry rates of approximately one-half gram per kg body weight per hour are almost double those found for overnight fasted rabbits by Katz *et al.* (1974), although the extent of glucose recycling which they reported (22–35%) is similar to those found in the present study. Net glucose entry rates are of greater significance to the energetic economy of the piglet, and the value of 0.27 g/kg per h determined in this study would

Table 89.3 GLUCOSE METABOLISM IN PIGLETS WEANED AFTER THREE HOURS, TWO DAYS OR 21 DAYS AND FASTED FOR 24 HOURS. ENTRY RATES CALCULATED FROM PLATEAU SPECIFIC ACTIVITIES ACHIEVED TWO HOURS AFTER BEGINNING INFUSION WITH ^{14}C AND ^{3}H LABELLED GLUCOSE. VALUES ARE MEANS FOR SEVEN PIGLETS PER GROUP

Age (d)	*Body weight* (kg)	*Plasma glucose concentration* (mg/dl)	*Glucose entry* (g/kg per h)		
			Total (^{3}H)	*Net* (^{14}C)	*Glucose recycled* (%)
1	1.1	74	0.43	0.27	36
3	1.4	74	0.63	0.45	29
22	4.6	87	0.37	0.26	31

supply almost half of the energy expended by the fasting newborn in the first part of the study. This entry rate would be sustained by the breakdown of 17 g glycogen in a 1.3 kg piglet fasting for 48 hours which compares well with the difference of 21.9 g glycogen between the newborn and 48 hour-fasted piglets observed in the second part of the study.

The net entry rate of glucose was 0.45 g/kg per h by three days of age, and the increase perhaps reflects the greater rate of energy metabolism in the three-day-old piglet. For the 22-day-old pigs the net entry rate was 0.26 g/kg per h which is consistent with the reduced dependence of these pigs on glucose as a metabolic fuel. The constancy of glucose recycling for the three age groups indicates that the one-day-old piglets were as capable as the older pigs in synthesizing glucose, at least from glucose metabolites.

These three studies have shown that although the glycogen stores of the neonatal piglet are a major part of its energy endowment, its ability to mobilize other compounds from its tissues to serve as fuels during starvation in the first few days of life is of great importance. Tissue protein catabolism occurs, but the results of these studies indicate that carbon residues of the amino acids must be catabolized directly as fuels, rather than being used extensively for gluconeogenesis. To investigate the utilization of amino acids by the piglet, single infusions of uniformly labelled ^{14}C-alanine and leucine were administered to piglets after a 24 hour fast. The activity released as ^{14}C-carbon dioxide by the animals was measured for 3 hours following the infusion: 30% of the activity administered as alanine to four-day-old piglets was recovered in comparison to only 11% of the activity administered as leucine. To investigate the incorporation of amino acid carbon into glucose, either labelled alanine or labelled leucine was administered to a further group of 24 hour fasted piglets. Blood samples were taken during the next hour and the total activity of the blood measured; this declined rapidly in the first 30 minutes and then remained relatively stable for the second 30 minute period. The contribution of the glucose to the total activity of the plasma was also measured (*Table 89.4*).

For both the alanine and the leucine approximately 5% of the administered activity could be recovered in the plasma of the four and 13-day-old piglets. Almost one-third of the plasma activity was accounted for by glucose in the

Table 89.4 RECOVERY OF ^{14}C ADMINISTERED EITHER AS ALANINE OR LEUCINE. VALUES SHOWN AS PERCENT OF DOSE IN BLOOD SAMPLES TAKEN BETWEEN 30 AND 60 MINUTES AFTER ADMINISTRATION OF DOSE. VALUES ARE MEANS OF SIX PIGLETS PER GROUP, PIGLETS FASTED 24 HOURS AFTER REMOVAL FROM THE SOW.

	Amino acid administered			
	[U-^{14}C] *alanine*		[U-^{14}C] *leucine*	
Age (d)	*Plasma* *	*Plasma*† *glucose*	*Plasma*	*Plasma glucose*
		(% of dose)		
4	7.0	2.1	5.8	0.1
13	6.4	1.9	5.7	0.3
18	13.4	4.0	5.1	0.1

* Total body plasma † Total plasma glucose

piglets which had received alanine, whereas there had been negligible incorporation of activity from the ketogenic leucine into glucose. The recovery of 30% of the activity from alanine in the expired carbon dioxide confirms the metabolic importance of tissue protein mobilization during a fast, and the incorporation of the label from alanine into plasma glucose shows that alanine can be used for gluconeogenesis even in the four-day-old piglet. The quantitative aspects of amino-acid catabolism in the fasting piglet deserve further investigation.

References

BAYLEY, H.S. and McDONALD, M.S. (1970). In *Energy Metabolism of Farm Animals*, pp. 157–160. Eds. A. Schürch and C. Wenk. EAAP Publication No. 13. Zurich: Juris Druck

GOODWIN, R.F.W. (1957). *J. Physiol., Lond.*, **136**, 208–217

KATZ, J., DUNN, A., CHENOWETH, M. and GOLDEN, S. (1974). *Biochem. J.*, **142**, 171–183

EFFECT OF TEMPERATURE AND ENERGY INTAKE ON HEAT LOSS AND ENERGY RETENTION OF EARLY-WEANED PIGS

K.J. McCRACKEN
Agricultural and Food Chemistry Research Division, Department of Agriculture, Northern Ireland and The Queen's University of Belfast, Newforge Lane, Belfast BT9 5PX, Northern Ireland
and
B.J. CALDWELL
Department of Crop and Animal Production, The Queen's University of Belfast, Newforge Lane, Belfast BT9 5PX, Northern Ireland

Summary

The basal metabolic rate (BMR) and fasting metabolic rate (FMR) of groups of pigs weaned at 10, 17 and 24 days were determined in an open-circuit respiration chamber. The mean FMR was 492 kJ/kg$^{0.75}$ per d and the activity increment over BMR averaged 15%. The efficiency of utilization for maintenance and production was 0.77, giving an estimate of 637 kJ/kg$^{0.75}$ per d for the maintenance requirement.

The mean rates of heat loss (H/ΔT, where ΔT is the rectal–environmental temperature difference) of fasting and fed pigs at the lower critical temperature (T_{cl}) were respectively 15.9 and 18.2 kJ/m^2 per h per °C ΔT. T_{cl} was 25.9°C for fasted pigs at 10 days and 20.3°C at 37 days. T_{cl} was 21.8, 18.3 and 18.6°C for *ad lib.* fed pigs at 17, 24 and 31 days respectively.

Introduction

There is an increased commercial interest in the weaning of pigs into cages as early as 14 days. While there is a considerable literature on the development of the thermoregulatory mechanisms in the neonatal pig (McCance and Widdowson, 1959; Mount, 1959, 1960, 1968; Curtis and Rogler, 1970; Mount and Stephens, 1970) there is a lack of information on the effect of temperature and nutrition on the heat production and heat loss of pigs between one and eight weeks of age. A series of studies has been conducted on various aspects of the heat production of pigs weaned at between 10 and 25 days and reared in groups in cages. The detailed results will be published elsewhere but this chapter summarizes the information available at present.

Materials and methods

Pigs were obtained from a hysterectomy-derived, Large White × Landrace herd and were weaned at 10, 17 or 24 days. They were removed from the sow at around 1000 h and transferred to the experimental unit within 1 h. Normally six pigs were selected from a litter to give the smallest live-weight range within the group. Measurements of fasting heat production were made for 48 h at a temperature of 29 °C on pigs weaned at 10 days. The pigs were subsequently given a pelleted diet containing (g/kg): dried skim milk, 360; flaked maize, 320; fish meal, 120; sucrose, 100; fat, 100. Amounts were increased by 25 g/pig per d, and estimates of the lower critical temperature (T_{cl}) of fed pigs were made at 17, 24 and 31 days. Measurements of fasting heat production were made for 72 h on pigs weaned at 17 or 24 days. BMR was estimated from the minimum values of heat production obtained between midnight and 08.00 h when the pigs were resting.

The rate of heat loss (H/m^2 per h per °C ΔT, subsequently abbreviated to H/ΔT) and T_{cl} was estimated by measuring heat production for 24 h at a temperature chosen to be below T_{cl} (H = total heat production, kJ; m^2 = surface area calculated as 0.097 kg$^{0.633}$ (Brody, 1945); °C ΔT = difference between rectal temperature, taken as 39 °C, and air temperature). The total H at a temperature chosen to be above T_{cl} divided by H/ΔT and subtracted from 39 °C then gave an estimate of T_{cl} (Kleiber, 1967).

H was measured by indirect calorimetry using an open-circuit respiration chamber of which the main features have been described (Gray and McCracken, 1976). Collections of waste food and excreta were made in order to obtain estimates of protein and energy retention.

Results and discussion

BMR (kJ/kg$^{0.75}$) at 10 days was significantly ($P<0.05$) higher than at 17 or 24 days (*Table 90.1*) but there were no significant differences in FMR, which averaged 492 kJ/kg$^{0.75}$ over the four ages. The energy increment due to activity averaged 15% of BMR. Assuming a value of 0.77 for efficiency of energy utilization for maintenance (*Table 90.2*) the maintenance requirement can be estimated as 680 kJ/kg$^{0.75}$ at 10 days and 610 kJ/kg$^{0.75}$ at 24 and 37 days with a mean value of 637 kJ/kg$^{0.75}$ for the period 10–37 days. This value is higher than that reported by Jordan and Brown (1970) but similar to the value of 644 obtained by multiple regression analysis of slaughter data for the period 8–32 days (McCracken, Eddie and Stevenson, 1980) and to the results of Close, Stanier and Sanz Sampelayo (1979).

The estimates of protein retention at 24 and 31 days (*Table 90.2*) are similar to those of Böhme, Gädeken and Oslage (1976) for pigs of equivalent weight,

Table 90.1 BASAL (BMR) AND FASTING (FMR) METABOLIC RATES OF PIGS REMOVED FROM THE SOW AT 10, 17, 24 DAYS OR 37 DAYS AND ESTIMATES OF THE ENERGY INCREMENT FROM ACTIVITY AND THE MAINTENANCE REQUIREMENT (ASSUMING k_m = 0.77)

	BMR (kJ/kg$^{0.75}$ per d)	*FMR* (kJ/kg$^{0.75}$ per d)	*Activity increment* (% BMR)	*Maintenance requirement* (kJ/kg$^{0.75}$ per d)
10 days	465*	527	13.3	680
17 days	434	501	15.4	650
24 days	411	469	14.3	610
37 days	422	470	16.7	610

* Significantly different ($P<0.05$) from 17 days and 24 days

Table 90.2 ENERGY RETENTION (ER) AND CALCULATED EFFICIENCY* OF ENERGY UTILIZATION FOR MAINTENANCE AND PRODUCTION OF WEANED PIGS GIVEN A MILK-BASED, LOW-FAT DIET *AD LIB.* AT AGE 24 AND 31 DAYS

Age (d)	*Litter*	*ME intake* (MJ/pig per d)	*Energy stored as protein* (MJ/pig per d)	*ER* (MJ/pig per d)	*ER/ME*	*Efficiency*
24	1	3.93	1.07	1.57	0.40	0.75
	2	4.86	1.14	2.27	0.47	0.79
31	1	6.53	1.46	3.00	0.46	0.77
	2	6.59	1.48	3.15	0.48	0.77

* Assuming FMR = 470 $kJ/kg^{0.75}$

and the efficiency of utilization of energy (0.77) is the same as found by them for pigs aged 15–22 days.

The experimental design did not permit statistical comparison of the values for $H/\Delta T$ in the fasting and fed states but the lower values obtained with fasting animals (*Table 90.3*) may be attributed to a reduction in activity due to the absence of feed. The value of 18.3 kJ/m^2 per h per °C ΔT at 10 days is not strictly comparable to the lower fasting values obtained at 17, 24 and 37 days since it relates to a 10-h period from 1200 h to 2200 h which has been found to correspond to the period of maximum activity whereas the others relate to the 24-h period. It is concluded that the minimum rate of heat loss in huddling pigs is around 15 kJ/m^2 per h per °C ΔT but that this should be increased by 20–30% to take account of extra heat loss associated with feeding, drinking and excretory functions. These values are similar to those calculated for older pigs (Holmes and Close, 1977).

Table 90.3 ESTIMATES OF RATE OF HEAT LOSS ($H/\Delta T$) AND LOWER CRITICAL TEMPERATURE (T_{cl}) OF GROUPS OF PIGS IN THE FED AND FASTING STATE BETWEEN 10 AND 37 DAYS OF AGE

Age (d)	$H/\Delta T$ (kJ/m^2 per h per °C ΔT)		T_{cl} (°C)	
	Fasting	*Fed*	*Fasting*	*Fed*
10	18.3	–	25.9	–
17	15.2	16.8	22.1	21.8
24	14.1	18.5	20.8	18.3
31	–	19.4	–	18.6
37	16.3	–	20.3	–

The difference between the estimates of T_{cl} for fasting and fed pigs at each age is less than would be expected on the basis of the extra heat production due to feeding because of the higher values of $H/\Delta T$ for the fed animals. The small difference at 17 days also reflects the fact that the energy intake of the pigs was only slightly above maintenance.

Acknowledgements

We wish to record our thanks to Dr R. Gray for help in the smooth running of the respiration chamber and for data-processing. One of us (BJC) acknowledges the award of a Postgraduate Scholarship from the Department of Agriculture for Northern Ireland.

References

BÖHME, H., GÄDEKEN, D. and OSLAGE, H.J. (1976). In *Energy Metabolism of Farm Animals*, pp. 165–168. Ed. M. Vermorel. EAAP Publication No. 19. Clermont-Ferrand: G. de Bussac

BRODY, S. (1945). *Bioenergetics and Growth.* New York: Reinhold Publishing Corp.

CLOSE, W.H., STANIER, M.W. and SANZ SAMPELAYO, M.R. (1979). *Proc. Nutr. Soc.* **38**, 47A

CURTIS, S.E. and ROGLER, J.C. (1970). *Am. J. Physiol.,* **218**, 149–152

GRAY, R. and McCRACKEN, K.J. (1976). In *Energy Metabolism of Farm Animals,* pp. 335–338. Ed. M. Vermorel. EAAP Publication No. 19. Clermont-Ferrand: G. de Bussac

HOLMES, C.W. and CLOSE, W.H. (1977). In *Nutrition and the Climatic Environment,* p. 51. Eds. W. Haresign, H. Swann and D. Lewis. London: Butterworths

JORDAN, J.W. and BROWN, W.O. (1970). In *Energy Metabolism of Farm Animals,* pp. 161–164. Eds. A. Schürch and C. Wenk. EAAP Publication No. 13. Zurich: Juris Druck and Verlag

KLEIBER, M. (1967). *Der Energiehaushalt von Mensl und Haustier,* p. 143. Hamburg: P. Parey

McCANCE, R.A. and WIDDOWSON, E.M. (1959). *J. Physiol.,* **147**, 124–134

McCRACKEN, K.J., EDDIE, S.M. and STEVENSON, W.G. (1979). *Br. J. Nutr.,* **43**, 305–319

MOUNT, L.E. (1959). *J. Physiol.,* **147**, 333–345

MOUNT, L.E. (1960). *J. agric. Sci., Camb.,* **55**, 101–105

MOUNT, L.E. (1968). *The Climatic Physiology of the Pig.* London: Arnold

MOUNT, L.E. and STEPHENS, D.B. (1970). *J. Physiol.,* **207**, 417–427

91

THE ENERGY COST FOR SHEEP OF WALKING ON GRADIENTS

J.M. BROCKWAY and A.W. BOYNE
Rowett Research Institute, Bucksburn, Aberdeen AB2 9SB, Scotland

Summary

Four sheep were exercised on a treadmill at 13 combinations of speed and gradient. Measurements were made of the energetic cost of walking and of the tension generated in the tether between the sheep and the treadmill. Dragging against the tether produced an energy saving for the sheep when exercise was strenuous and an increased energy expenditure during mild exercise. After adjustment of energy costs for the effects of tension in the tether the efficiency of work in raising body mass vertically averaged 30%.

Introduction

The efficiency of muscular work in raising body weight through unit vertical distance is estimated by measuring the energy expenditure when walking on a gradient and subtracting that when walking at the same speed on the level. On theoretical grounds it is to be expected that efficiency should be maximally 30–35%, and such values are commonly reported (Blaxter, 1962), but on occasion higher values have been found. For example Taylor, Caldwell and Rowntree (1972) quoted values of over 60% for chimpanzees and mice, Brockway and Gessaman (1977) 45% for red deer and Cohen, Robbins and Davitt (1978) over 70% for elk. These last authors made, but did not elaborate upon, two points which could lead to overestimation of efficiency: spuriously high estimates of the cost of horizontal locomotion and dragging of the animal by its tether.

The experiment described in this chapter was designed to investigate the effects of tension in the tether on the energy cost of walking on a treadmill by the sheep.

Materials and methods

ANIMALS

Four Suffolk half-bred wether sheep, aged 9 months old and weighing 60 kg at the beginning of the experiment, were used. They were fed on grass cubes and showed a mean gain in live weight of about 3 kg over the 5-month experimental period.

APPARATUS

The treadmill and open-circuit ventilated hood system for the measurement of oxygen consumption have been described previously (Brockway and Gessaman, 1977). The tethering system for the sheep on the treadmill consisted of a neck collar attached to the inside of the hood by a short chain. The hood was linked to the treadmill, parallel to the walking surface, by a line through a pulley wheel to a 20-kg capacity spring balance. The shaft of the balance actuated a potentiometer whose output voltage was applied to a digital integrator calibrated in counts/kg per min.

EXPERIMENTAL DESIGN

The treatment combinations of speed and gradient at which measurements were made are shown in *Table 91.1.* Each animal received the treatments in random order and the design was replicated four times. The experimental protocol at each treatment comprised 10 minutes standing followed by 10 minutes walking, but recordings were made only during the final 5 minutes of each activity. Each replicate took about one month to complete, with intervals of 2–3 weeks between replicates.

Results

COST OF WALKING

Analysis of the increased cost of walking, above that of standing, and disregarding any tension in the tether, showed significant differences between animals and between treatments. Virtually all the differences between treatments could be ascribed to the effects of gradients; effects of speed were not significant in this analysis.

EFFECT OF TENSION

A within-animal, within-treatment examination was carried out of the relationship between the energy cost of walking and the tension in the tether. For only one treatment 45 m/min, +3° gradient, were there significant differences in the regression coefficients, but it is not unexpected that one set out of 13 should show differences at the 5% level. For the more strenuous treatments the coefficients (*Table 91.1*) were negative, indicating that the sheep were deriving energetic assistance from dragging against their tethers, but, when they dragged during the less strenuous treatments the positive coefficients showed that the dragging represented an additional energy cost. The regression coefficients were statistically significant in only five out of the 13 treatments, but overall, taking tension into consideration reduced the standard deviation of the cost of walking from 0.464 J/kg per m to 0.414 J/kg per m, an effect which was highly significant ($P<0.001$).

Table 91.1 COEFFICIENTS AND THEIR SIGNIFICANCE, IN THE REGRESSIONS OF COST (J/kg per m) AGAINST TENSION (g/kg BODY WEIGHT)

Gradient	*Speed* (m/min)			
	45	55	65	75
9°	−0.024			
6°	−0.028	−0.034**		
3°	−0.007	−0.011	−0.011*	
0°	0.026b**	0.005	−0.006	−0.010
−5°	0.024*	0.012		
−10°	0.019*			

** $P<0.01$
* $P<0.05$

ADJUSTMENT OF COST TO ZERO TENSION

The regression coefficients from *Table 91.1* were then used to adjust the observed energy costs for each animal to zero tension. For these adjusted costs constants were estimated to represent the effects of speed and gradient. The residual mean square was reduced from 0.128 to 0.0717, and that representing animal differences from 0.83 to 0.38. The between-animal differences remained significant, but the adjustment did bring the cost for one sheep, which had shown consistently high tension values, into line with the other three. This effect is shown in *Table 91.2.* This analysis also revealed a small, but highly significant ($P<0.001$) effect of speed on the cost of walking unit distance. The results of the analysis are summarized in *Table 91.3.*

Table 91.2 MEAN TENSION AND ENERGY COST OF WALKING FOR EACH SHEEP

	Animal No.				*s.e. of means*
	1	*2*	*3*	*4*	
Tension (g/kg)	15.0	15.8	16.1	38.0	±2.50
Observed cost (J/kg per m)	2.98	3.21	3.41	2.84	±0.10
*Adjusted cost (J/kg per m)	3.15	3.33	3.56	3.28	±0.07

* Adjusted to zero tether tension

Simple quadratic expressions can be fitted to estimates of speed and gradient effects, which yield the equation:

$$\text{Cost (J/kg per m)} = 2.35 + 0.398G + 0.0286G^2 - 0.036S + 0.00052S^2 \qquad (1)$$

where G is the gradient measured in degrees and S is speed measured in m/min. The residual standard deviation for the equation is ± 0.24.

EFFICIENCY OF VERTICAL MOVEMENT

From equation (1) can be calculated the increased cost of walking on gradients, over and above that of walking on the level: the figures are 1.451, 3.418 and

Table 91.3 ESTIMATED INCREASED COST OF WALKING ABOVE STANDING, ADJUSTED TO ZERO TENSION

Speed (m/min)	*Cost of walking* (J/kg per m)
45	–0.198
55	0.039
65	0.196
75	0.641
Gradient	
9°	8.164
6°	5.962
3°	3.484
0°	2.296
–5°	1.208
–10°	1.130

5.899 J/kg per m for the 3°, 6° and 9° gradients respectively. The corresponding costs for raising unit body weight through 1 m vertically are 27.72, 32.70 and 37.71 J/kg, equivalent to efficiencies of 35%, 30% and 26% at 3°, 6° and 9°. Comparable efficiencies calculated from the costs before adjustment to zero assistance were approximately 11 percentage units higher.

Discussion

The results presented in this preliminary communication clearly demonstrate the need to take into account any tension which an animal imposes upon its tether when walking on a treadmill. For example in the treatment at 45 m/min, +6° gradient, the measured tension in the sheep's tether was 23 g/kg body weight, that is, less than 1.5 kg in total. The co-efficient in the regression of energy cost on assistance for this treatment was –0.028 J/m per g, which represents an energy saving of 0.644 J/kg per m, increasing the observed cost of 5.085 J/kg per m to 5.729 J/kg per m. On the other hand, in the treatment at 45 m/min, 0° gradient, the measured tension was only 11 g/kg, or less than 0.75 kg in total. The regression co-efficient for this treatment was 0.026 J/m per g representing an additional cost of 0.286 J/kg per m and reducing the observed cost from 2.405 J/kg per m to 2.199 J/kg per m. Thus, the unadjusted additional cost of walking at 45 m/min, +6° gradient, above that of walking on the level, was 2.680 J/kg per m, whereas after adjustment for tether tension it was 3.530 J/kg per m. The corresponding costs of moving through 1 m vertically (involving a distance of 9.567 m on the incline) are 25.64 and 33.77 J/kg: representing efficiencies of 38% and 29% respectively.

While it is not surprising that the co-efficients in the regressions of cost against tension had negative signs in the strenuous exercises, the finding of positive signs for the mild exercises is more difficult to explain. Tension in the tether has been assumed to represent a linear force acting to pull the sheep along the treadmill, but this is an oversimplification. Tension measured in this experiment was in fact the resultant of all forces acting within a large spherical angle centred on the pulley carrying the rope which linked the hood to the spring balance. Some of these forces may have been energetically expensive to generate in relation to their component in the direction of walking by the sheep. In the extreme case of an animal *standing* on the treadmill, any tension generated in the tether would represent an additional energy cost above that of standing.

References

BLAXTER, K.L. (1962). *The Energy Metabolism of Ruminants.* London: Hutchinson

BROCKWAY, J.M. and GESSAMAN, J.A. (1977). *Q. Jl exp. Physiol.*, **62**, 333–339

COHEN, Y., ROBBINS, C.T. and DAVITT, B.B. (1978). *Comp. Biochem. Physiol.*, **61A**, 43–48

TAYLOR, C.R., CALDWELL, S.L. and ROWNTREE, V.J. (1972). *Science,* **178**, 1096–1097

THE EFFECT OF DIFFERENT LIGHT REGIMENS ON FEED INTAKE, MOBILITY, WEIGHT GAIN AND ENERGY AND NITROGEN RETENTION OF GROWING CHICKENS

J.E. VOGT, K. BOUMA, J. VAN WEERD, H.A. BOEKHOLT, P.W.M. VAN ADRICHEM and A.J.H. VAN ES
Department of Animal Physiology, Wageningen, The Netherlands

Summary

In nitrogen and energy balance studies with groups of chickens lasting 5–7 weeks it was found that the animals did nearly as well under a regimen with two dark periods of 7 h and 5 h per day as under a regimen with one dark period of 4 h per day.

Introduction

Chickens are usually given a regimen of low-intensity light during the whole day. Often a daily dark period of one hour is included to accustom the animals to the dark in case of an electricity cut which can greatly disturb unaccustomed chickens. Including dark periods might be helpful also from the point of view of reducing the costs of electricity. However, in temperate climates such a saving might be completely offset by the chickens requiring more heat: in the dark their heat production is considerably lower, so they might no longer be within the zone of thermoneutrality. Finally, it is interesting to study the physiological reactions of chickens to various light regimens.

Energy and nitrogen balance studies using *ad lib.* fed chickens kept under different light regimens and including continuous measurements of respiratory gas exchange and physical activity seemed suited for this study.

Materials and methods

Four series of balance trials each lasting 5–7 weeks were performed using two respiration chambers of about 2 m^3 each. Each chamber housed a cage for one group of Starbro chickens. Females were used in the third series and males in the other three. The number of animals in each group was gradually reduced from 75 at an age of 2–4 days to about 20 at seven weeks. A commercial feed mixture with 49% milocorn, 10% corn, 27.5% soya bean meal, 4% fishmeal, 6% animal fat and 3.5% vitamin–mineral mixture, was given *ad lib.* as a meal; feed left was collected, weighed and analysed. Droppings were collected, weighed and analysed during six days of each week, and respiratory gas exchange was measured for two 48-h periods each week. Each group was weighed once a week. If the number of animals had to be reduced, care was taken that animals of similar weight were removed from each of the two groups. For each group in every experimental week averages were calculated that applied to one animal and one day.

After arrival of the one-day-old chickens, cage temperature was kept at 32 °C; thereafter in the first week the temperature was gradually lowered. In the second week the temperature was maintained at 25 °C, in the third at 21 °C and in later weeks at 19 °C. Relative humidity was kept at about 70%.

In the first series the two light treatments consisted of either 6 h dark (D) and 18 h light (L) per day or 2 h D, 2 h L, 2 h D, 2 h L, 2 h D and 14 h L per day. The experiment started when the chickens were aged 16 days. Treatments were changed after the third and fifth week, group A receiving 6 h dark continuously in the third, sixth and seventh experimental weeks, group B in the fourth and fifth weeks. In the other series, in which measurements started at age two days, the same light regimen was used throughout, either 4 h D and 20 h L (group A) or 7 h D, 7 h L, 5 h D and 5 h L (group B).

The same analytical and physical measurements were used as in earlier work (Wenk and van Es, 1976).

Initially the respiration chambers were covered with a black plastic sheet for darkening. The radiowaves of the activity meter, however, were partly absorbed or reflected by this sheet, especially if they had to cross several layers of it. Therefore, from the start of the fourth series, instead of the sheet black paint was used for darkening where the waves entered the chambers. Thus, reliable activity data were obtained only for the fourth series.

Results

FIRST SERIES: 3 x 2 h D vs 6 h D

The 6 h D treatment hardly influenced daily weight gain, feed conversion, metabolizable energy per gross energy (ME/GE), ME/$W^{0.75}$ and % protein-energy

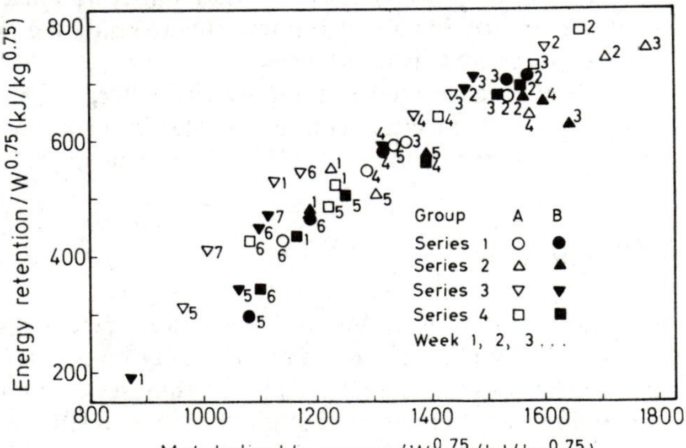

Figure 92.1 Metabolizable energy versus energy retention both scaled for metabolic weight; the values apply to the average animal within each week

Table 92.1 AVERAGES WITH MEAN ERROR, CALCULATED FROM THE EXPERIMENTAL WEEKS

	Series 1			*Series 2*		
	6 h D	*(3 x 2) h D*	*Difference 6–(3 x 2)*	*A*	*B*	*Difference A–B*
DLG (g)	47± 4	47±6	0±3	35±8	32±8	3±5
FC (g/g)	1.81±0.11	1.74±0.09	0.07±0.05	1.69±0.15	1.65±0.11	0.04±0.04
ME/GE	0.75±0.01	0.73±0.01	0.02±0.01	0.77±0.01	0.76±0.01	0.01±0.01
$ME/W^{0.75}$ ($kJ/kg^{0.75}$)	1372±75	1298±83	74±60	1515±109	1474±83	41±44
$ER/W^{0.75}$ ($kJ/kg^{0.75}$)	593±48	523±69	70±62	639±50	600±37	39±35
% PE	43±2	46±3	–3±4	51±1	51±2	0±3
n	5	5	5	5	5	5
	Series 3			*Series 4*		
	A	*B*	*Difference A–B*	*A*	*B*	*Difference A–B*
DLG (g)	31±5	30±5	1±2	38±6	37±7	0±1
FC (g/g)	1.84±0.16	1.82±0.12	0.02±0.07	1.67±0.17	1.63±0.17	0.05±0.03
ME/GE	0.77±0.01	0.74±0.02	0.02±0.01	0.76±0.01	0.75±0.01	0.01±0.01
$ME/W^{0.75}$ ($kJ/kg^{0.75}$)	1239±89	1200±84	38±49	1364±92	1327±77	37±22
$ER/W^{0.75}$ ($kJ/kg^{0.75}$)	552±59	489±71	63±51	596±58	536±56	61±17
% PE	45±2	52±5	–8±4	47±2	51±2	–4±1
Act.				537±69	523±64	14±21
n	7	7	7	6	6	6

DLG = daily live weight gain; FC = feed conversion; ME = metabolizable energy; GE = gross energy; ER = energy retention; % PE = % protein-energy in ER; Act. = physical activity.

(% PE) of the total energy retention (ER) (*Table 92.1*). From the regression of $ER/W^{0.75}$ on $ME/W^{0.75}$, combining all 10-week data sets, there appeared to be no significant difference in the relationship between these two variables between treatments within the same experimental week (*Figure 92.1*). Because of these results, a more extreme light treatment was used in the next three series.

SECOND, THIRD AND FOURTH SERIES: 4 h D (GROUP A) vs 7 h + 5 h D (GROUP B)

Initially group A grew slightly faster than B, but in the later weeks this was reversed so that final weights were only slightly lower for group B (*Figure 92.2*). Weekly feed conversion was usually slightly higher in group A (*Table 92.1*). Weekly ME/GE was also usually higher for A by 0.01–0.02. Relative ME intake ($ME/W^{0.75}$) tended to be higher for A initially, but lower finally. Compared to the regression line of $ER/W^{0.75}$ on $ME/W^{0.75}$, combining all 37-week data sets within weeks, group A did not behave differently from group B with respect to the relationship between these two variables (*Figure 92.1*). Relative energy retention ($ER/W^{0.75}$) did not differ significantly between groups. Group A tended to have a lower proportion of protein-energy in the daily energy retention than group B.

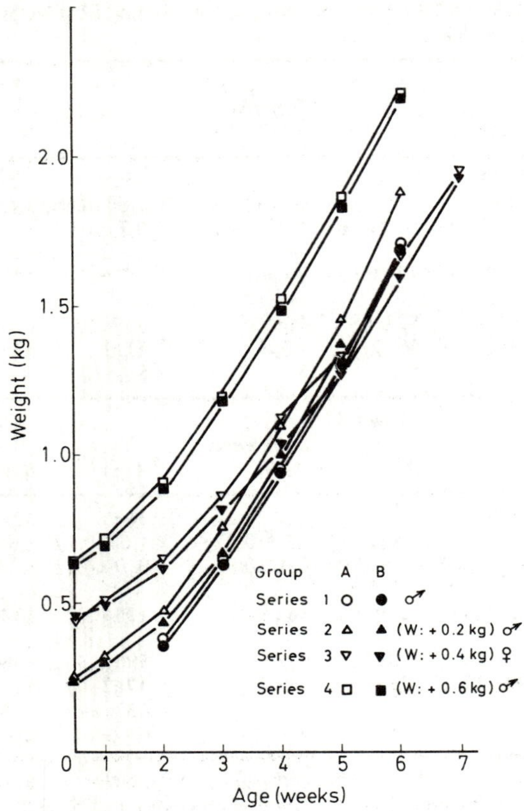

Figure 92.2 Growth curves

In the fourth series, in which physical activity was measured reliably, no significant differences in total activity could be found between the two groups but they showed different activity patterns. During both the dark and the light periods group B was more active than group A, nearly twice as active in the dark and about 20% more in the light. In both groups in the dark hourly heat production was about 6 $kJ/W^{0.75}$ lower (20–25%) than in the light.

Discussion

It might have been expected that chickens kept in the dark for a longer part of the day would eat less and show a lower activity. Obviously this is hardly the case. Their physical activity as measured in the fourth series suggests a more rapid or more frequent feed intake during the light and probably some feed intake during the dark. These higher activities offset the advantage with regard to energy needs of the much lower activity level in the dark compared to those in the periods with light. The results however are in accordance with the expectation that the younger animals would suffer somewhat more from the long dark periods: initially they ate and grew less than the animals of group A. It would be interesting to repeat the experiments using frequent short light periods as described by Gayner, Koon and Flood (1977) and Goodman (1978). In that way the chickens' feed intake would be ensured whereas their activity would probably be at its lowest.

References

GAYNER, R.M., KOON, J.L. and FLOOD, C.A. (1977). *Poult. Sci.*, **56**, 1381–1383

GOODMAN, B.L. (1978). *Poult. Sci.*, **57**, 1423–1428

WENK, C. and van ES, A.J.H. (1976). In *Energy Metabolism in Farm Animals*, pp. 189–192. Ed. M. Vermorel. EAAP Publication No. 19. Clermont-Ferrand: G. de Bussac.

93

DISCUSSION PAPER: METABOLIC RESPONSES IN THE GROWING ANIMAL

A.J.F. WEBSTER
Department of Animal Husbandry, University of Bristol, Langford House, Langford, Bristol BS18 7DU, England

Much of the discussion in this volume has been concerned with the factors that affect the heat production of an animal. This is by no means the whole of energy metabolism but it is a very large and economically important part of it. Differences in feed efficiency attributable to differences in body composition will not be discussed here.

It is appropriate to think of the differing approaches to this phenomenon as different, complementary modes of perception, like sight, hearing, touch, smell and taste, all of which contribute to our awareness of the world. The danger for the individual scientist, each wedded to his own experimental procedures, is that by concentrating too exclusively on one sense, he may allow others to atrophy. What alternative approaches are there to the analysis of the heat production of a growing animal?

Analysis according to external inputs

The following factors are taken into account:

1. Size
2. Physiological state
3. The quantity and quality of food ingested
4. Behaviour and activity
5. The thermal environment.

This is the classic approach of the nutritionist assessing energy flow in the whole animal by calorimetry or comparative slaughter. The majority of chapters in this book have been based on this approach and most of this discussion will be concerned with this type of analysis.

Analysis according to internal inputs

This is the 'lower level' of the modeller who uses his knowledge of stoichiometry and the power of his computer to derive a much more elegant and comprehensive picture of the flow of energy through an animal and of the likely consequences of some perturbation of the inputs to the system.

The approach has an enormous potential both to increase understanding of the mass of data that has been accumulated to date and to point not only to those experiments which need to be done in future but also, more especially, to those which *do not* need to be done again. It is to be hoped that at future energy symposia more attention can be given to this topic. However, this approach is, in essence, still that of the nutritionist; it assumes that the animal is what it eats, that is that the production of ATP and therefore heat by the animal is driven by the flow of substrates *into* the system. When such models become a little strange it is usual to postulate the existence of a 'futile cycle'. However, there is no convincing biological evidence for a genuinely futile cycle whose function is only to dissipate excess substrates in the form of heat.

It can be argued, at the risk of being called not blind to physiology but deaf to biochemistry, that an equally profitable, and more physiologically sound, approach is to model energy flow according to output or work done.

Analysis according to output

The philosophy behind this approach is that the metabolism of the animal is not so much driven by the energy it eats as pulled along by the work it has to do. In other words, the primary factor that dominates metabolic rate is the *requirement* of the various organs and tissues for energy substrates to regenerate (principally) ATP from ADP produced as a direct consequence of the work done by those tissues in support of maintenance and, for example, growth or lactation. This approach is consistent with calorimetric observations that for most simple-stomached animals the efficiency of utilization of digestible energy is predominantly a function of the physiological state of the animal rather than of the nature of the diet. This approach does not, of course, deny that different feeds, especially different feeds for ruminants, differ in the efficiency with which they generate ATP, but it does resurrect the suggestion that differences in efficiency of utilization between different ruminant feeds can be due, in major part, to differences in the amount of work done (ATP used) by tissues such as the epithelium of the gut, when presented with feeds differing in physical and chemical form (Webster, 1980).

This approach to analysis can be further subdivided into analysis by form and function. Analysis by form involves measurement *in vivo* or *in vitro* of heat production in different organs and tissues. The tissues that have been studied *in vivo* in the ruminant include the liver, the portal-drained viscera and the hind limb as indicative of the carcass. So far the metabolic rates of these components do not add up to more than about 60% of total heat production. Measurements of the metabolic rate of tissues are easy to make *in vitro* but are almost always very much lower than results obtained from intact animals. Measurements of the metabolic rate of individual small organs and tissues of intact animals in sufficient different nutritional and physiological states are almost impossible to contemplate. However, first principles dictate that the oxygen consumption and carbon dioxide production of tissues *in vivo* are determined primarily by capillary blood flow (rather than by Michaelis-Menton equations). It is possible, therefore, using techniques such as radioactive microspheres, to measure capillary flow *in vivo* and to achieve thereby a useful first approximation to the contribution to metabolic rate made by different organs and tissues.

The alternative approach to analysis of outputs has already been briefly discussed and consists in attempting to quantify the energy cost of the various synthetic and catabolic functions of the body. The calorimetric approach which attempts to partition energy between maintenance, protein and fat *deposition* in the growing animal is not sufficient to attack this problem since none of these is a specific work function – all, in fact, are differences between unknown rate terms. Protein synthesis, fat synthesis, cellular turnover and membrane pumping are all measurable components of the work done by the body. The trouble lies in distinguishing between them. The work done at the Rowett Institute with Zucker rats (Webster *et al.*, page 125) represents just one attempt to separate two of the variables, protein and fat synthesis, while disturbing other variables as little as possible.

Size

The effect of size on the heat production (H) of a growing animal is what is meant when maintenance fasting metabolism is referred to. Once this is appreciated, maintenance or basal metabolism need no longer be considered as having any profound physiological meaning in the growing animal, and can be seen instead simply as terms relating the residual component of heat production to some function of body weight. The precision of this term is undoubtedly improved if empty body weight is substituted for live weight, and almost certainly improved still further if the lean empty body weight is known with precision.

The expression, 'metabolic size', is only an arithmetic convenience to maintain the relationship between size and residual H in constant linear proportion. If the relationship between H and $W^{0.75}$ is not linear then no advantage is gained by using the exponent and $W^{1.0}$ would do. It is also worth remembering that Kleiber (1965) pointed out that 0.75 was the decimal form of the 'more approximate' term three-quarters, which described the relationship between basal metabolism and weight in mature animals of different species. The Agricultural Research Council's working parties on pigs (*see* Fowler *et al.*, page 151) and ruminants (unpublished, but *see* Webster, 1978) are at present recommending that the exponent of W needed to express the effect of size on H in the growing animal should be 0.66 and 0.69, respectively, both of which could be expressed more generally as two-thirds. The study of the effect of size on the growing animal has returned at last to the surface area law originally propounded by Sarrus and Rameaux in 1839.

Thermal environment

The effect of the thermal environment on energy metabolism is, in practice, much more marked in species like pigs and chickens than in ruminants. In the natural state, the former species usually produce heat in part to keep warm (to keep body temperature up) while ruminants dissipate by evaporation a proportion of the heat that they produce as an inevitable consequence of metabolism in order to keep body temperature down.

The cold tolerance of pigs and chickens is, as has been indicated often in this book, very dependent on food intake. It is, however, except in the neonatal period, probably less dependent on body size than has been formerly thought, since both residual heat production and sensible heat loss appear to be closely related to surface area. The critical temperature of the adult sow, for example, is about 18 °C. The interactions between temperature and plane of nutrition are, however, difficult to predict *a priori* and the best approach is probably that adopted by Close and Stanier (*see* page 399) and Le Dividich, Noblet and Aumaitre (*see* page 403), which is to investigate a matrix of temperature and nutritional treatments and unscramble the interactions retrospectively.

These studies suggest that the lower critical temperature of the three-week-old pig is about 25 °C, which is extremely high. However, air temperature is only one factor determining thermal exchange, especially in the young pig which loses so much heat by conduction. Verstegen, Hel and Willems (1977), Bruce (1979) and others have examined the extent to which floor insulation and bedding can improve the climatic environment for the young pig.

In these days of high energy costs the appropriate use of insulation assumes even greater importance. The practice of rearing piglets weaned at three weeks in batteries of expanded metal cages is one of several examples of intensive production that immediately repels many uneducated laymen on ethical grounds and appears finally to the informed scientist to be biologically unsound.

The net efficiency of utilization of metabolizable energy (ME) below the critical temperature increases to values above 0.8 but never approaches the value of 1.0 postulated for ruminants by Blaxter (1962). In other words, diet-induced thermogenesis in simple-stomached animals can never entirely replace cold thermogenesis because the heat is not produced at the required rate. However, when the interaction between nutrition and environment in pigs is such that animals are above their critical temperature at a high plane of nutrition but below it at (say) maintenance, it is not possible to derive efficiency terms that accurately describe the utilization of ME for growth (k_g).

Activity and behaviour

The two terms are not synonymous even in the restricted context of their effect on H since it is quite possible to imagine changes in vigilance or muscle tone that would alter H without being recorded on an activity meter. Halter, Wenk and Schürch (*see* page 395) quite correctly introduced activity as a factor confounding the relationship between ME and H in growing pigs and thus upsetting the estimation of k_g. Activity appeared to be more variable at about maintenance, when feed intake was restricted and, in their experiments, it certainly constituted a greater proportion of total H than at *ad lib.* intake. However, it is not yet clear whether the absolute amount of energy expended in activity was different at the two levels of intake. Vogt *et al.* (*see* page 455) described the effects of different light regimens on the activity of and utilization of ME by growing chickens. Here again problems of variable and alarming activity (such as cannibalism) were greatest at maintenance. These experiments raise questions about the wisdom of basing estimates of the efficiency of utilization of ME for rapid growth in pigs and poultry on experiments which include maintenance as an essential treatment.

Brockway and Boyne (*see* page 449), in resolving the problem of animals pulling on their tether while on a treadmill, have produced a very convincing case to suggest that the efficiency of utilization of ME for the work of vertical ascent has a maximum value of about 35%. Since this is rather close to the predicted efficiency of synthesis of ATP and creatine-phosphate from ME in ruminants, these results suggest that the efficiency of utilization of high-energy phosphate bonds for work by muscle is close to 100%, a conclusion which supports some fundamental studies done with isolated frog muscle (Curtin *et al.*, 1974).

Physiological state

Very few of the factors affecting H in the growing animal which have been discussed can be considered as functions of body size or of the quantity and quality of food intake. Frisch and Vercoe (*see* page 431) showed how selection for growth potential in a favourable environment was associated with an increase in fasting metabolism, whereas in a harsh, tropical environment the reverse was the case; the Brahman cattle or Hereford animals selected for optimal performance in Northern Australia had lower fasting H than unselected animals. Keller (*see* page 417), Wałach-Janiak, Kotarbińska and Kielanowski (*see* page 157) and van der Merwe and van Rooyen (*see* page 135) have also drawn attention to the relationship between growth potential and fasting (or residual) H.

A period of severe undernutrition is known to reduce that component of H usually described as maintenance requirement (Ledger and Sayers, 1977). However, when food intake is restored to that which will support normal growth, it appears that both 'maintenance' and k_g return rapidly to that of the normal animal. The experiments of Gädeken, Böhme and Oslage (*see* page 407) and Thomson *et al.* (*see* page 427) support this assumption and reinforce the conclusion that any improvement in food conversion efficiency noted during so-called 'compensatory growth' can be attributed almost entirely to a lower energy content of live weight gains. Ingram and Dauncey (*see* page 411) and Keller (*see* page 417) both discussed the extent to which metabolic rate might be affected by hormone activity. Inevitably, their contributions only touched on a vast and important subject. However, in assessing what value future endocrinological studies are likely to have in the analysis of H in the growing animal, it is necessary to point out first that no one hormone is likely to determine or even dominate H. Second, it is not certain to what extent individual or multiple hormones can be said to direct metabolism at all in a primary sense; it is equally possible to see changes in hormone secretion as secondary administrative adjustments required to sustain metabolism at a different rate as it is to see these changes as a result of some quite distinct external or internal stimulus. If this is the case, then hormone studies may not be particularly helpful in this context.

The conclusion that emerges as most important for the nutritionist is the inescapable fact that there are substantial differences in H between animals that cannot be described in terms of some function of body weight and the quantity and quality of food intake. Any feeding system that does not take these into account is not an adequate system. By this criterion there is no adequate feeding system for growing farm animals.

Turnover of metabolites

MacRae and Egan (*see* page 421) provided an elegant illustration of the kinetics of glucose in the ruminant. It appeared that even in pregnant and lactating sheep the production of glucose considerably exceeded the amount that was oxidized. There was considerable recycling of three-carbon units between pools of glucose and free non-essential amino acids. This turnover of three-carbon units may take its place alongside the turnover of nitrogen between proteins and amino acids as an essential contributor to metabolism. It is attractive to think of these functions in a teleological sense; the ruminant keeps these useful three-carbon units on standby rather like an army in peacetime. However, the cost of maintaining a standing army is high and similarly the turnover of useful metabolites must also be energetically expensive since the one thing the body cannot recycle is energy. The experiments of Edmunds, Buttery and Fisher (*see* page 129) and Webster *et al.* (*see* page 125) suggested that the energy cost of protein turnover was substantial but even so could only account for about 20–25% of H in the growing animal. MacCrae and Egan's chapter (*see* page 421) has indicated another aspect of turnover which may make a significant contribution to metabolic rate. Similar studies in the future may quantify other aspects of turnover whose energy costs may then be predicted by stoichiometry. This approach should permit a reasonably complete analysis of metabolic rate according to output by function and a biologically sound interpretation of changes in metabolic rate resulting from changes in the composition and rate of supply of nutrients to the tissues of the body.

References

BLAXTER, K.L. (1962). *The Energy Metabolism of Ruminants*, pp. 116–148. London: Hutchinson

BRUCE, J.M. (1979). *Farm Building Progress 55.* Scottish Farm Buildings Investigation Unit 1–4

CURTIN, N.A., GILBERT, C., KRETZSCHMAR, K.M. and WILKIE, D.R. (1974). *J. Physiol.*, **238**, 455–472

KLEIBER, M. (1965). In *Energy Metabolism of Farm Animals*, pp. 427–432. Ed. K.L. Blaxter. EAAP Publication No. 11. New York: Academic Press

LEDGER, H.P. and SAYERS, A.R. (1977). *J. agric. Sci., Camb.*, **88**, 11–26

VERSTEGEN, M.W.A., HEL, W. van der and WILLEMS, G.E.J.M. (1977). *Anim. Prod.*, **24**, 253–260

WEBSTER, A.J.F. (1978). *Wld Rev. Nutr. Diet*, **30**, 189–226

WEBSTER, A.J.F. (1980). *Proc. Vth Symp. Digestion and Metabolism in the Ruminant.* Eds. Y. Ruckebusch and P. Thivend, pp. 423–438. Lancaster: MTP Press

94

GENERAL DISCUSSION

The discussion was organized under four headings:

1. Biochemical aspects of metabolism
2. Effects of activity and behaviour
3. Effects of climate
4. Effects of the physiological state of an animal.

Biochemical aspects of metabolism

It was pointed out that 'protein turnover rate' is deduced from measurement of rates of synthesis and degradation of a mixture of proteins which includes enzymes with a very short half-life and body muscle with a very long half-life. The precise physiological interpretation of these data is not simple.

Discussion of the energetic costs of maintaining reserves or pools of substrates suggested that this was not expensive for reserves which are stored, as in fat, but may be expensive if such substrates have to be continually resynthesized and remobilized into and out of reserves. In this context it is probable that although the contributions of individual organs to total body heat production may be small, when taken together these become important.

Effects of activity and behaviour

Physical activity can be measured readily, but the exact significance of such measurements in relation to the physiology of the animal body is not clear. The amount of physical activity and the behaviour of animals can exert very important influences on their rates of heat production, and their behaviour can in turn be influenced by management. For example, the incidence of fighting and cannibalism was higher in chickens fed on a low level of feeding than in those fed on a higher level. Another example was given by the observation that when sheep were first brought indoors from pasture, there were large differences in heat production between them when measured under controlled conditions; these differences decreased considerably as the sheep became accustomed to the conditions and there was a decrease in the average rate of production. It was reassuring to find that, when the energy requirements of sheep grazing on pasture were calculated by adding together the costs of such activities as eating and walking, which were measured calorimetrically, these requirements were generally in reasonable agreement with those actually measured in the field.

Effects of climate

Young piglets grow very rapidly so that their lower critical temperature (T_c) probably changes rapidly with age; this creates problems in relation to the design of experiments which are intended to measure the effects of conditions below the T_c on the energy metabolism of early weaned piglets. Nevertheless, reliable estimations of energy requirements do require balance periods of several days' duration, and such measurements are likely to provide information which is of practical value.

Effects of the physiological state

It is important to consider the influence of genotype on the energy metabolism of livestock, although energy balance must be analysed in more appropriate ways than at present before geneticists will be able to cooperate with nutritionists and physiologists to any extent. There appears to be no evidence to show that 'genetic improvement' has resulted in changes in the maintenance requirement for metabolizable energy (ME) nor in the net efficiency with which ME is converted into animal products. The data for piglets which were in negative fat balance but in positive protein balance when fed at maintenance accentuated the problems of interpreting the term 'maintenance'.

LIST OF PARTICIPANTS

Alderman, Mr G.	Ministry of Agriculture, Fisheries and Food, Room 324, Great Westminster House, Horseferry Road, London SW1P 2AE, U.K.
Annison, Prof. E.F.	Department of Animal Husbandry, University of Sydney, Camden, New South Wales 2570, Australia
Apfelbaum, Prof. M.	Hôpital Bichat, Exploration Fonctionelle Nutritionelle, 170 bd Ney, 75018 Paris, France
Baldwin, Dr R.L.	Department of Animal Science, University of California, Davis, California 95616, U.S.A.
Bascuñan, Dr G.	Departamento de Zootecnia, Facultad de Agronomia, Universidad Catolica de Chile, Casilla 114-D, Chile
Bayley, Dr H.S.	Department of Nutrition, University of Guelph, Guelph, Ontario N1G 2W1, Canada
Beever, Dr D.E.	Grassland Research Institute, Hurley, Maidenhead, Berkshire, U.K.
Bell, Dr A.W.	School of Agriculture, La Trobe University, Bundoora, Victoria 3083, Australia
Berschauer, Dr F.	Universität Hohenheim, Abteilung Tierernährung, 7000 Stuttgart 70, P.O. Box 106 – 06100, Federal Republic of Germany
Bickel, Prof. H.	Institut für Tierproduktion, Gruppe Ernährung, ETH-Zentrum, 8092 Zurich, Switzerland

Blaxter, Sir Kenneth, FRS	The Rowett Research Institute, Bucksburn, Aberdeen AB2 9SB, U.K.
Braude, Dr R.	National Institute for Research in Dairying, Shinfield, Reading, Berks., U.K.
Brockway, Dr J.M.	The Rowett Research Institute, Bucksburn, Aberdeen AB2 9SB, U.K.
Bull, Dr L.S.	Department of Animal and Veterinary Sciences, Hitchner Hall, University of Maine, Orono, Maine 04469, U.S.A.
Burlacu, Dr G.	Institute of Animal Nutrition, Balotesti, Ilfov, Romania
Buttery, Dr P.J.	Department of Applied Biochemistry and Nutrition, University of Nottingham School of Agriculture, Sutton Bonington, Loughborough LE12 5RD, Leicestershire, U.K.
Byers, Dr F.M.	Department of Animal Science, Ohio Agricultural Research and Development Center, Wooster, Ohio 44691, U.S.A.
Campbell, Dr L.D.	Department of Animal Science, University of Manitoba, Winnipeg, Manitoba, Canada R3T 2N2
Cañas, Dr R.	Departamento de Zootecnia, Facultad de Agronomia, Universidad Catolica de Chile, Casilla 114-D, Chile
Chamberlain, Dr D.G.	Hannah Research Institute, Ayr KA6 5HL, U.K.
Charlet-Lery, Dr G.	Laboratoire de Physiologie de la Nutrition, CNZR, Jouy-en-Josas 78350, France
Cho, Dr C.Y.	Ontario Ministry of Natural Resources, Fisheries Branch, Fish Nutrition Laboratory, University of Guelph, Guelph, Ontario, N1G 2W1, Canada
Christensen, Dr. K.	The Royal Veterinary and Agricultural University, Department of Animal Physiology, Rolighedsvej 25, DK-1958 Copenhagen V, Denmark
Chwalibog, Dr A.	National Institute of Animal Science, Rolighedsvej 25, DK-1958 Copenhagen V, Denmark

Close, Dr W.H.	ARC Institute of Animal Physiology, Babraham, Cambridge CB2 4AT, U.K.
Cole, Dr D.J.A.	Department of Agriculture and Horticulture, School of Agriculture, Sutton Bonington, Loughborough LE12 5RD, U.K.
Corbett, Dr J.L.	CSIRO, Division of Animal Production, Pastoral Research Laboratory, Armidale, New South Wales 2350, Australia
Czerkawski, Dr J.W.	The Hannah Research Institute, Ayr KA6 5HL, U.K.
Dauncey, Dr M.J.	ARC Institute of Animal Physiology, Babraham, Cambridge CB2 4AT, U.K.
Decuypere, Dr E.	Laboratorium voor ecofysiologie der huisdieren, Kardinaal Mercierlaan 92, 3030 Heverlee, Belgium
Egan, Dr A.R.	Waite Agricultural Research Institute, University of Adelaide, Glen Osmond, South Australia 5064
Ekern, Dr A.	Agricultural University of Norway, Institute of Animal Nutrition, 1432 Ås-NLH, Norway
Ellis, Dr W.C.	Animal Science Department, Texas A – M University, College Station, TX 77843, U.S.A.
Es, Dr A.J.H. van	Institute for Livestock Feeding and Nutrition, Research Hoorn, Runderweg 2, Postbox 160, Lelystad, The Netherlands
Evans, Dr P.J.	Unilever Research Ltd., Colworth House, Sharnbrook, Bedford MK44 1LQ, U.K.
Farrell, Dr D.J.	Department of Biochemistry and Nutrition, University of New England, Armidale, New South Wales 2351, Australia
Fisher, Dr C.	ARC Poultry Research Centre, King's Buildings, West Mains Road, Edinburgh EH9 3JS, U.K.
Fontaine, Dr G.	Rijksstation voor Kleinveeteelt, Burgemeester van Gansberghelaan, 92, 9220 Merelbeke, Belgium

Foot, Dr J.Z.	Pastoral Research Station, Box 180, Hamilton, Victoria 3300, Australia
Fowler, Dr V.R.	The Rowett Research Institute, Bucksburn, Aberdeen AB2 9SB, U.K.
Freeman, Dr B.	Houghton Poultry Research Station, Houghton, Huntingdon PE17 2DA, U.K.
Fuller, Dr M.F.	The Rowett Research Institute, Bucksburn, Aberdeen AB2 9SB, U.K.
Gädeken, Dr D.	Institut für Tierernährung der Forschungsanstalt f. Landwirtschaft, 3300 Braunschweig, Bundesallee 50, Federal Republic of Germany
Garrett, Dr W.N.	Department of Animal Science, University of California, Davis, California 95616, U.S.A.
Geay, M.Y.	Centre de Recherche Zootechnique et Vétérinaire de Theix, 63110 Beaumont, France
Glem-Hansen, Dr N.	National Institute of Animal Science, Department of Furbearing Animals, Trollesminde, Roskildevej 48H, DK-3400 Hilleroed, Denmark
Gray, Dr R.	Department of Agriculture, Agriculture and Food Science Centre, Newforge Lane, Belfast BT9 5PX, N. Ireland
Guillaume, Dr J.	CNEXO-COB, Boite Postale 337, F29273 Brest Cedex, France
Hel, Dr W. van der	Agricultural University, Department of Animal Husbandry, Zodiac, Marijkeweg 40, 6700 AH Wageningen, The Netherlands
Henkel, Dr H.	Institut für Tierernährung, Universität Kiel, 2300 Kiel, Federal Republic of Germany
Henry, M.Y.	Station de Recherches sur l'Elevage des Porcs, CNRZ, Jouy-en-Josas 78530, France
Holmes, Dr C.W.	Department of Dairy Husbandry, Massey University, Palmerston North, New Zealand

Honing, Dr Y. van der	Institute for Livestock Feeding and Nutrition, Research Hoorn, Runderweg 2, Postbox 160, Lelystad, The Netherlands
Ingram, Dr D.L.	ARC Institute of Animal Physiology, Babraham, Cambridge CB2 4AT, U.K.
Johnson, Dr D.E.	Metabolic Laboratory, Colorado State University, Fort Collins, Colorado 80523, U.S.A.
Jordan, Dr J.W.	Department of Agriculture, Agricultural and Food Chemistry Research Division, Newforge Lane, Belfast BT9 5PX, N. Ireland
Just, Dr A.	National Institute of Animal Science, Rolighedsvej 25, 1958 Copenhagen V, Denmark
Kallay, Dr K.	European Association for Animal Production, Corso Trieste 67, 00198 Rome, Italy
Kameoka, Dr K.	National Institute of Animal Industry, Tsukuba 305, Japan
Kampen, Dr M. van	Vakgreop Veterinaire Fysiologie, Alexander Numankade 93, Utrecht 2502, The Netherlands
Keller, Prof. J.	Polish Academy of Sciences, Institute of Genetics and Animal Breeding, Jastrzebiec, 05-551 Mroków, Poland
Kirchgeßner, Prof. M.	Technische Hochschule Munchen, Institut für Ernährungsphysiologie, 8050 Freising-Weihenstephan, Federal Republic of Germany
Kleimenov, Prof. N.I.	All-Union Institute of Animal Sciences, Moscow, Podolsk, Dubrovizi, U.S.S.R.
Koong, Dr L.	Department of Animal Science, University of California, Davis, California, 95616, U.S.A.
Kotarbińska, Dr M.	Polish Academy of Sciences, Institute of Animal Physiology and Nutrition, 05-110 Jablonna, near Warsaw, Poland

Kubasek, Dr F.O.T.	Department of Applied Nutrition, School of Agriculture, University of Nottingham, Sutton Bonington, Loughborough, Leicestershire LE12 5RD, U.K.
Kushima, Mr M.	Laboratory of Animal Nutrition, Faculty of Agriculture, Nagoya University, Nagoya 464, Japan
Le Dividich, M.J.	Station de Recherches sur l'Elevage des Porcs, CNRZ, Jouy-en-Josas 78350, France
Lister, Dr D.	ARC Meat Research Institute, Langford, Bristol BS18 7DY, U.K.
Ludvigsen, Dr J.B.	National Institute of Animal Science, Rolighedsvej 25, 1958, Copenhagen V, Denmark
Lundy, Dr H.	ARC Poultry Research Centre, King's Buildings, West Mains Road, Edinburgh EH9 3JS, U.K.
McCracken, Dr K.J.	Agricultural and Food Chemistry Department, Newforge Lane, Belfast BT9 5PX, N. Ireland
McLean, Dr J.A.	The Hannah Research Institute, Ayr KA6 5HL, U.K.
MacLeod, Dr M.G.	ARC Poultry Research Centre, King's Buildings, West Mains Road, Edinburgh EH9 3JS, U.K.
MacRae, Dr J.C.	The Rowett Research Institute, Bucksburn, Aberdeen AB2 9SB, U.K.
Margan, Mr D.	CSIRO, Division of Animal Production, P.O. Box 239, Blacktown, New South Wales 2148, Australia
Meissner, Dr H.H.	Research Institute for Animal and Dairy Science, Private Bag X 2, Irene 1675, South Africa
Menke, Prof. K.H.	Universität Hohenheim, Abteilung Tierernährung, Postfach 106 - 06100, 7 Stuttgart 70, Federal Republic of Germany

Merwe, Prof. F.J. van der	Department of Animal Science, Faculty of Agriculture, University of Stellenbosch, Stellenbosch 7600, South Africa
Miller, Dr E.L.	Department of Applied Biology, University of Cambridge, Pembroke Street, Cambridge CB2 3DX, U.K.
Millward, Dr B.J.	Clinical Nutrition and Metabolism Unit, Hospital of Tropical Diseases, 4 St. Pancras Way, London NW1 2PE, U.K.
Milne, Dr J.A.	Hill Farming Research Organization, Bush Estate, Penicuik, Midlothian EH26 0PY, U.K.
Misson, Mr B.H.	Houghton Poultry Research Station, Houghton, Huntingdon PE17 2DA, U.K.
Moe, Dr P.W.	Ruminant Nutrition Laboratory, US Department of Agriculture, Beltsville Agricultural Research Center, Maryland 20705, U.S.A.
Mount, Dr L.E.	ARC Institute of Animal Physiology, Babraham, Cambridge CB2 4AT, U.K.
Müller, Dr H.L.	Technische Hochschule München, Institut für Ernährungsphysiologie, 8050 Freising-Weihenstephan, Federal Republic of Germany
Nadalyak, Dr E.A.	All-Union Institute of Physiology, Biochemistry and Nutrition of Domestic Animals, Kalujskaia Oblast, Borovsk, U.S.S.R.
Neergaard, Dr L.	National Institute of Animal Science, Rolighedsvej 25, 1958 Copenhagen V, Denmark
Nichelmann, Dr M.	Sektion Tierproduktion und Veterinärmedizin, Humboldt-Universität zu Berlin, Bereich Physiologie, 104 Berlin, Reinhardstrasse 4, German Democratic Republic
Nicol, Dr A.M.	Department of Animal Science, University of Alberta, Edmonton, Alberta, Canada T6H 4M5
Noblet, M.J.	Station de Recherches sur l'Elevage des Porcs, CNRZ, Jouy-en-Josas 78530, France

Nordén, Prof. Å.	Forskn. avd. 2 E-bl. Lasarettet, S-221 85 Lund, Sweden
Ørskov, Dr E.R.	The Rowett Research Institute, Bucksburn, Aberdeen AB2 9SB, U.K.
Oslage, Prof. H.J.	Institut für Tierernährung, Forschungsanstalt für Landwirtschaft, 3300 Braunschweig, Bundesallee 50, Federal Republic of Germany
Peaker, Dr M.	The Hannah Research Institute, Ayr KA6 5HL, U.K.
Petry, Dr P.	Institut für Physiologie, Physiologische Chemie und Ernährungsphysiologie der Tierärztlichen Fakultät der Universität München, Veterinarstrasse 13, 8000 München 22, Federal Republic of Germany
Pullar, Dr J.D.	The Rowett Research Institute, Bucksburn, Aberdeen AB2 9SB, U.K.
Rattray, Dr P.V.	Ruakura Agricultural Research Centre, Private Bag, Hamilton, New Zealand
Reichl, Prof. Dr J.R.	Universität Hohenheim, Abteilung Tierer-nährung, Postfach 106 - 06100, 7000 Stuttgart 70, Federal Republic of Germany
Reynolds, Dr P.J.,	Ruminant Nutrition Laboratory, Building 200, BARC-East, Beltsville, Maryland 20705, U.S.A.
Robb, Dr. J.	Unitrition Int. Ltd., Springpark House, Basing View, Basingstoke, Hampshire, U.K.
Rogers, Dr G.L.	The Dairy Research Institute, Ellinbank, Via Warragul, Victoria 3820, Australia
Rosochacki, Dr S.	Clinical Nutrition and Metabolism Unit, Hospital for Tropical Diseases, 4 St. Pancras Way, London NW1 2PE, U.K.
Rothwell, Dr N.	Department of Physiology, St. George's Hospital Medical School, Tooting, London SW17 0RE, U.K.
Roux, Dr C.Z.	Research Institute for Animal and Dairy Science, Private Bag X 2, Irene 1675, South Africa

Schnyder, Mr W.	Institut für Tierproduktion, Gruppe Ernährung, ETH-Zentrum, 8092 Zurich, Switzerland
Schulz, Dr A.R.	Department of Biochemistry, School of Medicine, Indiana University, Indianapolis, Indiana 46202, U.S.A.
Schürch, Prof. A.	Institut für Tierproduktion, Gruppe Ernährung, ETH-Zentrum, 8092 Zurich, Switzerland
Sibbald, Dr I.R.	Animal Research Institute, Central Experimental Farm, Ottawa, Ontario K1A 0C6, Canada
Smith, Mr J.S.	The Rowett Research Institute, Bucksburn, Aberdeen AB2 9SB, U.K.
Smith, Dr N.E.	Department of Animal Science, University of California, Davis, California 95616, U.S.A.
Sørensen, Prof. P.H.	The Royal Veterinary and Agricultural University, Bülowsvej 13, DK-1870, Copenhagen V, Denmark
Stanier, Dr M.W.	ARC Institute of Animal Physiology, Babraham, Cambridge CB2 4AT, U.K.
Stock, Dr M.J.	Department of Physiology, St. George's Hospital Medical School, Tooting, London SW17 0RE, U.K.
Stombaugh, Dr D.P.	Department of Agricultural Engineering, Ohio State University, 2073 Neil Avenue, Columbus, Ohio 43210, U.S.A.
Sundstøl, Dr F.	Agricultural University of Norway, Institute of Animal Nutrition, 1432 Ås-NLH, Norway
Sutton, Dr J.D.	National Institute for Research in Dairying, Shinfield, Reading RG2 9AT, U.K.
Tasaki, Prof. I.	Nagoya University, Laboratory of Animal Nutrition, Nagoya 464, Japan
Taylor, Dr A.J.	Unilever Research, Colworth House, Sharnbrook, Bedford, U.K.
Thomas, Dr P.C.	The Hannah Research Institute, Ayr KA6 5HL, U.K.

Thomson, Dr D.J.	The Grassland Research Institute, Hurley, Maidenhead, Berkshire SL6 5LR, U.K.
Thomson, Dr E.F.	Institute of Animal Production, ETH-Zentrum, CH-8092 Zurich, Switzerland
Thorbek, Dr G.	National Institute of Animal Science, Rolighedsvej 25, 1958 Copenhagen V, Denmark
Tissier, M.M.	Centre de Recherches Zootechniques et Vétérinaires, Theix, 63110 Beaumont, France
Todorov, Dr N.A.	Agricultural University, Department of Nutrition, Stara Zagora, 6000, Bulgaria
Trigg, Dr T.E.	Ruakura Agricultural Research Centre, Private Bag, Hamilton, New Zealand
Tyrrell, Dr H.F.	Ruminant Nutrition Laboratory, Building 162, BARC-East, Beltsville, Maryland 20705, U.S.A.
Ulyatt, Dr M.J.	Grassland Research Institute, Hurley, Maidenhead, Berkshire SL6 5LR, U.K.
Valli, Miss S.	Butterworth & Co. (Publishers) Ltd., Borough Green, Sevenoaks, Kent TN15 8PH, U.K.
Vercoe, Dr J.E.	Joint FAO/IAEA Division, IAEA, P.O. Box 542, A-1011, Vienna, Austria
Vermorel, M.M.	Institut National de la Recherche Agronomique, CRZV, Theix, 63110 Beaumont, France
Verstegen, Dr M.W.A.	Department of Animal Husbandry, Zodiac, Marijkeweg 40, 6700 AH Wageningen, The Netherlands
Vogt, Dr H.	Institut für Kleintierzucht, Doernbergstrasse 25/27, 3100 Celle, Federal Republic of Germany
Wainman, Mr F.W.	The Rowett Research Institute, Bucksburn, Aberdeen AB2 9SB, U.K.
Walker, Dr D.M.	Department of Animal Husbandry, University of Sydney, Sydney, New South Wales 2006, Australia

Webster, Prof. A.J.F.	Department of Animal Husbandry, University of Bristol, Langford House, Langford, Bristol BS18 7DU, U.K.
Wenk, Dr C.	Institut für Tierproduktion, Gruppe Ernährung, Eidg. Techn. Hochschule, ETH-Zentrum, 8092 Zurich, Switzerland
Wilde, Dr R. de	Laboratory of Animal Nutrition, University of Gent, Heidestraat 19, 9220 Merelbeke, Belgium

INDEX